GLASGOW COLLEGE OF TECHNOLOGY

This book should be returned by the last date stamped below

CONCANAVALIN A

ADVANCES IN EXPERIMENTAL MEDICINE AND BIOLOGY

Editorial Board:

Nathan Back	State University of New York at Buffalo
N. R. Di Luzio	Tulane University School of Medicine
Bernard Halpern	Collège de France and Institute of Immuno-Biology
Ephraim Katchalski	The Weizmann Institute of Science
David Kritchevsky	Wistar Institute
Abel Lajtha	New York State Research Institute for Neurochemistry and Drug Addiction
Rodolfo Paoletti	University of Milan

Recent Volumes in this Series

Volume 46
PARENTERAL NUTRITION IN INFANCY AND CHILDHOOD
Edited by Hans Henning Bode and Joseph B. Warshaw • 1974

Volume 47
CONTROLLED RELEASE OF BIOLOGICALLY ACTIVE AGENTS
Edited by A. C. Tanquary and R. E. Lacey • 1974

Volume 48
PROTEIN-METAL INTERACTIONS
Edited by Mendel Friedman • 1974

Volume 49
NUTRITION AND MALNUTRITION: Identification and Measurement
Edited by Alexander F. Roche and Frank Falkner • 1974

Volume 50
ION-SELECTIVE MICROELECTRODES
Edited by Herbert J. Berman and Normand C. Hebert • 1974

Volume 51
THE CELL SURFACE: Immunological and Chemical Approaches
Edited by Barry D. Kahan and Ralph A. Reisfeld • 1974

Volume 52
HEPARIN: Structure, Function, and Clinical Implications
Edited by Ralph A. Bradshaw and Stanford Wessler • 1975

Volume 53
CELL IMPAIRMENT IN AGING AND DEVELOPMENT
Edited by Vincent J. Cristofalo and Emma Holečková • 1975

Volume 54
BIOLOGICAL RHYTHMS AND ENDOCRINE FUNCTION
Edited by Laurence W. Hedlund, John M. Franz, and
  Alexander D. Kenny • 1975

Volume 55
CONCANAVALIN A
Edited by Tushar K. Chowdhury and A. Kurt Weiss • 1975

CONCANAVALIN A

Edited by
Tushar K. Chowdhury and A. Kurt Weiss
Department of Physiology and Biophysics
University of Oklahoma Health Sciences Center
Oklahoma City, Oklahoma

PLENUM PRESS • NEW YORK AND LONDON

Library of Congress Cataloging in Publication Data

International Symposium on Concanavalin A, Norman, Okla., 1974.
   Concanavalin A: [proceedings]

  (Advances in experimental medicine and biology; v. 55)
  Includes bibliographies and index.
  1. Concanavalin A—Congresses. I. Chowdhury, Tushar K., 1936-   ed. II. Weiss, Adolph Kurt, 1923-   ed. III. Title. IV. Series. [DNLM: 1. Concanavalin A—Congresses. W1 AD559 v. 55 1974 / QU55 I642c 1974]
QP935.C6I58   1974   574.1'9245   75-4528
ISBN-13: 978-1-4684-0951-2   e-ISBN-13: 978-1-4684-0949-9
DOI: 10.1007/978-1-4684-0949-9

Proceedings of an International Symposium on Concanavalin A held at the Oklahoma Center for Continuing Education, Norman, Oklahoma, April 19-20, 1974.

© 1975 Plenum Press, New York
Softcover reprint of the hardcover 1st edition 1975

A Division of Plenum Publishing Corporation
227 West 17th Street, New York, N.Y. 10011

United Kingdom edition published by Plenum Press, London
A Division of Plenum Publishing Company, Ltd.
4a Lower John Street, London W1R 3PD, England

All rights reserved

No part of this book may be reproduced, stored in a retrieval system, or transmitted, in any form or by any means, electronic, mechanical, photocopying, microfilming, recording, or otherwise, without written permission from the Publisher.

PREFACE

Concanavalin A (Con A), a plant lectin, has become an object of extensive research not only for the biochemist, but also for the biologist, biophysicist, pathologist, immunologist and others.

On April 19-20, 1974 a group of scholars from four continents met on the campus of the University of Oklahoma for an International Symposium on Con A. This volume contains all lectures presented by the invited speakers on this occasion, as well as the abstracts for all shorter technical papers which were presented. Further, the editors invited additional contributions from a few selected laboratories to cover the areas not covered in the symposium itself. This volume then reports the present status of research with Con A. In a sense it is encyclopedic and should be a useful reference tool to the worker in this field. The newcomer to this area will find that a careful study of this book will bring him up-to-date with the latest techniques, approaches and the "state of the art" in this specialized area of research, as well as with traps and ensnarements in which he need not get entangled again. Starting at the molecular level and advancing from there to the organismic level as well as to a number of highly specialized areas, this volume presents the most complete survey of research findings with Con A available today.

The editors wish to acknowledge the generous financial support provided by the University of Oklahoma through its President, Dr. Paul F. Sharp, the S. R. Noble Foundation of Ardmore, Oklahoma, and the Smith, Kline and French Laboratory of Philadelphia, Pennsylvania. Without this help no Symposium could have been held and no Proceedings of the Symposium could have been published. The encouragement, professional skill and assistance of Mr. Seymour Weingarten of the Plenum Publishing Company, New York, was of inestimable value in assembling the written papers presented here. The expert typing and patient attention to details of Mrs. Ruth McKillen is especially appreciated as is the editorial assistance of Mrs. Sharon Dana, Mrs. Eileen Spain, Dr. Q. S. Kapoor, and Dr. S. V. K. N. Murthy. Our warm gratitude is expressed to the scientific contributors who have prepared their manuscripts promptly and have made every effort to present to the reader the very recent status in the various areas of Con A research. A special note of appreciation is extended to Drs. Irwin J. Goldstein,

Garth Nicolson and Richard D. Berlin for the stimulus which led to the initiation of this project and their cooperation throughout the planning of the symposium.

Tushar K. Chowdhury, Ph.D.
A. Kurt Weiss, Ph.D.

Oklahoma City, Oklahoma
1975

CONTENTS

Chapter
1   CONCANAVALIN A: AN INTRODUCTION
    T. K. Chowdhury

      I. Background Information. 1
     II. Physical and Chemical Properties of
          Concanavalin A. 3
    III. Biological Properties of Concanavalin A . . . 4
    References. 7

2   STRUCTURE AND FUNCTION OF CONCANAVALIN A
    G. N. Reeke, Jr., J. W. Becker, B. A. Cunningham,
    J. L. Wang, I. Yahara, and G. M. Edelman

    Abstract. 13
      I. Introduction. 14
     II. Structure of Concanavalin A 15
   III. Saccharide Binding Function of
         Concanavalin A. 19
    IV. Multivalent Nature of Con A 21
     V. A Model for Interactions between Cell
         Surface Receptors and Cytoplasmic
         Structures. 25
    References. 29

3   STUDIES ON THE COMBINING SITES OF CONCANAVALIN A
    I. J. Goldstein

    Abstract. 35
      I. Introduction. 36
     II. Methods, Results, and Discussion. 36
    References. 49

Chapter

4   ^{13}C NMR STUDIES OF THE INTERACTION OF CONCANAVALIN A WITH SACCHARIDES
C. F. Brewer, H. Sternlicht, D. M. Marcus, and
A. P. Grollman

    Abstract. 55
       I. Introduction. 55
      II. Methods 56
     III. Results and Discussion. 56
    References. 69

5   SELF-ASSOCIATION, CONFORMATION AND BINDING EQUILIBRIA OF CONCANAVALIN A
W. H. Sawyer

    Abstract. 71
       I. Introduction. 72
      II. Methods, Results, and Discussion. 73
     III. Conclusion. 89
    References. 89

6   STUDIES ON THE INTERACTION OF CONCANAVALIN A WITH GLYCOPROTEINS
A. Surolia, S. Bishayee, A. Ahmad,
K. A. Balasubramanian, D. Thambi-Dorai,
S. K. Podder, and B. K. Bachhawat

    Abstract. 95
       I. Introduction. 96
      II. Methods 96
     III. Results 101
    References. 111

7   INTERACTION OF CONCANAVALIN A WITH THE SURFACE OF VIRUS - INFECTED CELLS
G. Poste

    Abstract. 117
       I. Introduction. 117
      II. Methods, Results, and Discussions 118
    References. 145

CONTENTS ix

Chapter
8   CONCANAVALIN A AS A QUANTITATIVE AND ULTRASTRUCTURAL
     PROBE FOR NORMAL AND NEOPLASTIC CELL SURFACES
     G. L. Nicolson

        Abstract. 153
          I. Introduction. 154
          II. Methods 156
          III. Results and Discussion. 157
        References. 166

9   MICROTUBULAR PROTEINS AND CONCANAVALIN A RECEPTORS
     R. D. Berlin

        Abstract. 173
          I.   Introduction. 173
          II. Methods 174
          III. Results 174
          IV. Discussion. 183
        References. 185

10   EFFECTS OF CONCANAVALIN A ON CELLULAR DYNAMICS
     AND MEMBRANE TRANSPORT
     T. K. Chowdhury

        Abstract. 187
          I.   Introduction. 188
          II. Methods of Procedure. 189
          III. Results 191
          IV. Discussion. 201
        References. 204

11   THE CHARACTERISTICS OF SUCCINYLATED CON A INDUCED
     GROWTH INHIBITION OF 3T3 CELLS IN TISSUE CULTURE
     R. J. Mannino, Jr., and M. M. Burger

        Abstract. 207
          I.   Introduction. 207
          II. Methods 209
          III. Results 211
          IV. Discussion. 214
        References. 218

Chapter
12   CELL CYCLE DEPENDENT AGGLUTINABILITY, DISTRIBUTION
     OF CONCANAVALIN A BINDING SITES AND SURFACE
     MORPHOLOGY OF NORMAL AND TRANSFORMED FIBROBLASTS
     J. G. Collard, J. H. M. Temmink, and L. A. Smets

     Abstract. 221
          I.   Introduction. 222
          II. Materials and Methods 223
          III. Results 225
          IV. Discussion. 238
     References. 241

13   CONCANAVALIN A AND OTHER LECTINS IN THE STUDY OF
     TUMOR CELL SURFACE ORGANIZATION
     S. Friberg and S. Hammarström

     Abstract. 245
          I.   Introduction. 246
          II. Methods 247
          III. Results 247
          IV. Discussion. 258
     References. 259

14   MODIFICATION OF THE BIOLOGICAL ACTIVITIES OF
     CONCANAVALIN A BY ANTI-CONCANAVALIN A
     M. A. Leon

     Abstract. 261
          I.   Introduction. 261
          II. Methods, Results, and Discussion. 263
     References. 270

15   CONCANAVALIN A AS A PROBE FOR STUDYING THE MECHANISM
     OF METABOLIC STIMULATION OF LEUKOCYTES
     D. Romeo, G. Zabucchi, M. Jug, N. Miani, and
       M. R. Soranzo

     Abstract. 273
          I.   Introduction. 274
          II. Materials and Methods 275
          III. Results 276
          IV. Discussion. 283
     References. 285

Chapter
16 ENHANCED CYTOTOXICITY IN MICE OF COMBINATIONS
OF CONCANAVALIN A AND SELECTED ANTITUMOR DRUGS
S. G. Bradley, N. M. Marecki, J. S. Bond,
A. E. Munson, and D. T. John

    Abstract. 291
      I. Introduction. 292
      II. Materials and Methods 293
      III. Results 295
      IV. Discussion. 305
    References. 306

17 EFFECT OF CONCANAVALIN A AND PHYTOHEMAGGLUTININ ON
THE MODIFICATION OF IMMUNOGENICITY OF CANINE
KIDNEY ALLOGRAFTS
L. H. Toledo-Pereyra, C. O. Callender, P. K. Ray,
J. S. Najarian, and R. L. Simmons

    Abstract. 309
      I. Introduction. 309
      II. Materials and Methods 310
      III. Results 312
      IV. Discussion. 314
    References. 317

ABSTRACTS

EVIDENCE FOR CONFORMATIONAL CHANGES IN CONCANAVALIN A UPON
  BINDING OF SACCHARIDES AS DETERMINED FROM SOLVENT WATER
  PROTON MAGNETIC RELAXATION RATE DISPERSION MEASUREMENTS
  R. D. Brown, III, C. F. Brewer, and S. H. Koenig. 323

MAGNETIC RESONANCE STUDIES OF CONCANAVALIN A: LOCATION
  OF THE BINDING SITE OF METHYL-D-MANNOPYRANOSIDE
  B. H. Barber, A. Quirt, and J. P. Carver. 325

THE METAL ION REQUIREMENTS OF CONCANAVALIN A
  A. D. Sherry and A. Newman. 327

THE KINETICS OF CELLULAR COMMITMENT DURING STIMULATION
  OF LYMPHOCYTES BY CONCANAVALIN A
  G. R. Gunther, J. L. Wang, and G. M. Edelman. 329

ISOLATION OF A GLYCOPROTEIN RECEPTOR FOR CONCANAVALIN
A FROM THE OUTER SURFACE OF MOUSE L CELLS
R. C. Hunt, J. C. Brown, and C. M. Bullis 331

ELECTRON MICROSCOPIC STUDY ON INTERACTION OF CONCANAVALIN
A WITH MOUSE LYMPHOSARCOMA CELLS IN TISSUE CULTURE AND
IN ASCITES FORM
C. A. Feltkamp. 333

THE EFFECT OF GLUTARALDEHYDE FIXATION ON THE AGGLUTINATION
OF HUMAN ERYTHROCYTES BY CONCANAVALIN A AND SOYBEAN
AGGLUTININ
J. A. Gordon and M. D. Marquardt. 335

ALTERED NET CATION TRANSFER ACROSS THE EHRLICH MOUSE
ASCITES TUMOR CELL DURING EXPOSURE TO CONCANAVALIN A
F. Aull, M. S. Nachbar, and J. D. Oppenheim 337

EFFECTS OF CON A ON FROG NERVE AND MUSCLE
R. J. Person. 339

MODIFICATION OF THE SURFACE CHARACTERISTICS OF DEVELOPING
HEMOPOIETIC CELLS FROM NORMAL HUMAN BONE MARROW REVEALED
ULTRASTRUCTURALLY BY THE CONCANAVALIN A-PEROXIDASE-
DIAMINOBENZIDINE TECHNIQUE
G. A. Ackerman and S. D. Waksal 341

EFFECTS OF SUCCINYL-CON A ON THE GROWTH OF NORMAL AND
TRANSFORMED MOUSE FIBROBLASTS
D. A. Hilborn and I. S. Trowbridge. 343

ENDOTOXIN-LIKE ACTIVITIES IN CONCANAVALIN A PREPARATIONS
K. W. Brunson and D. W. Watson. 345

CONCANAVALIN A INDUCED INFLAMMATION
W. T. Shier 347

ENHANCED IMMUNOGENICITY OF CON A COATED EL-4 LEUKEMIA
CELLS
W. J. Martin, E. Esber, and J. R. Wunderlich. 349

List of Contributors 351

Index 357

1

CONCANAVALIN A: AN INTRODUCTION

Tushar K. Chowdhury
University of Oklahoma Health Sciences Center
Oklahoma City, Oklahoma, U. S. A.

I. BACKGROUND INFORMATION

During the last quarter of the nineteenth century and in the beginning of the twentieth century, bacteriologists were searching for agglutinins which could be used as antibacterial agents. Field and Teague (1907), for example, were working with typhoid agglutinins. Kolmer (1923) was working with various other types of antibacterial agglutinins. Sumner and his associates were the first group to isolate an agglutinin in relatively pure form (1936a). They were able to crystalize a lectin from the jack bean *(Canavalia ensiformis)* and referred to it as concanavalin fraction A. This particular fraction of the jack bean later became known as concanavalin A or simply as Con A. This Con A indeed had the property to agglutinate certain bacteria, such as *Mycobacterium* and *Actinomyces*.

What is it that makes this lectin so important that an international symposium was arranged to discuss its actions, which in turn led to the printing of the proceedings in this book? In the first place, during the past ten years this globular lectin was found to possess numerous interesting properties. Con A, for example, forms insoluble complexes with many polysaccharides (Goldstein *et al.*, 1965) and glycoproteins (Clarke and Denborough, 1971). This precipitation reaction of Con A has been utilized to assay many dextrans, lipopolysaccharides and other carbohydrate-containing substances (Goldstein *et al.*, 1965; Goldstein and Staub, 1970; So and Goldstein, 1968). Con A is a potent mitogen (Douglas *et al.*, 1969; Novogrodsky and Katchalski, 1971; Powell and Leon, 1970) and it stimulates the blast formation of lymphocytes (Perlman *et al.*, 1970). Con A also inhibits the migration of tumor cells in *in vitro* chambers (Friberg *et al.*, 1972). In some skin allograft experiments (Markowitz *et al.*,

1968) low dosages of Con A have been found to be immunosuppressive, thus prolonging allograft survival. This lectin has also been reported to inhibit both fertilization (Lallier, 1972) and phagocytic activity (Allan et al., 1971; Berlin, 1972). Interestingly enough, there have been several reports that Con A inhibits tumor cell growth (Burger and Noonan, 1970; Dent, 1971; Shoham et al., 1972). A recent review article by Nicolson (1974) lists some of the other effects of Con A as well as those of other lectins.

It is not possible to cover in a single volume all aspects of Con A research. Consequently, a limited number of topics has been selected for this volume. In selecting these topics, however, particular attention has been given not to exclude those topics over which some controversies have developed. After all, controversies in scientific research actually make science more interesting and challenging. As a matter of fact, there are several controversial issues in Con A research. The purpose of this symposium is not necessarily to resolve these controversies; rather, by assembling opinions from experts from various countries, our aim is to find out where we stand and in which direction we are heading in Con A research.

Some of the fundamental questions pertaining to Con A and posed in this volume are stated below.

1. What is the structure of Con A? How are the various properties of Con A related to its structure (in the crystalline phase, as well as in the solution phase)?

2. What is the nature of the specific interaction between Con A molecules and the cell membrane?

3. What are the effects of Con A on the physiological processes, e.g., cellular transport, synthesis and growth?

4. Are Con A effects specifically related to certain specific phases of the cell cycle?

5. What is its immunologic effect?

6. How can Con A be utilized best in understanding various fundamental processes in cell biology?

7. How does Con A affect cancer cells?

While some of these questions relate primarily to basic research, there are others which relate to more applied, health-related research. In understanding the effects of this lectin on biological systems, it is important that the physicochemical properties of this substance are well characterized. Consequently, the first few chapters of this book

INTRODUCTION

are devoted to a discussion of various physicochemical properties of Con A. The other portion of this book deals with the biological effects of Con A, with particular emphasis on its effect on cancer cells. It is hoped that the succeeding 16 chapters along with the abstracts of additional papers at the end of this book will provide welcome information regarding the current status of research with Con A.

II. PHYSICAL AND CHEMICAL PROPERTIES OF CONCANAVALIN A

Studies of the physical structure of the Con A molecule have been complicated by the pH-dependent dissociation and re-association of its subunits (McCubbin and Kay, 1971; Olson and Liener, 1967; Wang et al., 1971). As early as 1936, Sumner and Howell showed that the agglutinating power of Con A was lost in acid medium, but that the activity could be restored by addition of Ca^{++} and Mn^{++} at neutral pH. More recently it has been shown that treatment of Con A with ethylenediamine tetrachloroacetate (EDTA) destroys its saccharide-binding activity which can be regained by the addition of $MnCl_2$ and $CaCl_2$ (Uchida and Matsumoto, 1972; Yariv et al., 1968). The metal ion requirement of Con A has itself been extensively studied with the aid of nuclear magnetic resonance spectroscopy (Brewer et al., 1973; Sherry and Cottam, 1973). Several groups of investigators have been engaged in a study of the three-dimensional structure of Con A (Hardman et al., 1971; Quiocho et al., 1971). These X-ray diffraction studies of crystalline Con A have shown that it is a tetramer. The basic subunit has a molecular weight of 25,500 daltons and associates to form predominantly dimers in solution at pH 5 or less and tetramers above pH 7 (Kalb and Lustig, 1968; McCubbin and Kay, 1971). Each monomer binds one Mn^{++}, one Ca^{++} and one saccharide molecule. These metals are essential for saccharide-binding activity.

The second chapter in this volume is a contribution from the laboratory of Reeke et al., where the amino acid sequence and a three-dimensional structure of Con A are presented. These authors have discussed not only the interactions between the subunits, but also the possible loci of the saccharide-specific binding sites.

The saccharide-binding behavior of Con A in the crystalline state does not necessarily represent what actually happens in the solution phase. Indeed there is some experimental evidence suggesting that Con A behaves differently in solution phase than in crystalline phase (Chapters 2 through 4). The saccharide-binding specificity of Con A in solutions has been extensively investigated by Goldstein and his co-workers (Goldstein et al., 1965; Agrawal and Goldstein, 1968). They had demonstrated earlier that the reaction between Con A and carbohydrate-containing molecules is specific for α-D-mannopyranosyl and α-D-glucopyranosyl residues. Some of the most recent findings of Goldstein are presented in Chapter 3 where he discusses the specificity of interaction between Con A and carbohydrate containing sub-

stances. On the basis of some extensive physical and chemical
studies, Goldstein has suggested that the saccharide-specific binding
sites identified in Con A in the crystalline phase (reported in
Chapter 2) may not represent the actual site responsible for the
interaction between Con A and biological systems. This viewpoint is
somewhat strengthened by the results of some nuclear magnetic resonance studies (Brewer *et al.*, 1973; Hardman and Ainsworth, 1973)
which are discussed by Brewer *et al.*, in Chapter 3. Further behavior
of Con A in solution is being discussed by Sawyer in Chapter 4. This
chapter deals with the characteristics of various associated forms of
Con A. Such discussion is particularly useful when the action of Con
A on biological systems may depend upon the molecular forms and stability of this lectin. Chapter 6 is the last in the group of physicochemical studies of Con A and deals with the application of Con A in
the chemical purification of several glycoprotein enzymes. In summarizing their work on the binding specificity between Con A and glycoprotein Surolia and his colleagues have suggested that such specificity is not only dependent on the nature of the sugar present in
the glycoprotein, but also on the nature of the protein moiety.

III. BIOLOGICAL PROPERTIES OF CONCANAVALIN A

One of the biological properties of Con A which has been investigated more extensively than others is its cell-agglutinating ability.
This particular property was first recognized in 1936 when Sumner and
Howell reported that Con A can agglutinate erythrocytes in certain
animal species. Recent interest has been centered on its reactions
with other cells, particularly transformed cells and normal cells *in
vitro*. Unfortunately, not all studies of cell agglutination were
conducted under a common set of physicochemical conditions. Firstly,
the form of Con A was not the same in these studies. In some experiments the tetrameric form of Con A was used, whereas in other experiments the dimeric or monomeric form or even a fragmented, reaggregated
or a derivatized form of Con A was used. Secondly, the media in which
the cells were allowed to react with Con A ranged from culture medium
to simple salt solution. The ionic strength, temperature and pH of
the medium are likely to affect cell agglutination. We have indeed
shown that agglutination of some leukemic cells by Con A is significantly affected by the ionic strength and ionic contents of the solution (Killion and Chowdhury, 1972). Possible effects of some other
physical and chemical parameters of the reaction media have also been
implicated in Chapters 2 through 6. One additional factor deserves
some consideration in an evaluation of cell agglutination by Con A.
This is the fact that in any agglutination study the cells must be in
suspension. While a suspension preparation may be quite a suitable
environment for cells like erythrocytes and leukemic cells, for epithelial cells and fibroblasts the process of preparation of the
cellular suspension itself may be quite injurious to such cells,
particularly when the cells are detached from the substratum by

INTRODUCTION

treatment with trypsin. Variations in some of the above factors may thus add significantly to the complication of the phenomenon of cell agglutination as well as to the production of some variation in the experimental results.

Several studies have shown that cells transformed by RNA- or DNA-containing tumor viruses are agglutinated by Con A at concentrations which fail to agglutinate nontransformed cells (Inbar and Sachs, 1969a,b). However, following a brief trypsinization, nontransformed cells can be made susceptible to agglutination with low concentrations of Con A (Burger, 1969). There are, however, other reports indicating that there exist many normal cells which agglutinate as readily as transformed cells (Liske and Frank, 1968; Moscona, 1971). In addition there are several reports which suggest that normal cells infected with a wide variety of nononcogenic enveloped viruses become susceptible to agglutination with low concentrations of Con A (Becht et al., 1972; Birdwell and Strauss, 1973; Poste, 1972; Tevethia et al., 1972). In Chapter 7 Poste discusses the similarities in Con A-agglutination between transformed cells and normal cells infected with, but not transformed by, nononcogenic viruses. Some possible explanation of this Con A-agglutination is also offered. Finally, modification of the viral cytopathogenicity by Con A is also discussed.

Let us next turn to those cell lines where the transformed cells have been found to agglutinate more than their nontransformed counterparts (Burger, 1969; Inbar and Sachs, 1969). One obvious question which can be asked is this: Is there a quantitative difference in Con A binding between the transformed and nontransformed cells? This particular question has generated significant interest among scientists engaged in Con A research. At first it was thought that transformed cells have more Con A binding sites on their surface than normal cells (Inbar and Sachs, 1969b; Noonan and Burger, 1973). In a number of subsequent investigations, ^{3}H- or ^{125}I-labelled Con A was used to estimate the Con A binding sites. The results indicate very little or no difference between the number of Con A binding sites on the surfaces of normal and transformed cells (Arndt-Tovin and Berg, 1971; Cline and Livingston, 1971; Inbar et al., 1971; Nicolson, 1973; Phillips et al., 1974; Rosenblith et al., 1973). At this juncture, an obvious question must be posed: Is there a difference in the distribution of Con A binding sites between normal and tumor cell surfaces? Nicolson has conducted an extensive investigation of this particular question and his findings strongly indicate that, while the number of Con A binding sites (CABS) may be the same for the normal and transformed cells, the surface distribution of these binding sites is different in each of the two categories of cells studied. The CABS on the transformed cell surface are arranged in clusters, while those on the normal cell surface are randomly dispersed. A detailed discussion of Nicolson's own results, as well as those of other investigators, is presented in Chapter 8. The following chapter also concerns itself with the topographical distribution of the CABS. Here Berlin concerns himself more with the mechanisms which may control the distribution of CABS. He has suggested that cellular microtubular proteins may play some important role in the topographical distribution of CABS.

This leads directly to Chowdhury's chapter where the primary emphasis has been placed on studying transient as well as long-term effects of Con A in normal as well as tumor cells. The specific parameters covered in Chapter 10 are cellular motility, transport activity and some associated electrophysiological properties of cells. It has been shown that, even though Con A-treated tumor cells do not acquire the morphological and physiological properties of normal cells, the effects of Con A are specifically exhibited by tumor cells and these effects are not of a permanent nature. The mediator of these effects is postulated to be a Con A-specific contractile membrane component.

Another interesting aspect of Con A research has centered around the question of whether or not this lectin can affect cellular growth, particularly of tumor cells. One of the growth characteristics of cells grown *in vitro* which has drawn special interest is the so-called contact-inhibited movement or contact-inhibited growth. There are numerous reports indicating that normal cells, which can grow by attaching themselves to a plastic or glass substratum, proliferate until a confluent monolayer is reached. On the other hand, tumor cells in general lack this contact-inhibited movement and contact-inhibited growth. Consequently, when compared with normal cells, tumor cells have a higher saturation density, i.e., a greater number of cells per unit area of the culture vessel on which the cells are grown. There are, of course, some exceptions to this finding and these have been discussed in Chapters 8 and 10.

Interestingly enough, it has been reported that binding to certain transformed cells of monovalent Con A, obtained by limited tryptic digestion of Con A, results in the regaining of contact-inhibited growth (Burger and Noonan, 1970). Moreover, Pollack and Burger (1969) have claimed a good correlation between the *in vitro* saturation density of cells and their agglutinability by the lectin. The higher the saturation density, the higher the agglutinability. Aaronson and Todaro (1969), on the other hand, have shown that the higher the *in vitro* saturation density, the greater the *in vivo* tumorigenicity. Thus it appears that a Con A-detectable cell surface configuration is associated both with uncontrolled cell growth and with tumorigenicity. In addition, Fox *et al.* (1971) have reported that nontransformed cells in interphase bind very little fluorescent-labelled Con A, while transformed cells are highly fluorescent. Nontransformed cells in mitosis, however, bind the same amounts of Con A as do the transformed cells (Noonan and Burger, 1973). Moreover, the Con A-agglutinability of these nontransformed mitotic cells is similar to that of the

INTRODUCTION

transformed cells. The Con A-detectable cell surface configuration is not a peculiar property of transformed cells, since it is also present in nontransformed mitotic cells. Burger and his colleagues have observed that mild proteolytic treatment of the surfaces of nontransformed cells in a contact-inhibited monolayer results in an additional round of cell division and also in a susceptibility to agglutination by Con A (Burger, 1969, 1970). In Chapter 11, Mannino and Burger review some of their recent data on the growth inhibitory effect of a derivatized form of Con A, i.e., succinylated Con A. These authors propose a lectin-sensitive regulatory mechanism of cell division. In Chapter 12 Collard et al. add further details to the discussion of the regulation of cell division in both normal and tumor cells. Furthermore, they show that agglutinability of cells with Con A is minimal during the G_2 phase of the cell cycle, but maximal during mitosis and the G_1 phase. Chapters 13 through 15 illustrate some of the uses of Con A in a wide variety of problems, ranging from an assay of biochemical reactions to the nature of the specificity of antigen-antibody type reactions. In Chapter 16 Bradley and his colleagues discuss the cytotoxicity of Con A in animals. The final chapter of this book deals with the immunosuppressive effect of Con A.

The findings presented here obviously do not constitute complete or final answers to the questions posed in the beginning of this chapter. Nevertheless, it is quite clear that significant progress has been made toward achieving the answers. Most importantly, several aspects of the Con A research discussed in this book open new approaches toward unravelling some of the fundamental processes associated with the growth of a living cell, be it normal or tumorous.

REFERENCES

1. Aaronson, S., and Todaro, G. (1968). "Basis for the acquisition of malignant potential by mouse cells cultivated in vitro." *Science* 162,1024

2. Agrawal, B. B. L., and Goldstein, I. J. (1968). "Protein-carbohydrate interaction. VII. Physical and chemical studies on concanavalin A, the hemagglutinin of the jack bean." *Arch. Biochem. Biophys.* 124,218.

3. Allan D., Auger, J., and Crumpton, M. J. (1971). "Interaction of phytohemagglutinin with plasma membranes of pig lymphocytes and thymus cells." *Exptl. Cell Res.* 66,362.

4. Arndt-Jovin, D. J., and Berg, P. (1971). "Quantitative binding of ^{125}I-concanavalin A to normal and transformed cells." *J. Virol.* 8,716.

5. Becht, H. A., Rott, R., and Klenk, H. D. (1972). "Effect of concanavalin A on cells infected with enveloped RNA viruses." *J. Gen. Virol.* 14,1.

6. Berlin, R. D. (1972). "Effect of concanavalin A on phagocytosis." *Nature New Biol.* 235,44.

7. Birdwell, C. R., and Strauss, J. H. (1973). "Agglutination of *Sindbis* virus and of cells infected with *Sindbis* virus by plant lectins." *J. Virol.* 11,502.

8. Brewer, C. F., Sternlicht, H., Marcus, D. M., and Grollman, A. P. (1973). "Binding of ^{13}C-enriched α-methyl-D-glycopyranoside to concanavalin A as studied by carbon magnetic resonance." *Proc. Nat. Acad. Sci. U. S. A.* 70,1007.

9. Burger, M. M. (1969). "A difference in the architecture of the surface membrane of normal and virally transformed cells." *Proc. Nat. Acad. Sci. U. S. A.* 62,994.

10. Burger, M. M. (1970). "Proteolytic enzymes initiating cell division and escape from contact inhibition of growth." *Nature* 227,170.

11. Burger, M. M., and Noonan, K. D. (1970). "Restoration of normal growth by covering of agglutinin sites on tumour cell surface." *Nature* 228,512.

12. Clarke, A. E., and Denborough, M. A. (1971). "The interaction of concanavalin A with blood-group-substance glycoproteins from human secretions." *Biochem. J.* 121,811.

13. Cline, M. J., and Livingston, D. C. (1971). "Binding of ^{3}H-concanavalin A by normal and transformed cells." *Nature New Biol.* 232,155.

14. Dent, P. B. (1971). "Inhibition by phytohemagglutinin of DNA synthesis in cultured mouse lymphomas." *J. Natl. Cancer Inst.* 46,763.

15. Douglas S. D., Kamin, R. M., and Fudenberg, H. H. (1969). "Human lymphocyte response to phytomitogens *in vitro*: normal, agammaglobulinemic and paraproteinemic individuals." *J. Immunol.* 103, 1185.

16. Field, C. W., and Teague, O. (1907). "On the electrical charge of the native proteins and the agglutinins." *J. Exptl. Med.* 9, 222.

INTRODUCTION

17. Fox, T., Sheppard, J., and Burger, M. (1971). "Cyclic membrane changes in animal cells: Transformed cells permanently display a surface architecture detected in normal cells only during mitosis." *Proc. Nat. Acad. Sci. U. S. A.* $\underline{68}$,244.

18. Friberg, S., Golub, S. H., Lilliehook, B., and Cochran, A.J. (1972). "Assessment of concanavalin A reactivity to murine ascites tumours by inhibition of tumour cell migration." *Exptl. Cell Res.* $\underline{73}$,101.

19. Goldstein, I. J., and Staub, A. M. (1970). "Interaction of concanavalin A with polysaccharides of Salmonellae." *Immunochemistry* $\underline{7}$,315.

20. Goldstein, I. J., Hollerman, C. E., and Merrick, J. M. (1965). "Protein-carbohydrate-interaction. I. The interaction of polysaccharides with concanavalin A." *Biochim. Biophys. Acta* $\underline{97}$,68.

21. Hardman, K. D., and Ainsworth, C. F. (1973). "Binding of nonpolar molecules by crystalline concanavalin A." *Biochemistry* $\underline{12}$,4442.

22. Hardman, K. D., Wood, M. K., Schiffer, M., Edmundson, A. B., and Ainsworth, C. F. (1971). "Structure of concanavalin A at 4.25 Ångstrom resolution." *Proc. Nat. Acad. Sci. U. S. A.* $\underline{68}$,1393.

23. Inbar, M., and Sachs, L. (1969a). "Interaction of the carbohydrate-binding protein concanavalin A with normal and transformed cells." *Proc. Nat. Acad. Sci. U. S. A.* $\underline{63}$,1418.

24. Inbar, M., and Sachs, L. (1969b). "Structural difference in sites on the surface membrane of normal and transformed cells." *Nature* $\underline{223}$,710.

25. Inbar, M., Ben-Bassat, H., and Sachs, L. (1971). "A specific metabolic activity on the surface membrane in malignant cell transformation." *Proc. Nat. Acad. Sci. U. S. A.* $\underline{68}$,2748.

26. Kalb, A. J., and Lustig, A. (1968). "The molecular weight of concanavalin A." *Biochim. Biophys. Acta* $\underline{168}$,366.

27. Killion J. J., and Chowdhury, T. K. (1972). "Some physical parameters affecting the agglutination of L-5178Y leukemic cells with concanavalin A." Unpublished data.

28. Kolmer, J. A. (1923). *Infection, immunity and biologic therapy.* 3rd ed., W. B. Saunders Company, Philadelphia, Pa.

29. Lallier, R. (1972). "Effects of concanavalin A on the development of sea urchin egg." *Exptl. Cell Res.* $\underline{72}$,157.

30. Lin, H., and Bruce, W. R. (1971). "Selective action of concanavalin A on transplanted tumors of the mouse." *Proc. Amer. Assoc. Cancer Res.* <u>12</u>,66.

31. Liske, R., and Franks, D. (1968). "Specificity of the agglutinin in extracts of wheat germ." *Nature* <u>217</u>,860.

32. Markowitz, H., Person, D. A., Gitnick, G. L., and Ritts, R. E. (1969). "Immunosuppressive activity of concanavalin A." *Science* <u>163</u>,476.

33. McCubbin, W. D., and Kay, C. M. (1971). "Molecular weight studies on concanavalin A." *Biochem. Biophys. Res. Commun.* <u>44</u>, 101.

34. Moscona, A. A. (1971). "Embryonic and neoplastic cell surface: Availability of receptors for concanavalin A and wheat germ agglutinin." *Science* <u>171</u>,905.

35. Nicolson, G. L. (1974). "The interactions of lectins with animal cell surfaces." *Int'l Rev. Cytol.* In Press.

36. Nicolson, G. L. (1973). "Neuraminidase unmasking and the failure of trypsin to unmask β-D-galactose-like sites on erythrocytes, lymphoma and normal and virus-transformed fibroblast cell membranes." *J. Natl. Cancer Inst.* <u>50</u>,1443.

37. Noonan, K. D., and Burger, M. M. (1973). "Binding of ^{3}H-concanavalin A to normal and transformed cells." *J. Biol. Chem.* <u>248</u>,4286.

38. Novogrodsky, A., and Katchalski, E. (1971). "Lymphocyte transformation induced by concanavalin A and its reversion by methyl-alpha-D-mannopyranoside." *Biochim. Biophys. Acta* <u>228</u>,579.

39. Olson, M. O. J., and Liener, I. E. (1967). "The association and dissociation of concanavalin A, the phytohemagglutinin of the jack bean." *Biochemistry* <u>6</u>,3801.

40. Perlmann, P., Nilsson, H., and Leon, M. A. (1970). "Inhibition of cytotoxicity of lymphocytes by concanavalin A *in vitro*." *Science* <u>168</u>,1112.

41. Phillips, P. G., Furmanski, P., and Lubin, M. (1974). "Cell surface interactions with concanavalin A: Location of bound radiolabeled lectin." *Exptl. Cell Res.* <u>86</u>,301.

42. Pollack, R., and Burger, M. (1969). "Surface-specific characteristics of a contact-inhibited cell line containing the SV40 genome." *Proc. Nat. Acad. Sci. U. S. A.* <u>62</u>,1074.

43. Poste, G. (1972). "Changes in susceptibility of normal cells to agglutination by plant lectins following modification of cell coat material." *Exptl. Cell Res.* 73,319.

44. Powell, A. E., and Lion, M. A. (1970). "Reversible interaction of human lymphocytes with the mitogen concanavalin A." *Exptl. Cell Res.* 62,315.

45. Quiocho, F. A., Reeke, G. N., Becker, J. W., Lipscomb, W. N., and Edelman, G. M. (1971). "Structure of concanavalin A at 4 Å resolution." *Proc. Nat. Acad. Sci. U. S. A.* 68,1853.

46. Rosenblith, J. F., Ukena, T. E., Yin, H. H., Berlin, R. D., and Karnowsky, M. J. (1973). "A comparative evaluation of the distribution of concanavalin A binding sites on the surfaces of normal, virally-transformed and protease-treated fibroblasts." *Proc. Nat. Acad. Sci. U. S. A.* 70,1625.

47. So, L. L., and Goldstein, I. J. (1968). "Protein-carbohydrate interaction. XIII. The interaction of concanavalin A with α-mannans from a variety of microorganisms." *J. Biol. Chem.* 243, 2003.

48. Sharon, N., and Lis, H. (1972). "Lectins: Cell-agglutinating and sugar-specific proteins." *Science* 177,949.

49. Sherry, A. D., and Cottam, G. L. (1973). "Protein relaxation rate and fluorometric studies of manganese and rare earth binding to concanavalin A." *Arch. Biochem. Biophys.* 156,665.

50. Shoham, J., Inbar, M., and Sachs, L. (1970). "Differential toxicity on normal and transformed cells *in vitro* and inhibition of tumour development *in vivo* by concanavalin A." *Nature* 227, 1244.

51. Sumner, J. B., and Howell, S. F. (1936). "The role of divalent metals in the reversible inactivation of jack bean hemagglutinin." *J. Biol. Chem.* 115,583.

52. Sumner, J. B., and Howell, S. F. (1936a). "The identification of the hemagglutinin of the jack bean with concanavalin A." *J. Bacteriol.* 32,227.

53. Tevethia, S. S., Lowry, S., Rawls, W. E., Melmick, J. L., and McMillan, V. (1972). "Detection of early cell surface changes in *Herpes simplex* virus-infected cells by agglutination with concanavalin A." *J. Gen. Virol.* 15,93.

54. Uchida, T., and Matsumoto, T. (1972). "Heterogeneity of commercially available concanavalin A with respect to carbohydrate-

binding ability." *Biochim. Biophys. Acta* 257,230.

55. Wang, J. L., Cunningham, B. A., and Edelman, G. M. (1971). "Unusual fragments in the subunit structure of concanavalin A." *Proc. Nat. Acad. Sci. U. S. A.* 69,608.

56. Yariv, J., Kalb, A. J., and Levitake, A. (1968). "The interaction of concanavalin A with methyl alpha-D-glucopyranoside." *Biochim . Biophys. Acta* 165,303.

2

STRUCTURE AND FUNCTION OF CONCANAVALIN A*

George N. Reeke, Jr., Joseph W. Becker, Bruce A.
  Cunningham, John L. Wang, Ichiro Yahara and Gerald
  M. Edelman
The Rockefeller University
New York, New York, U. S. A.

ABSTRACT

Lectins have been extensively used to analyze a variety of fundamental processes in cell biology. In conjunction with our studies on the cell surface and mitosis, we have determined the amino acid sequence and three-dimensional struction of concanavalin A (Con A), the mitogenic lectin from the jack bean. Knowledge of the structure has been helpful in interpreting experiments on lymphocyte mitogenesis and the effects of Con A on cell surface receptor mobility.

Con A subunits of molecular weight 25,500 are folded into dome-like structures of maximum dimensions 42 x 40 x 39 Å. The domes are related by 222 symmetry to form roughly tetrahedral tetramers. Each subunit contains two large antiparallel pleated sheets, and subunits are joined to form dimers and tetramers by interactions involving one of these pleated sheets. We have examined the binding of a variety of carbohydrates to Con A and have obtained preliminary data which suggest that there are differences in the saccharide-binding behavior of Con A in solution and in the crystalline state.

Dimeric chemical derivatives of Con A have been prepared and shown to have biological activities different from those of the

---

*This work was supported by grants from the National Institutes of Health, the National Science Foundation, and a Teacher-Scholar Grant from the Camille and Henry Dreyfus Foundation. J. L. Wang is a Fellow of the Damon Runyon Memorial Fund for Cancer Research.

native tetrameric protein. Under different conditions, native Con A exhibits two antagonistic activities on the lymphoid cell surface: the induction of cap formation by its own receptors and the inhibition of the mobility of a variety of receptors, including its own receptors. The dimeric derivative, succinyl-Con A, is just as effective a mitogen as the native lectin, but it lacks the ability to modulate cell surface receptor mobility. The data suggest that neither extensive immobilization of cell surface receptors nor cap formation is required for cell stimulation. Further studies on modulation of receptor translocation suggest the hypothesis that there exists a connecting network of colchicine-sensitive proteins that links receptors of different kinds and mediates their rearrangement. The degree of connectivity of this postulated network appears to be altered by changes in the state of attachment of various surface receptors to the network. Thus the network might provide the cell with a means of transmitting signals such as the stimulus for mitosis by lectins or antigens.

I. INTRODUCTION

Concanavalin A (Con A) is the most widely used of the plant agglutinins for research in cell biology. Con A was first isolated from the jack bean by Sumner (1919), and was subsequently shown to agglutinate erythrocytes (Sumner and Howell, 1935) and to precipitate various polysaccharides (Sumner and Howell, 1936a). Since the observation that transformed fibroblasts are more readily agglutinated by Con A than the corresponding untransformed cells (Inbar and Sachs, 1969), there have been extensive studies of the differences in number, distribution, and mobility of Con A receptors on the surfaces of normal and transformed cells in culture (Ozanne and Sambrook, 1961; Nicolson, 1971; Inbar *et al.*, 1972; Rosenblith *et al.*, 1973; DePetris *et al.*, 1973), and attempts have been made to correlate the properties of the transformed cell membrane with those of the normal cell during mitosis (Fox *et al.*, 1971).

Our interest in Con A has centered on its interactions with lymphoid cells. Con A binds to the lymphocyte surface (Powell and Leon, 1970), where it can be visualized, for example, by fluorescence microscopy of fluorescein-Con A conjugates (Unanue *et al.*, 1972) or electron microscopy of ferritin-labelled Con A (Nicholson and Singer, 1971). Under appropriate conditions, Con A in low doses first forms patchy distributions, and then so-called caps (Unanue *et al.*, 1972; Gunther *et al.*, 1973). Higher concentrations of Con A inhibit formation of patches and caps, resulting in a diffuse distribution of Con A or immunoglobulin receptors (Yahara and Edelman, 1972; 1973a). Lymphocytes cultured with Con A undergo blast transformation and mitosis (Powell and Leon, 1970; Beckert and Sharkey, 1970), and the stimulation of individual cells can be assayed by autoradiography of incorporated radioactive thymidine (Gunther et al., 1974). Con A-induced mitogenesis resembles the stimulation of immune cells by specific antigens, and Con A, unlike antigen, stimulates a large percentage of lymphocytes. These properties make it possible to study stimulation independent of many of the complications of the immune system.

The biological properties of Con A can be inhibited by specific saccharides such as α-methyl-mannoside, and these properties are thus assumed to involve binding of Con A to glycoprotein or glycolipid receptors on the cell surface. The saccharide-binding specificity of Con A has been extensively investigated by Goldstein and coworkers (Goldstein et al., 1965; So and Goldstein, 1967, 1969; Poretz and Goldstein, 1970; Goldstein et al., 1974), who have shown that Con A binds to non-reducing terminal D-mannopyranosyl and D-glucopyranosyl residues, and to certain nonterminal mannopyranosyl residues.

In order to make the best use of Con A as a reagent in biological experiments, one would like to know what specific features of the molecule are responsible for its various activities. It is therefore necessary to know the structure of the lectin in some detail. Accordingly, we have determined the complete amino acid sequence and three-dimensional structure of Con A (Edelman et al., 1972). In addition, we have chemically derivatized the protein and determined the effects of the resulting structural modifications on the activities of Con A (Gunther et al., 1973). These studies have led us to hypothesize the presence of an intracellular network of colchicine-binding proteins to link cell surface receptors of different kinds and to mediate their mobility and redistribution (Edelman et al., 1973).

II. STRUCTURE OF CONCANAVALIN A

Con A is a subunit protein composed of identical subunits of molecular weight 25,500 (Wang et al., 1971). The subunits associate to form predominantly dimers below pH 6 and tetramers about pH 7 (Kalb and Lustig, 1968). Low-resolution crystallographic studies (Reeke et al., 1971; Quiocho et al., 1971; Hardman et al., 1971a, b) showed that the protomers are dome or gumdrop-shaped and are approximately 42 Å high x 40 Å wide x 39 Å thick (Figure 1). They are a bit thinner - about 25 Å - at the bases. Monomers are paired base-to-base by an exact twofold symmetry axis, perpendicular to the plane of the page in Figure 1, to form ellipsoidal dimers of about 84 Å x 40 Å. The dimers are in turn paired by additional twofold axes to form roughly tetrahedral tetramers, as shown schematically in Figure 1. Each monomer binds one Mn^{++} ion, one Ca^{++} ion, and one saccharide molecule, and the metals are necessary for saccharide-binding activity (Sumner and Howell, 1936b; Kalb and Levitski, 1968). In

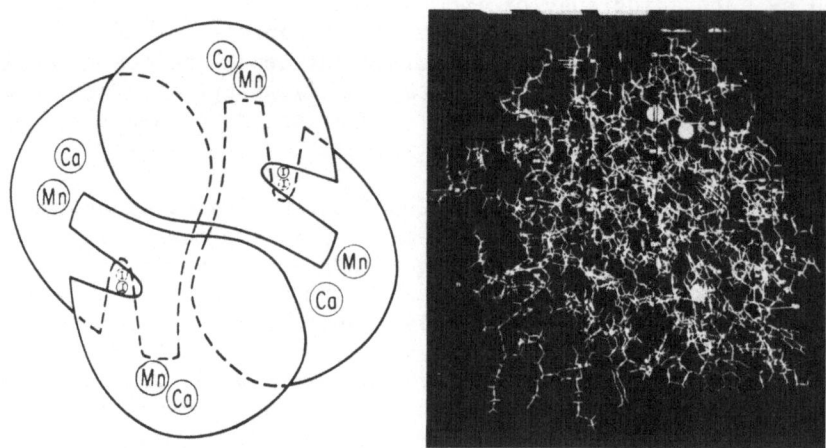

Fig. 1. Composite view of Con A as seen at low and high resolution. (Left) Schematic view of the Con A tetramer, illustrating 222 symmetry of the molecule. Binding sites for Ca^{++}, Mn^{++}, and β-IPG are indicated by Ca, Mn, and I respectively. Largest dimension of the molecule is about 84 Å. (Right) Kendrew atomic model of a single protomer. Binding sites for Ca^{++} and Mn^{++} (top) and β-IPG (lower right) are represented by large white spheres.

our low-resolution electron density maps, the molecular surface appeared to be relatively smooth and uninterrupted except for a large depression in the surface of each protomer, and it was surmised that these depressions might contain the carbohydrate-binding sites (Reeke et al., 1971; Quiocho et al., 1971). Later it was shown by difference Fourier techniques that an inhibitor of Con A activity, β-(o-iodophenyl)-D-glucopyranoside (β-IPG), is in fact bound in these cavities (Becker et al., 1971) as indicated by the I's in Figure 1. The question has been raised whether β-IPG is bound primarily by its saccharide moiety or by its phenyl group, and thus there is currently some doubt whether the cavities are the actual saccharide-specific binding sites of Con A. This question will be discussed in some detail below.

The detailed structure (Edelman et al., 1972; Hardman and Ainsworth, 1972) was worked out from the X-ray data and from the amino acid sequence (Figure 2). The intact protomer contains 237 amino acid residues. However, as usually isolated, some Con A molecules contain subunits with an internal cleavage (Wang et al., 1971), indicated by the wavy arrow between residues 118 and 119 in Figure 2. Interestingly, there are no obvious differences between the three-dimensional structures or biological activities of Con A tetramers made up entirely of intact subunits and of tetramers con-

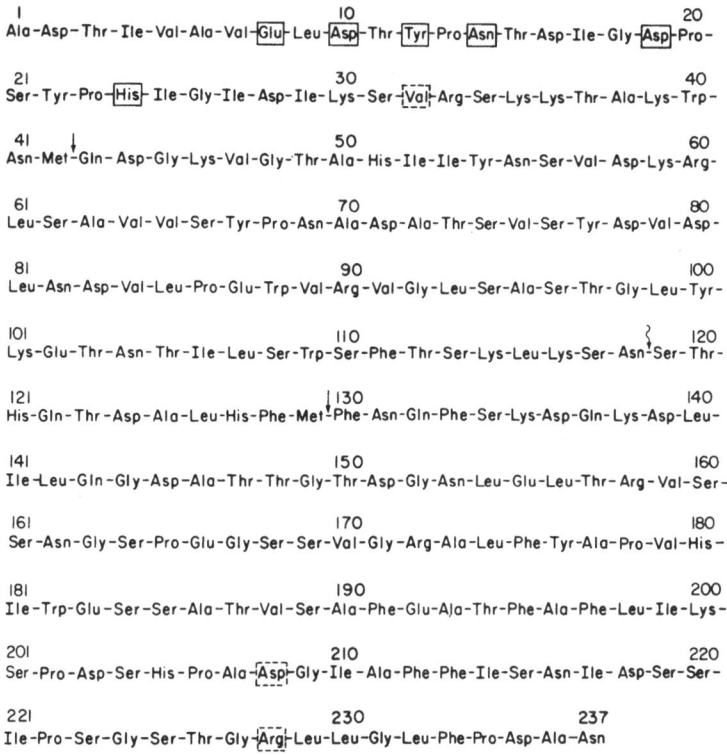

Fig. 2. Amino-acid sequence of Con A. Straight vertical arrows indicate the positions of methionine residues; wavy arrow denotes the position at which some molecules are naturally cleaved. Residues in boxes serve as direct ligands for metal ions; residues in dashed boxes are hydrogen-bonded to water molecules which are metal ligands.

taining the fragments. The amino-terminal portion of the polypeptide chain contains an acidic region in which most of the ligands for the metal ions, indicated by boxes in Figure 2, are located. The residues in dashed boxes appear to be hydrogen-bonded to water molecules that in turn serve as ligands for the metal ions.

The geometry of the metal-protein complex is shown in Figure 3. The metals are bound near the top of the protomer in the orientation seen in Figure 1. Each metal is surrounded by six ligands which form an octahedral coordination shell. In each case, four of the ligands are from the protein and two are water molecules. Two of the protein ligands, aspartic acid 10 and aspartic acid 19, are common to both metals. The whole arrangement thus forms a binuclear complex of two octahedra sharing a common edge. One of the water

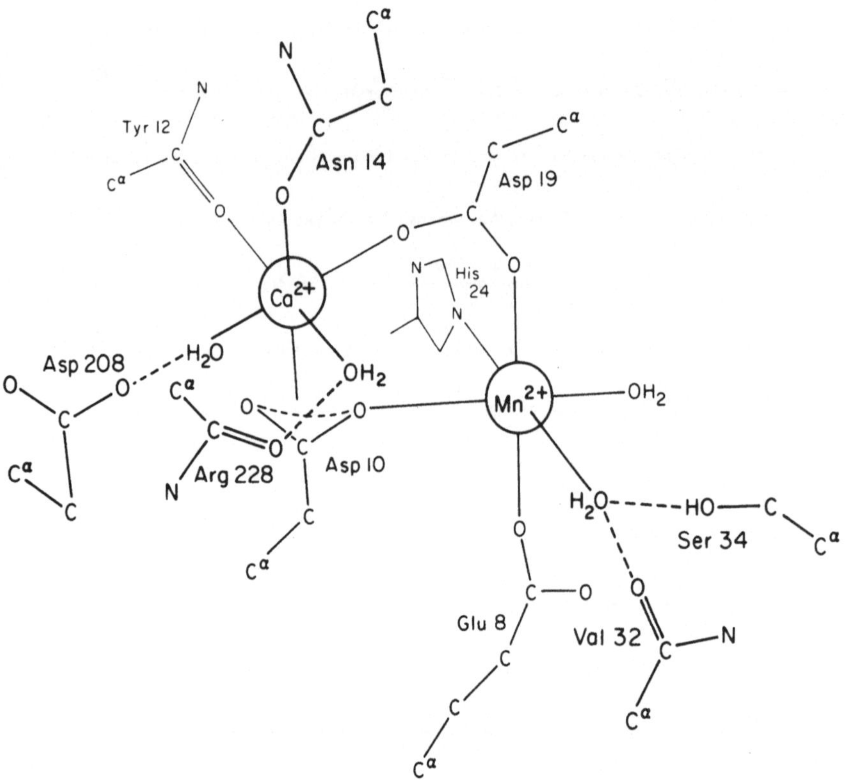

Fig. 3. Schematic drawing of the metal-binding region of Con A. Mn^{++} and Ca^{++} ions are both approximately octahedrally coordinated by four protein ligands and two water molecules. Mn^{++} ligands are Glu 8, Asp 10, Asp 19, and His 24; Ca^{++} ligands are Asp 10, the backbone carbonyl of Tyr 12, Asn 14, and Asp 19. Asp 10 and Asp 19 are ligands of both metals. Reproduced by permission of the New York Academy of Sciences.

molecules bound to Mn^{++} is hydrogen-bonded to the protein, and the other is in a shallow channel exposed to the surface. The Ca^{++} coordination is less symmetrical than the Mn^{++}, but it is still definitely octahedral. As with the Mn^{++}, all the protein ligands come from the acidic part of the sequence near the amino-terminus, but in the case of the Ca^{++}, the two water molecules are hydrogen-bonded to groups near the carboxyl-terminus.

The geometry of the metal-binding complex appears to explain some aspects of the chemistry of metal-binding to Con A. First, treatment of Con A with acid removes the metals (Sumner and Howell, 1936b) and this behavior is an expected consequence of protonating the carboxylic acid ligands. Second, it is known that Mn^{++} must be present in order for Ca^{++} to bind (Kalb and Levitzki, 1968). This requirement may be related to the fact that all the Mn^{++} ligands come from near the amino-terminus of the chain, whereas Ca^{++} is bound via two waters to the opposite end of the chain. Perhaps binding of Mn^{++} is necessary to bring the amino-terminal part of the chain, especially the shared ligands 10 and 19, into the appropriate conformation to create part of the Ca^{++} binding site. Calcium binding would in turn stabilize the native conformation of the carboxyl terminal portion of the chain.

III. SACCHARIDE BINDING FUNCTION OF CONCANAVALIN A

The saccharide-binding activity of Con A, which appears to be involved in the biological effects of the protein, is perhaps the most interesting feature of the molecule. Observation of the binding interactions in atomic detail would permit an explanation of the observed saccharide-binding specificity, and a more precise analysis of the interaction of Con A with cell surfaces than has yet been possible. Unfortunately, saccharide binding has proven to be very resistant to crystallographic analysis. Treatment of crystals with high concentrations of inhibitory sugars results in either dissolution of the crystals or loss of the diffraction pattern. This effect is probably related to a conformational change associated with saccharide binding, and such a change has, in fact, been observed by spectroscopic methods (Pflumm *et al.*, 1971).

We have made high resolution difference electron density maps from crystals treated with the highest concentrations of several saccharides that leave measurable diffraction patterns, but only one of these compounds, β-IPG, yields a map that shows anything that can be positively identified as a bound molecule. Figure 4 shows the opening in the surface of the molecule that leads to the β-IPG binding site. The opening is extremely narrow, e.g., only about 3.5 to 6 Å wide, 7.5 Å high at the narrowest point, and 18 Å deep, widening somewhat at the back. The iodine atoms of β-IPG, represented by the large ball in Figure 4, binds at the back of the cavity, about 15 Å from the surface of the protein and 20 Å from the Mn^{++} ion. The iodine-binding part of the cavity is a highly hydrophobic region which is capable in the crystal of binding such compounds as iodophenol. The saccharide part of β-IPG is probably bound farther out toward the opening of the cavity, where the side chains of the protein form a more hydrophilic subsite. Such a picture is consistent with the protein structure in model-building experiments, and is also consistent with equilibrium binding studies (Becker *et al.*, 1974) which show that Con A has one binding site per monomer for β-IPG in solution.

There is, however, some evidence which indicates that the β-IPG

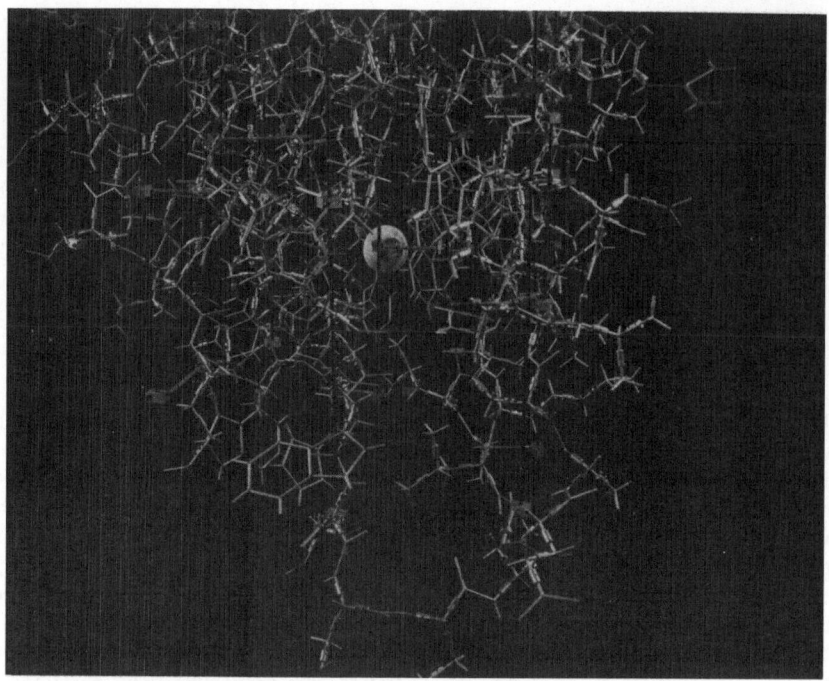

Fig. 4. View of the Kendrew model showing the cavity where β-IPG is bound. View is from the right of the model as seen in Fig. 1. The large ball represents the iodine atom of β-IPG. Back β-structure is at the right (see Fig. 5).

binding site observed in the crystal may not be the same as the inhibitory saccharide-binding site in solution. ^{13}C-nuclear magnetic resonance studies have placed the saccharide ring carbons of bound α-methyl-glucoside and β-IPG in nearly identical positions, 10 to 12 Å from the Mn^{++} ion of Con A (Brewer et al., 1973a, b; Villafranca and Viola, 1974; Brewer et al., 1975), in conflict with the crystallographic results. Barber et al. (1975) on the other hand, have concluded from proton magnetic resonance studies that their data are only consistent with the site located crystallographically for β-IPG. Another relevant evidence is the finding that in addition to β-IPG, various analogous compounds such as o-iodophenol (Becker et al., 1974) and β-(o-iodophenyl)-D-galactopyranoside (β-IPGal) (Hardman and Ainsworth, 1973) are bound in the β-IPG-binding cavity. Inasmuch as these compounds do not contain the glucoside or mannoside moiety usually associated with Con A binding, it appears that the binding specificity of the protein is less selective in the crystalline state. In addition, preliminary experiments (Becker et al., 1974) suggest that these compounds are not detectably bound by Con A in solution. This fact implies that there are differences in the binding behavior of

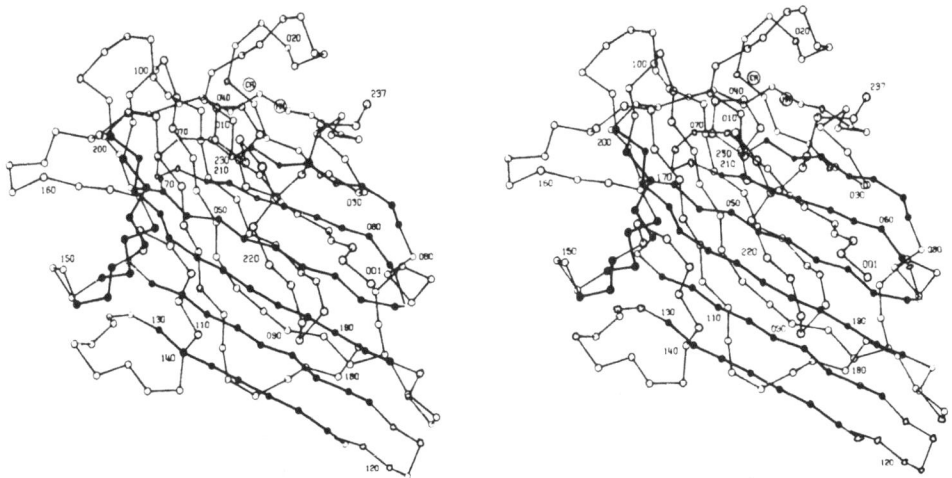

Fig. 5. Stereo drawing of the polypeptide chain of a Con A protomer, oriented as in Fig. 1 (right). Small circles represent α-carbons. The back β-structure, which participates in the formation of dimers and tetramers is emphasized by heavy black lines. The loop between the two lowest chains at bottom right contains the site of the natural cleavage in Con A between residues 118 and 119.

Con A in solution and in the crystal. Either the true binding site is in the cavity, but its specificity is modified on crystallization, or else the solution binding site does not exist in the crystal, and the cavity seen in the crystal is not active in solution. The present evidence is not sufficient to distinguish these two possibilities, but it seems more likely that the site is in the cavity, and that its activity is modulated, but not entirely changed, by the process of crystallization. Until the location of the site can be definitely established by location of the individual atomic positions of the saccharide, great care should be exercised in interpreting studies of the binding carried out by various physical methods.

IV. MULTIVALENT NATURE OF CON A

Some studies (Gunther et al., 1973) have indicated the importance of the multiple valence of Con A in its biological activities. Multiple valence derives from the tetrameric structure of the protein, which in turn can be understood in terms of certain basic features of the protomer structure. The most striking structural feature of the molecule is the presence of extensive β-structure or pleated sheets. Each monomer contains two large entirely antiparallel pleated sheets, comprising in all more than half of the residues in the molecule. The first of these sheets has seven chains

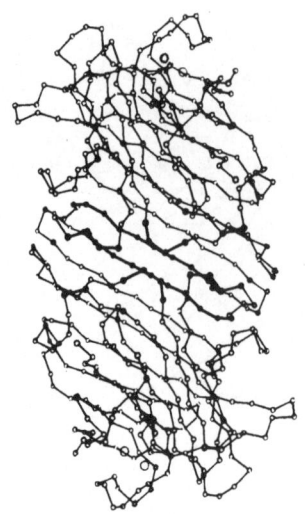 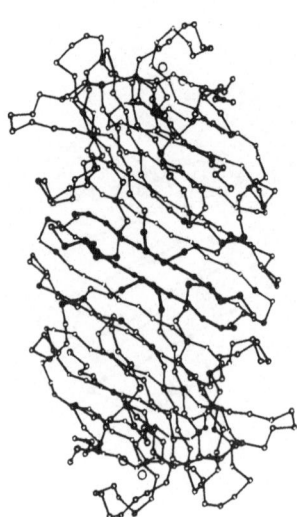

Fig. 6. Stereo backbone drawing of a Con A dimer. Residues participating in non-covalent interactions between the two protomers are emphasized in heavy black.

running through the molecule from back to front. The plane of the sheet is twisted, somewhat like the β-structures in carboxypeptidase A and carbonic anhydrase, and it divides the rest of the molecule into unequal randomly coiled regions on the left and right. The other β-structure is located at the back of the molecule (Figure 5), where it plays a major role in the interactions giving rise to the formation of dimers and tetramers. This sheet contains 6 segments of polypeptide chain. It is not twisted like most pleated sheets, but is somewhat bowed out of plane at the top. The naturally-occurring cleavage between residues 118 and 119 of some molecules is in a loop between the bottom two strands of this β-structure (Figure 5), and it seems reasonable to expect that the loss of one covalent bond in such a highly stabilized part of the molecule would not disrupt the overall structure in any major way.

In the dimer, the second monomer is related to the first by a twofold axis (bottom center, Figure 5) to give the arrangement shown in Figure 6. The main-chain nitrogen and carbonyl oxygen atoms of residues Ala 125, His 127, and Met 129 in the bottom chain of the back β-structure are hydrogen bonded to the corresponding atoms in the other monomer, forming a single large pleated sheet of twelve antiparallel chains which forms the entire back of the dimer. There are at least eight other hydrogen bonds as well as a large number of non-polar interactions holding the dimer together across the contact area.

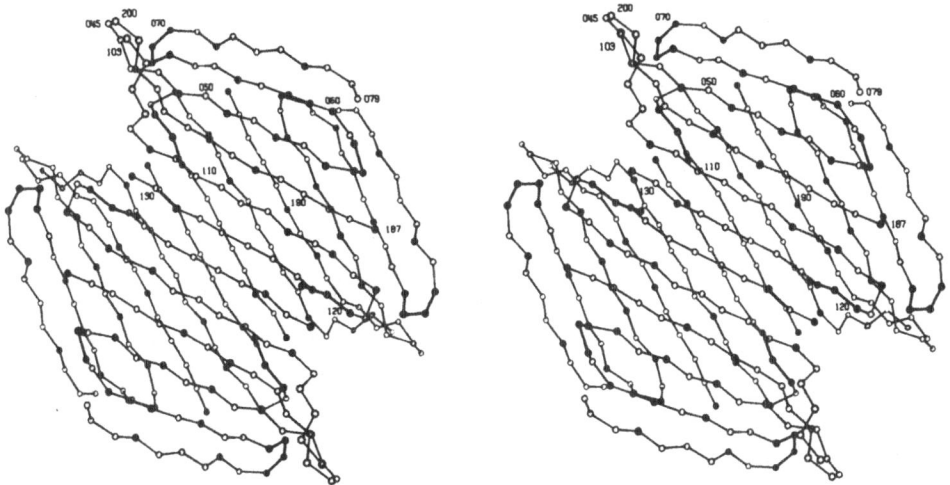

Fig. 7. Stereo backbone drawing of a Con A tetramer, showing only the back β-structures of the two constituent dimers. Residues participating in interactions between the two dimers are indicated by filled circles (●). At the center of the figure, surrounding the point of 222 symmetry, is a cavity where the two dimers are not in direct contact.

The pleated sheet at the back of each dimer is crucial in the formation of the tetramer, as shown in Figure 7. This figure is a stereo view of just the two large back pleated sheets in the two dimers of a single tetramer. The sheets fit each other very closely, but they are slightly concave, leaving a solvent-filled cavity at the center. Half of the side chains from each pleated sheet project out into the region of contact, where they provide the interactions that hold the tetramer together. In all, there are about 12 salt links, 36 hydrogen bonds, and several hundred (344) atoms in van der Wall's contact between the two dimers. Among the significant interactions between dimers are four pairs of salt links between lysines 114 and 116 on one dimer and glutamic acid 192 on the other dimer. It seems possible that chemical modification of these lysines could prevent tetramer formation and yield a dimeric form of Con A. Indeed, it has been found (Gunther et al., 1973) that succinylation or acetylation of Con A splits the molecule into dimers that remain stable over the range of physiological pH. A similar result was obtained upon maleylation of Con A (Young, 1974). These chemical modifications do not alter the carbohydrate-binding properties of the protein as measured by equilibrium dialysis.

Both native Con A and succinyl-Con A are equally potent mitogens at the optimal dose of about 3 µg/ml (Gunther et al., 1973). At

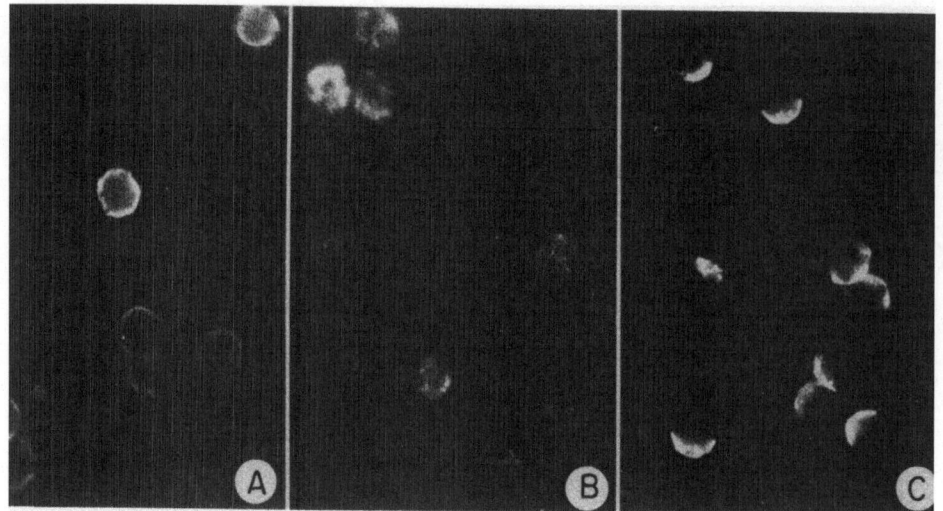

Fig. 8. Distribution of fluorescein-labelled anti-immunoglobulin on mouse splenic lymphocytes, illustrating diffuse staining (A), patches (B), and caps (C). Cap formation is prevented by Con A, but not by succinyl-Con A.

higher doses, the response to native Con A decreases much more rapidly than that to succinyl-Con A. This difference seems to be correlated with decreased cell viability at high doses of native Con A.

Succinylation of Con A also alters its immediate effects on the lymphocyte surface, as assayed by the modulation of cell-surface receptor mobility. Interaction of lymphocytes with fluorescence-labelled anti-immunoglobulin results in formation of patches and caps (Taylor et al., 1971) (Figure 8). Yahara and Edelman (1972, 1973a) have observed that patch and cap formation induced by anti-immunoglobulin as well as by a variety of other ligands including Con A itself is inhibited by Con A. Under different conditions, however, Con A can form caps with its own glycoprotein receptors (Unanue et al., 1972). Succinyl-Con A lacks both of these activities (Table I). In contrast to native Con A, succinyl-Con A neither forms caps nor inhibits caps. Inasmuch as succinyl-Con A is just as mitogenic as the native lectin, it appears, therefore, that neither immobilization of cell surface receptors nor cap formation is required for lymphocyte stimulation.

We have also found that addition of anti-Con A antibodies to cells with bound succinyl-Con A restores both cap formation and the

TABLE I

COMPARISON OF THE BIOLOGICAL ACTIVITIES OF CON A AND SUCCINYL-CON A

Assay	Con A	Succinyl-Con A
(1) Number of lectin sites per cell		
(a) sheep erythrocytes	1.1×10^6	2.8×10^6
(b) mouse spleen cells	1.4×10^6	4.4×10^6
(2) Mitogenesis		
(a) lectin (5 μg/ml)	+	+
(b) lectin (50 μg/ml)	−	+
(3) Inhibition of anti-Ig capping		
(a) lectin (100 μg/ml)	100%	0
(b) lectin (50 μg/ml) + anti-Con A (100 μg/ml)	100%	40%
(4) Formation of lectin-receptor caps		
(a) lectin (5 μg/ml, 37°C)	0.2-6%	0
(b) lectin (20 μg/ml) + anti-Con A (100 μg/ml)	18%	82%

inhibition of receptor mobility. On the assumption that the antibodies act primarily to cross-link the Con A dimers, and since anti-Con A itself has no effect on the cells, this restoration of activity appears to indicate the critical role of the valence of the bound lectin and cross-linkage of receptors in these experiments. This conclusion is further supported by the observation that the addition of monovalent anti-Con A Fab' fragments (i.e., fragments from pepsin digest of anti-Con A) does not restore the activities of the native lectin.

V. A MODEL FOR INTERACTIONS BETWEEN CELL SURFACE RECEPTORS AND CYTOPLASMIC STRUCTURES

Several possible mechanisms for the Con A-induced inhibition of anti-immunoglobulin patching and capping are consistent with the above observations. First, binding of Con A, but not of succinyl-Con A,

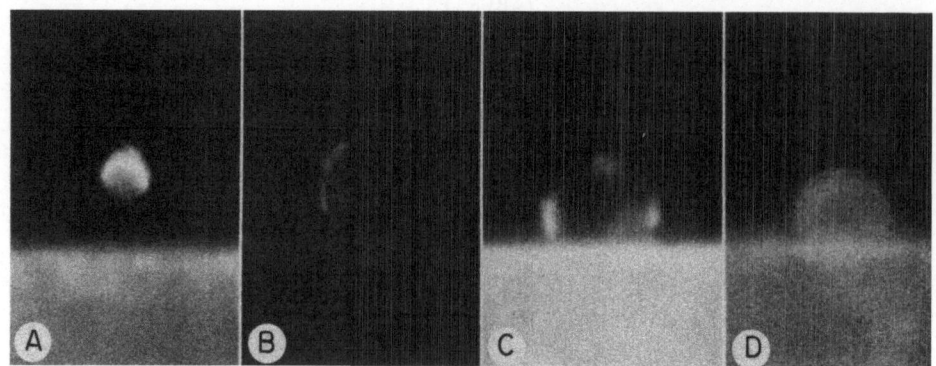

Fig. 9. Staining patterns of fluorescein-labelled anti-immunoglobulin on mouse splenic lymphocytes bound to nylon fibers. Cells in (A), (B), and (C) are bound to a dinitrophenyl-bovine serum albumin-derivatized fiber and stained with fluorescent anti-immunoglobulin. A cap is shown in (A), and patchy distributions in (B) and (C). In (D), the fiber is derivatized with Con A, which binds the cell via its Con A surface receptors. Distribution of the fluorescent anti-Ig stain is diffuse, indicating that contact of Con A with a local region of the cell surface is sufficient to prevent redistribution of surface components over the entire cell surface.

---

might induce a phase transition in the lipid component of the plasma membrane, leading to a decrease of membrane fluidity and hence of receptor mobility. However, no evidence has been found for such a phase transition, and spin label data suggest that, if anything, membrane fluidity is increased slightly by Con A binding (Barnett et al., 1974). Secondly, Con A might act directly by forming a cross-linked network of Con A-receptor complexes on the cell surface which would physically impede the mobility of other receptor systems. This explanation is shown to be an unlikely one by the results of an experiment using cells bound to chemically derivatized nylon fibers (Figure 9) (Rutishauser et al., 1974). When lymphocytes are bound to nylon fibers derivatized with the antigen, dinitrophenyl bovine serum albumin, treatment with fluorescein-labelled anti-immunoglobulin results in formation of caps and patches similar to those formed by cells in suspension (Figure 9B). In contrast, cells bound to fibers derivatized with Con A are inhibited from forming patches and caps when stained with fluorescent anti-immunoglobulin. The distribution of the fluorescent label is diffuse, as shown in Figure 9C. Note that although Con A is in contact with the cell only over a small area, aggregation of Ig receptors is prevented over the entire cell surface, and Ig receptors remain uniformly distributed. Inhibition of patch and cap formation by Con A therefore cannot be attributed to cross-linkage among various receptors by Con A. This leaves the

TABLE II

EFFECT OF DRUGS ON THE CON A INHIBITION OF CAP FORMATION

Drug	Conc (M)	Caps (%) with fl-Con A	Caps (%) with fl-anti-Ig and Con A
None	-	1	1
Colchicine	10^{-4}	24	30
Colcemid	10^{-4}	25	20
Vinblastine	10^{-4}	51	51
Vincristine	10^{-4}	19	15
Lumicolchicine	5×10^{-4}	3	1
Cytochalasin B	4×10^{-5}	1	1

intriguing possibility that Con A binding might induce some alteration of structures in the cell which are attached to and can influence the mobility of other receptors.

In considering this possibility further, one is led to ask what cellular structure might be involved. A pertinent observation is the fact that Con A inhibits cap formation when cells are preincubated with Con A at 37° C but not at 4° C (Gunther et al., 1973; Yahara and Edelman, 1973a). Furthermore, incubation of cells with colchicine, colcemid or vinca alkaloids suppresses the inhibitory effect of Con A on cap formation (Table II) (Yahara and Edelman, 1973b). For example, less than 1% of the cells form caps in the presence of Con A alone while more than 20% of the cells form caps in the presence of Con A and colchicine. This effect is more pronounced in the case of the vinca alkaloid vinblastine (50% caps). Lumicolchicine, a photo-inactivated derivative of colchicine (Wilson and Friedkin, 1967), has no effect. Inasmuch as cellular microtubules are disrupted either by colchicine or by incubation in the cold, the data suggest that inhibition of receptor mobility by Con A is correlated with the state of microtubules or related colchicine-binding proteins (CBP) in the cell.

On the basis of these observations, Edelman, Yahara and Wang (1973) have suggested the model depicted schematically in Figure 10 for interactions between cell surface receptors and cytoplasmic structures. In this model, it is assumed that the cell surface has several types of receptor molecules attached to or through the mem-

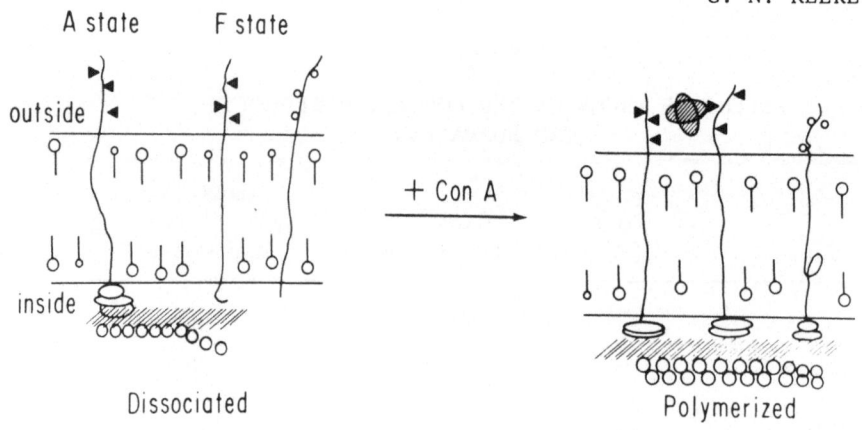

Fig. 10. Schematic model illustrating the features and events which might play an important role in the modulation of cell surface receptor mobility by Con A. (Left) Various cell surface receptors are in an equilibrium between anchored (A) and free (F) states with respect to a submembranous assembly of colchicine-binding proteins. The CBP are, in turn, in equilibrium between dissociated and polymerized states. (Right) Binding of Con A to its glycoprotein receptors can alter the state of assembly of the CBP network, thus altering the mobility of other receptors, such as immunoglobulin receptors, which also interact with the network. The effect of Con A is abolished by colchicine, which prevents polymerization of the CBP network.

brane. Surface receptors for Con A and other substances exist in either an anchored (A) or free (F) state, with respect to their interactions with a system of colchicine-binding proteins. Thus, the distribution of the receptors can be influenced by the state of this CBP network. A similar proposal has recently been made by Berlin et al. 1974. Conversely, we also proposed that the organization and mobility of the CBP are in turn affected by binding of ligands to the associated surface receptors. The CBP network and the equilibria between anchored and free receptors provide a communication link among the various kinds of receptors. Alteration of the CBP network by binding to one set of receptors, such as the Con A receptors, can affect the mobility of another set of receptors, such as immunoglobulin receptors. Finally, the equilibria among the various receptors and CBP states are of course affected by such factors as temperature and drugs. This hypothesis provides a framework for further investigations into the nature of signalling at the cell surface, and perhaps for intercellular communications as well. It promises to unify a number of phenomena which have until now been viewed as distinct. Although the hypothesis is framed in terms of the properties of lymphocytes, the postulated system for communication among receptors may well be a general feature of other cell systems as well.

In conclusion, our structural studies have revealed the details of the subunit interactions and features of the binding sites of the protein. The primary and three-dimensional structures of Con A should serve as references with which to compare the structures of other mitogenic and nonmitogenic lectins, when these are determined. Such comparisons may reveal secondary sites or other features that may not be apparent from examination of a single structure. Extension of these approaches by the use of other lectins, chemically modified lectins, and antibody reagents as cell surface probes promises to contribute to our understanding of lymphoid cell stimulation and also of a variety of other cellular phenomena, including specific cell-cell interactions, contact inhibition, and cellular mobility.

REFERENCES

1. Barber, B. H., Quirt, A., and Carver, J. P. (1975). "Magnetic resonance studies of concanavalin A: Location of the binding site of α-methyl-D-mannopyranoside." This book, p. 325.

2. Barnett, R. E., Scott, R. E., Furcht, L. T., and Kersey, J. H. (1974). "Evidence that mitogenic lectins induce changes in lymphocyte membrane fluidity." *Nature* 249, 465.

3. Becker, J. W., Reeke, G. N., and Edelman, G. M. (1971). "Location of the saccharide-binding site of concanavalin A." *J. Biol. Chem.* 246, 6123.

4. Becker, J. W., Reeke, G. N., Wang, J. L., Cunningham, B. A., and Edelman, G. M. (1974). "The covalent and three-dimensional structure of concanavalin A. III. Structure of the monomer and its interactions with metals and saccharides." *J. Biol. Chem.*, submitted for publication.

5. Beckert, W. H., and Sharkey, M. M., Sr. (1970). "Mitogenic activity of the jack bean (*Canavalia ensiformis*) with rabbit peripheral blood lymphocytes." *Int. Arch. Allergy and Appl. Immunol.* 39, 337.

6. Berlin, R. D., Oliver, J. M., Ukena, T. E., and Yin, H. H. (1974). "Control of cell surface topography." *Nature* 247, 45.

7. Brewer, C. F., Sternlicht, H., Marcus, D. M., and Grollman, A. P. (1973a). "Binding of ^{13}C-enriched α-methyl-D-glucopyranoside to concanavalin A as studied by carbon magnetic resonance." *Proc. Nat. Acad. Sci. U.S.A.* **70**, 1007.

8. Brewer, C. F., Sternlicht, H., Marcus, D. M., and Grollman, A. P. (1973b). "Interactions of saccharides with concanavalin A. Mechanism of binding of α- and β-methyl-D-glucopyranoside to concanavalin A as determined by ^{13}C nuclear magnetic resonance." *Biochemistry* **12**, 4448.

9. Brewer, C. F., Sternlicht, H., Marcus, D. M., and Grollman, A. P. (1975). "^{13}C NMR studies of the interaction of concanavalin A with saccharides." This book, p. 55.

10. DePetris, S., Raff, M. C., and Mallucci, L. (1973). "Ligand-induced redistribution of concanavalin A receptors on normal, trypsinized, and transformed fibroblasts." *Nature New Biol.* **244**, 276.

11. Edelman, G. M., Cunningham, B. A., Reeke, G. N., Becker, J. W., Waxdal, M. J., and Wang, J. L. (1972). "The covalent and three-dimensional structure of concanavalin A." *Proc. Nat. Acad. Sci. U.S.A.* **69**, 2580.

12. Edelman, G. M., Yahara, I., and Wang, J. L. (1973). "Receptor mobility and receptor-cytoplasmic interactions in lymphocytes." *Proc. Nat. Acad. Sci. U.S.A.* **70**, 1442.

13. Fox, T. O., Sheppard, J. R., and Burger, M. M. (1971). "Cyclic membrane changes in animal cells: Transformed cells permanently display a surface architecture detected in normal cells only during mitosis." *Proc. Nat. Acad. Sci. U.S.A.* **68**, 244.

14. Gunther, G. R., Wang, J. L., Yahara, I., Cunningham, B. A., and Edelman, G. M. (1973). "Concanavalin A derivatives with altered biological activities." *Proc. Nat. Acad. Sci. U.S.A.* **70**, 1012.

15. Gunther, G. R., Wang, J. L., and Edelman, G. M. (1974). "The kinetics of cellular commitment during stimulation of lymphocytes by lectins." *J. Cell Biol.*, in press.

16. Goldstein, I. J., Hollerman, C. E., and Smith, E. E. (1965). "Protein-carbohydrate interaction. II. Inhibition studies on the interaction of concanavalin A with polysaccharides." *Biochemistry* **4**, 876.

17. Goldstein, I. J., Reichert, C. M., and Misaki, A. (1974).

"Interaction of concanavalin A with model substrates." *Ann. N.Y. Acad. Sci.*, in press.

18. Hardman, K. D., Wood, M. K., Schiffer, M., Edmundson, A. B., and Ainsworth, C. F. (1971a). "X-ray crystallographic studies of concanavalin A." *Cold Spring Harbor Symp. Quant. Biol.* 36, 271.

19. Hardman, K. D., Wood, M. K., Schiffer, M., Edmundson, A. B., and Ainsworth, C. F. (1971b). "Structure of concanavalin A at 4.25-Ångstrom resolution." *Proc. Nat. Acad. Sci. U.S.A.* 68, 1393.

20. Hardman, K. D., and Ainsworth, C. F. (1972). "Structure of concanavalin A at 2.4 Å resolution." *Biochemistry* 11, 4910.

21. Hardman, K. D. and Ainsworth, C. F. (1973). "Binding of nonpolar molecules by crystalline concanavalin A." *Biochemistry* 12, 4442.

22. Inbar, M., and Sachs, L. (1969). "Interaction of the carbohydrate-binding protein concanavalin A with normal and transformed cells." *Proc. Nat. Acad. Sci. U.S.A.* 63, 1418.

23. Inbar, M., Ben-Bassat, H., and Sachs, L. (1972). "Membrane changes associated with malignancy." *Nature New Biol.* 236, 3, 16.

24. Kalb, A. J., and Levitzki, A. (1968). "Metal-binding sites of concanavalin A and their role in the binding of α-methyl-D-glucopyranoside." *Biochem. J.* 109, 669.

25. Kalb, A. J., and Lustig, A. (1968). "The molecular weight of concanavalin A." *Biochim. et Biophys. Acta* 168-366.

26. Nicolson, G. L. (1971). "Different distribution of ferritin-conjugated concanavalin A on surfaces of normal and tumor cell membranes." *Nature New Biol.* 233, 244.

27. Nicolson, G. L., and Singer, S. J. (1971). "Ferritin-conjugated plant agglutinins as specific saccharide stains for electron microscopy: Application to saccharides bound to cell membranes." *Proc. Nat. Acad. Sci. U.S.A.* 68, 942.

28. Ozanne, B., and Sambrook, J. (1971). "Binding of radioactively labelled concanavalin A and wheat germ agglutinin to normal and virus-transformed cells." *Nature New Biol.* 232, 156.

29. Pflumm, M. N., Wang, J. L., and Edelman, G. M. (1971). "Conformational changes in concanavalin A." *J. Biol. Chem.* 246, 4369.

30. Poretz, R. D., and Goldstein, I. J. (1970). "An examination of

the topography of the saccharide binding sites of concanavalin A and of the forces involved in complexation." *Biochemistry* 9, 2890.

31. Powell, A. E., and Leon, M. A. (1970). "Reversible interaction of human lymphocytes with the mitogen concanavalin A." *Exptl. Cell Res.* 62, 315.

32. Quiocho, F. A., Reeke, G. N., Becker, J. W., Lipscomb, W. N., and Edelman, G. M. (1971). "Structure of concanavalin A at 4 Å resolution." *Proc. Nat. Acad. Sci. U.S.A.* 68, 1853.

33. Reeke, G. N., Becker, J. W., and Quiocho, F. A. (1971). "The structure of concanavalin A at 4 Å resolution." *Cold Spring Harbor Symp. Quant. Biol.* 36, 277.

34. Rosenblith, J. Z., Ukena, T. E., Yin, H. H., Berlin, R. D., and Karnovsky, M. J. (1973). "A comparative evaluation of the distribution of concanavalin A binding sites on the surfaces of normal, virally-transformed, and protease-treated fibroblasts." *Proc. Nat. Acad. Sci. U.S.A.* 70, 1625.

35. Rutishauser, U., Yahara, I., and Edelman, G. M. (1974). "Morphology, motility, and surface behavior of lymphocytes bound to nylon fibers." *Proc. Nat. Acad. Sci. U.S.A.* 71, 1149.

36. So, L. L., and Goldstein, I. J. (1967). "Protein-carbohydrate interaction. IX. Application of the quantitative hapten inhibition technique to polysaccharide-concanavalin A interaction. Some comments on the forces involved in concanavalin A-polysaccharide interaction." *J. Immunol.* 99, 158.

37. So, L. L., and Goldstein, I. J. (1969). "Protein-carbohydrate interaction. XXI. The interaction of concanavalin A with D-fructans." *Carbohyd. Res.* 10, 231.

38. Sumner, J. B. (1919). "The globulins of the jack bean, *Canavalia ensiformis*." *J. Biol. Chem.* 37, 137.

39. Sumner, J. B., and Howell, S. F. (1935). "The non-identity of jack bean agglutinin with crystalline urease." *J. Immunol.* 29, 133.

40. Sumner, J. B., and Howell, S. F. (1936a). "The identification of the hemagglutinin of the jack bean with concanavalin A." *J. Bacteriol.* 32, 227.

41. Sumner, J. B., and Howell, S. F. (1936b). "The role of divalent metals in the reversible inactivation of jack bean hemagglutinin." *J. Biol. Chem.* 115, 583.

42. Taylor, R. B., Duffus, P. H., Radd, M. C., and DePetris, S. (1971). "Redistribution and pinocytosis of lymphocyte surface immunoglobulin molecules induced by anti-immunoglobulin antibody." *Nature New Biol.* 233, 225.

43. Unanue, E. R., Perkins, W. D., and Karnovsky, M. J. (1972). "Ligand-induced movement of lymphocyte membrane macromolecules. I. Analysis by immunofluorescence and ultrastructural radiography." *J. Exptl. Med.* 136, 885.

44. Villafranca, J. J., and Viola, R. E. (1974). "The use of ^{13}C spin lattice relaxation times to study the interaction of α-methyl-D-glucopyranoside with concanavalin A." *Arch. Biochem. Biophys.* 160, 465.

45. Wang, J. L., Cunningham, B. A., and Edelman, G. M. (1971). "Unusual fragments in the subunit structure of concanavalin A." *Proc. Nat. Acad. Sci. U.S.A.* 68, 1130.

46. Wilson, L., and Friedkin, M. (1967). "The biochemical events of mitosis. II. The *in vivo* and *in vitro* binding of colchicine in grasshopper embryos and its possible relation to inhibition of mitosis." *Biochemistry* 6, 3126.

47. Yahara, I., and Edelman, G. M. (1972). "Restriction of the mobility of lymphocyte immunoglobulin receptors by concanavalin A." *Proc. Nat. Acad. Sci. U.S.A.* 69, 608.

48. Yahara, I., and Edelman, G. M. (1973a). "The effects of concanavalin A on the mobility of lymphocyte surface receptors." *Exptl. Cell Res.* 81, 143.

49. Yahara, I., and Edelman, G. M. (1973b). "Modulation of lymphocyte receptor redistribution by concanavalin A, anti-mitotic agents, and alterations of pH." *Nature* 246, 152.

50. Young, N. M. (1974). "The effects of maleylation on the properties of concanavalin A." *Biochim. et Biophys. Acta* 336, 46.

3

STUDIES ON THE COMBINING SITES OF CONCANAVALIN A

I. J. Goldstein
University of Michigan
Ann Arbor, Michigan, U. S. A.

ABSTRACT

The initial event in the biological activity of concanavalin A (Con A) involves binding of the protein to cell surface receptors. The nature and mechanism whereby such binding may occur is described in terms of cell surface carbohydrates and the demonstrated specificity of the protein. Although considerable latitude is tolerated at the C-2 position of the α-D-hexopyranose ring system, the carbohydrate binding site of Con A appears to be complementary to α-D-mannopyranosyl residues. Hapten inhibition studies indicate that each of the hydroxyl groups of this sugar is probably involved in the binding mechanism. Of the common sugars present on cell surfaces (D-glucose, D-mannose and N-acetyl-D-glucosamine), it is probably α-D-mannopyranosyl residues which react with Con A. Since the latter units are generally located in the core region of cell surface glycoproteins, it is necessary to postulate that 2-o-glycosyl-α-D mannopyranosyl units are primary receptors for Con A. Evidence supporting this view includes hapten inhibition studies with model oligosaccharides and precipitin studies with macromolecules containing internal 2-o-substituted α-D-mannopyranosyl residues. The binding to Con A of a series of oligosaccharides containing α-(1→2)-linked D-mannosyl units appears to increase up to the tetraose and then decreases; several possible explanations are considered. Acetylated Con A, although retaining its specificity, is about 50% as active as the native protein. Some biological properties of the modified protein are described. Data suggesting that Con A behaves differently in the solution phase than in the crystalline state are presented in terms of UV difference displacement studies. It is suggested that the so-called carbohydrate binding site reportedly identified in Con A crystals may not be correct.

I. INTRODUCTION

A complete understanding of the diverse and intriguing biological properties of concanavalin A (Con A), the jack bean lectin, will depend to a large measure on our knowledge of the physical and chemical properties of this unique protein. Several investigators have made significant contributions in this direction (Edelman et al., 1971, 1972; Hardman et al., 1971; Hardman and Ainsworth, 1972; Quiocho et al., 1971; Wang et al., 1971 and Waxdal et al., 1971). In the second chapter Reeke et al. have elucidated the X-ray crystal structure of Con A at a resolution of 2.0 Å. However, the solution properties of this protein must also be examined critically because there may well be important differences between the properties and behaviour of the Con A molecules in the crystalline phase and that in the solution phase. In fact, such differences have already been observed (Brewer et al., 1973; Villafranca and Viola, 1974; Bessler et al., 1974) and will be discussed below.

Our own early interest in the jack bean lectin stems from observations of Sumner and Howell (1935, 1936), made over 40 years ago that Con A interacted with certain polysaccharides and plant and animal cells. Intrigued by these reports, we explored the generality of the reaction of Con A with a wide variety of carbohydrate-containing substances, both macromolecular and low molecular weight (Goldstein et al., 1965; So and Goldstein, 1967, 1968a, 1968b, 1969a, 1969b; Goldstein and Poretz, 1968; Smith and Goldstein, 1967; Poretz and Goldstein, 1970; Goldstein and So, 1965; Iyer and Goldstein, 1973). In these investigations we have employed the techniques of immunochemistry because we believe the Con A-polysaccharide interaction displays many of the properties of an idealized antibody-antigen system, an observation that was hinted at by Sumner and Howell (1936), and Hehre (1960) sometime later. As a model antibody system, Con A also has the distinctive feature of possessing homogeneous, noninteracting combining sites (Yariv et al., 1968; So and Goldstein, 1968b).

Inasmuch as an important feature of this symposium is its concern with the interaction of Con A with animal cells, I would like to direct my own comments to this point. Specifically, I would like to answer the questions: With what chemical groupings on the cell surface does Con A react? Further, is it possible to rationalize the interaction of Con A with cell surfaces on the basis of what we know regarding its carbohydrate-binding specificity?

II. METHODS, RESULTS AND DISCUSSIONS

Animal cells contain bound to their surface a number of carbohydrate-containing species (Cook et al., 1973). These include glycoprotein, glycolipids and mucopolysaccharides. To be complete one should also include nucleic acids, sugar nucleotides and other, per-

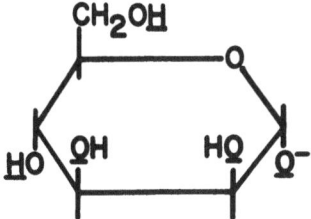

Fig. 1. An α-D-mannopyranosyl residue. Underlined atoms are believed to be involved in hydrogen bonding with the protein.

haps, unidentified species (e.g., polysaccharides). The simple sugars which occur as building units of these substances include the pentoses, D-xylose, D-ribose, and perhaps L-arabinose; the hexoses, D-glucose, D-mannose and D-galactose; the deoxy sugars, L-fucose (6-deoxy-L-galactose) and 2-deoxy-D-ribose; the hexosamines, N-acetyl-D-glucosamine and N-acetyl-D-galactosamine and the uronic acids, D-glucuronic acid and L-iduronic acid. Of course this list would be incomplete without citing the important sialic acids which are constituents of all living cells. Of all these sugars, only D-glucose, D-mannose and N-acetyl-D-glucosamine have been shown to react with the protein. This is known from hapten inhibition studies in which the ability of several hundred naturally occurring and chemically modified sugars to inhibit Con A-polysaccharide interaction was examined (Goldstein et al., 1965; So and Goldstein, 1967, 1968; Smith and Goldstein, 1967; Poretz and Goldstein, 1970; So and Goldstein, 1969a; Duke et al., 1972; Poretz and Goldstein, 1971). The results of such investigations indicate that the carbohydrate binding sites of Con A are most complementary to the α-D-mannopyranosyl residue (Fig. 1). In fact, our binding studies indicate that each of the hydroxyl groups in this molecule probably makes contact with an amino acid residue(s) in the protein combining site. Using deoxy, fluoro and o-methyl derivatives of D-glucose and D-mannose (So and Goldstein, 1967; Poretz and Goldstein, 1970) we have been able to analyze whether it is the oxygen atom or the hydrogen atom of each hydroxyl group which may be involved in hydrogen bonding with the protein. A specific example is the C-4 hydroxyl group of D-glucose (Poretz and Goldstein, 1970; Goldstein et al., 1974).

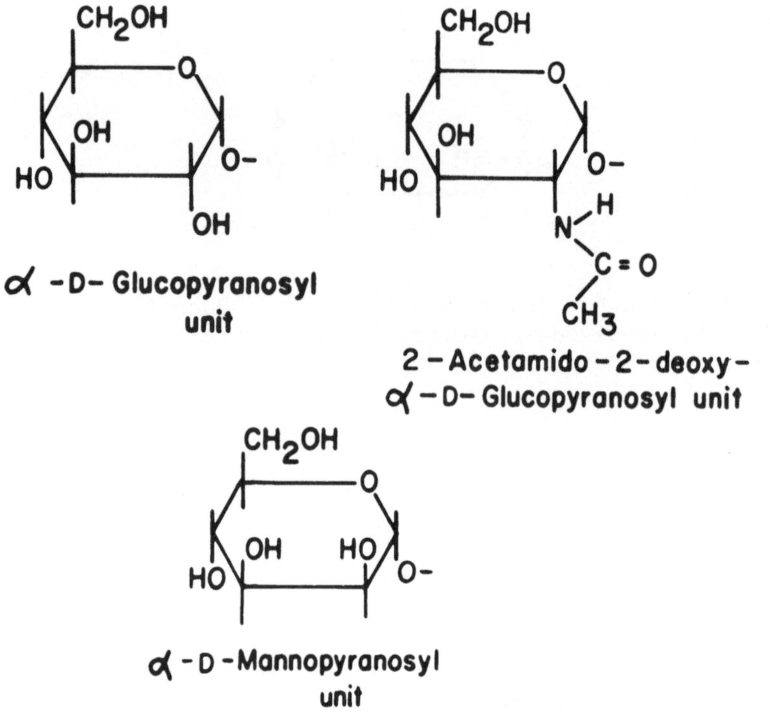

Fig. 2. Structure of cell surface carbohydrates capable of reacting with Con A.

Replacement of the equatorial C-4 hydroxyl group of methyl α-D-glucopyranoside by an axial hydroxyl group (methyl α-D-galactopyranoside) or by H (methyl 4-deoxy-α-D-xylo-hexopyranoside) abolishes completely the capacity of the resulting sugar to bind to Con A. This is also true of the 4-o-methyl derivative. These data suggest a definite role for the C-4 hydroxyl group, and indicate a very close fit between it and the Con A binding site. Furthermore, substitution of the C-4 hydroxyl function by a fluorine atom renders the resulting derivative (4-deoxy-4-fluoro-D-glucose) inactive as an inhibitor of dextran-Con A precipitation. The latter finding is particularly interesting; the highly electronegative fluorine atom which is isosteric with the hydroxyl group may participate as an electron donor in H-bond formation. The fact that this substance is inactive suggests that it is the hydrogen atom of the C-4 hydroxyl group which is involved in H-bonding with the protein.

Con A will tolerate considerable variation at the C-2 position (Poretz and Goldstein, 1970). The 2-deoxy derivative (2-deoxy-D-arabino-hexopyranose) binds to the protein and D-glucose, which possesses an equatorial hydroxyl group in the C-2 position, also binds, although with less affinity than the D-manno-configuration. N-acetyl-D-glucosamine which possesses an equatorial C-2 acetamido group also binds to Con A but N-acetyl-D-mannosamine does not.

The commonly occurring sugar moieties which react with Con A are shown in Fig. 2. They all are depicted in their α-anomeric form, the configuration in which as components of oligosaccharides, polysaccharides, etc., they bind to the protein (So and Goldstein, 1967; Smith and Goldstein, 1967; Poretz and Goldstein, 1970). As far as the cell surfaces are concerned, D-glucose is probably not an important constituent. It occurs only infrequently in glycoproteins, and when present in glycolipids, it generally occurs as a component of lactosyl residues (Gottschalk, 1972) in a form which is unavailable for binding to Con A. Non-reducing, terminal 2-acetamido-2-deoxy-α-D-glucopyranosyl units were identified by Kabat and his coworkers (Lloyd et al., 1969) as receptors for Con A in several blood group substances isolated from hog and human stomach linings. In most glycoproteins that have been described, however, N-acetyl-D-glucosamine generally occurs in its β-anomeric configuration (Gottschalk, 1972), a form which does not bind to Con A. This leaves α-D-mannopyranosyl residues of cell surfaces and animal glycoprotein as the most likely candidate to interact with Con A. But this possibility immediately raises a paradox. Although α-D-mannopyranosyl units are common constituents of animal glycoproteins (D-mannose has not yet been reported to be a constituent of glycolipids), this sugar is usually found in the interior or core region of the oligosaccharide side chains of glycoproteins, a position previously believed unavailable to Con A. Where then are the Con A receptor sites located?

The answer, I believe, is inherent in the specificity of the protein. The C-3, C-4 and C-6 hydroxyl groups of the D-arabino-configuration are the minimal, essential configurational features which a molecule must possess in order to interact with Con A (Goldstein et al., 1965; So and Goldstein, 1967; Smith and Goldstein, 1967; Poretz and Goldstein, 1970). We have also shown that 2-o-methyl-D-mannose (Poretz and Goldstein, 1970) binds as well as the parent sugar to Con A. It would appear therefore, that Con A can interact with α-D-mannopyranosyl units bearing a glycosyl residue in the C-2 position as depicted in Fig. 3. This possibility has been considered by several investigators (Andersen, 1969; Chase and Miller, 1973) studying glycoproteins as inhibitors of various biological activities of Con A. A number of glycoproteins and glycopeptides con-

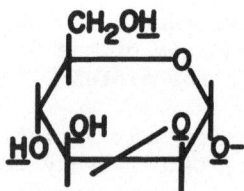

Fig. 3. Structure of a 2-o-substituted α-D-mannopyranosyl residue.

taining internally linked α-D-mannopyranosyl residues has been reported (Andersen, 1969; Chase and Miller, 1973; Toyoshima et al., 1972; Kornfeld et al., 1972). These include glycopeptides isolated from erythrocyte membranes, fibrinogen, thyroglobulin, and immunoglobulins IgG and IgM. A particularly good example is a glycopeptide (Fig. 4) isolated from an IgG_2 myeloma protein which was shown by Chase and Miller (1973) to inhibit Con A-stimulated ^{3}H-thymidine incorporation in cultured lymphocytes. It possesses a branched oligosaccharide structure with terminal neuraminic acid and N-acetyl-D-glucosamine units, the latter linked by a β-D-glycosidic bond. Neither of these sugar units will interact with Con A. On the other hand, this molecule contains two α-D-mannopyranosyl residues substituted at the C-2 position. These are the chemical groupings which I believe to be the receptor sites for Con A. Furthermore, the fact that there are two receptor sites in the same oligosaccharide side chain could result in an enhancement of the binding properties of this molecule to Con A.

The recently reported evidence (Goldstein et al., 1973, 1974) which supports the fact that Con A interacts with internal 2-o-linked α-D-mannopyranosyl units is of such critical importance that it will be summarized here.

Hapten inhibition data demonstrate the capacity of several oligosaccharides to inhibit the Con A-dextran precipitation system. The trisaccharide o-α-D-galactopyranosyl-(1→2)-o-α-D-mannopyranosyl-(1→2)-D-mannose is a very potent inhibitor. Its terminal α-D-galactopyranosyl unit is unreactive with the protein, but the two internal

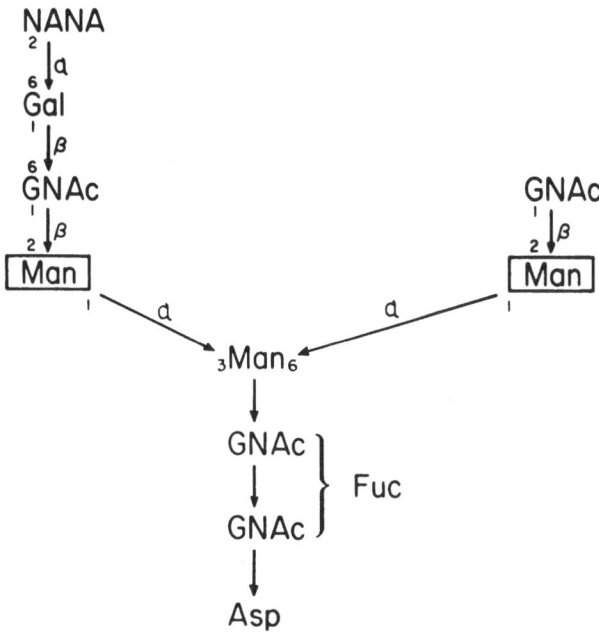

Fig. 4. Proposed structure of IgG$_2$ glycopeptide inhibiting Con A-stimulated incorporation of (^{3}H)-thymidine in cultured lymphocytes (Chase and Miller, 1973).

α-(1→2)-D-mannopyranosyl residues do indeed possess the reactive groupings necessary for interaction with Con A making it more than twice as effective as methyl α-D-mannopyranoside as an inhibitor of Con A-dextran precipitation. Borohydride reduction of this trisaccharide affords the corresponding trisaccharide alditol which, although less potent than the parent trisaccharide, still inhibits to the same extent as methyl α-D-mannopyranoside. A second trisaccharide (o-α-D-galactopyranosyl-(1→6)-o-α-D-mannopyranosyl-(1→2)-D-mannose) with a nonreactive α-D-galactopyranosyl end unit also inhibits the Con-A-dextran precipitation reaction by virtue of its subterminal 2-o-linked D-mannopyranosyl component.

TABLE I

INHIBITION OF CONCANAVALIN A-DEXTRAN INTERACTION BY α-(1→2)-D-MANNODEXTRINS

Inhibitor	μMoles for 50% Inhibition
Methyl α-D-mannopyranoside	0.70
Mannobiose (M_2)	0.12
Mannotriose (M_3)	0.035
Mannotetraose (M_4)	0.025
Mannohexaose (M_6)	0.054
Mannoheptaose (M_7)	0.08

Further support for assessing that potential of 2-o-substituted α-D-mannopyranosyl residues as receptor sites for Con A is forthcoming from precipitation studies employing macromolecules of known constitution. The structure of several *Klebsiella* species - specific polysaccharides has been elucidated. A few of these complex biopolymers possess internal 2-o-linked α-D-mannopyranosyl units as structural features. Two of these, K-24 and K-57, have been shown to form precipitin bands with Con A in Ouchterlony plates and to afford typical precipitin-type curves. These experiments indicate that 2-o-substituted α-D-mannopyranosyl units (and almost certainly 2-o-linked α-D-glucopyranosyl units), when accessible to the protein, may serve as receptor sites for Con A.

Pertinent to this discussion are some experiments which were initiated several years ago (So and Goldstein, 1968a) and are now being continued. These involve the capacity of a series of α-(1→2)-linked mannodextrins to interact with Con A. Previously we had shown that the α-(1→2)-linked mannobiose and mannotriose were respectively four and twenty times more potent inhibitors than methyl α-D-mannopyranoside. We have just extended this series with some interesting results (Reichert *et al.*, 1974). Table I lists these oligosaccharides and their inhibition potency. It will be observed that the mannotetraose is a somewhat more potent inhibitor than the mannotriose, but then the higher homologues of the series display less affinity for Con A, at least by this assay. These results are in contrast to our previous studies (Smith and Goldstein, 1967) with the higher homologues of maltose (maltodextrins) and isomaltose (isomaltodextrins) in which it was found that, on a molar basis, the individual members of each of these homologous series inhibited the Con A-dextran precipitation reaction to the same extent. Interestingly, Allen *et al.* (1973) reported that the capacity of a series of β-(1→4)-linked *N*-acetyl-D-glucosamine oligosaccharides (chitodextrins) to inhibit hemagglutination by wheat germ agglutinin increased several thousand fold in proceeding from N-acetyl-D-glucosamine to N-acetylated chitopentaose. The classic experiments of Kabat and his coworkers (1966) may also be cited in which it was shown that the combining sites of several antidextran antibodies were complementary to a sequence of about six α-(1→6)-linked D-glucopyranosyl units. Several explanations for these observations suggest themselves. These include the possibility that two or more lectin molecules may become associated with each oligosaccharide, or the more likely possibility suggested by Allen et al. (1973) and by us (So and Goldstein, 1968a) that perhaps the combining sites of lectins may be more extensive than previously believed, encompassing a sequence of two or more properly linked glycosyl units. A third, very intriguing possibility, which is discussed by Brown and his coworkers (Brown et al., 1975) involves statistical considerations (So and Goldstein, 1969a).

With this introduction to the carbohydrate-binding specificity of Con A, I would now like to consider the sequence of events which may be involved in the binding of lectins to animal cells. The primary event is almost certainly the interaction of the carbohydrate-binding sites of the lectins with a complementary chemical grouping on the cell surface. In the case of Con A it may well be α-(1→2)-D-mannopyranosyl units. This is shown in Fig. 5 as the first stage of the reaction of Con A with an animal cell. At this point it is probably possible to reverse lectin-receptor site complexation to a rather large extent by the addition of 0.1 M α-methyl-D-mannopyranoside. Further reaction of unsaturated lectin combining sites probably occurs next, and might also involve the formation of intramolecular cross linkages of sites on the same cell surface (e.g., patch and cap formation), as well as uniting receptor sites from two different cell surfaces to establish bridges between two or more different cells. This stage might be somewhat more difficult to reverse but one would still expect that saturation of the system with α-methyl mannoside would result in the displacement of Con A from the cell surface and resuspension of cellular aggregates.

The last stage in this sequence is that one about which we know least but which may be of very great importance so far as the biological effects of lectins are concerned. Further secondary, noncovalent bonds may now be established between the lectin protein and the cell surface. These interactions may involve electrostatic (including hydrogen bonding) and nonpolar forces. The latter hydrophobic interactions are especially important in stabilizing protein structures. These noncovalent interactions cannot be reversed by the simple addition of hapten inhibitors. To dislodge the protein from the cell surface at this stage could require the addition of detergents, high concentrations of salt, the application of heat, or alteration of pH to low or high values.

The current literature on the biological activity of lectins makes frequent reference to the fact that Con A cannot be recovered

1. Primary binding of concanavalin A to carbohydrate receptors (reversible with α-methyl mannoside)

2. Further saturation of concanavalin A sites (reversible with α-methyl mannoside)

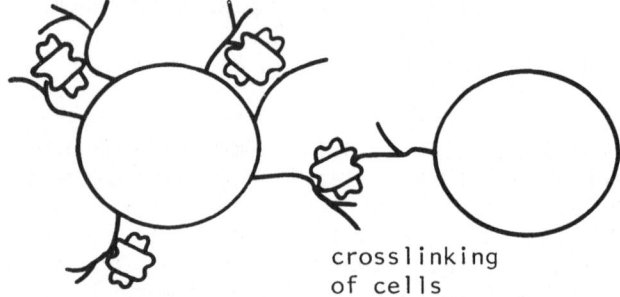

crosslinking of cells

3. Additional noncovalent, nonspecific interactions (not reversible with α-methyl mannoside)

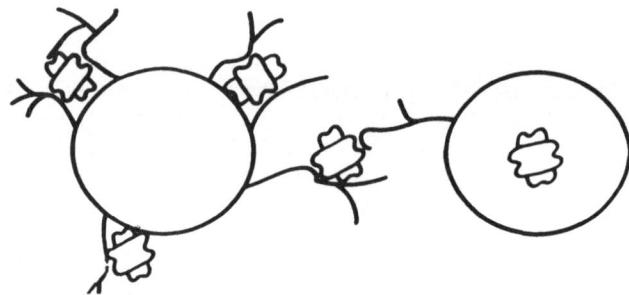

Fig. 5. Sequence of events involved in the binding of Con A to animal cells. ▧, tetravalent Con A; ⋎, cell surface glycoproteins.

quantitatively from, for example, a lymphocyte preparation after a certain time period, despite repeated washings with α-methyl mannoside (Powell and Leon, 1970; Pauli et al., 1973). In view of the above consideration, perhaps such reports are not surprising. One further possibility is endocytosis, the engulfing or internalization of Con A molecules by the cell, in which case the protein would be immune from release (Pauli et al., 1973). There is good evidence that this does indeed occur (Pauli et al., 1973).

In this regard I would like to cite one experimental approach which might be worthy of consideration: the use of chemically modified lectins. Six years ago we reported on the preparation and properties of acetylated Con A (Agrawal et al., 1968). About 85% of all available amino groups were masked by acetyl functions. Although the protein retained its specificity it was only 50% as active as the native preparation. As a mitogen (Reichert et al., 1973), acetylated protein was also about one half as active as native Con A. Not unexpectedly, the mitogenic activity of acetylated Con A was more readily inhibited by α-methyl mannoside than the natural protein. Acetylated Con A is an example of a modified lectin which should have less affinity for glycoproteins and as such could be useful in studying the mechanism of lectin binding to cell surfaces. Preliminary experiments with fluorescein-labelled acetylated Con A indicate that this may indeed be the case (Hellström, 1974).

Finally, I would like to refer to some recent studies which indicate that the site identified by Edelman and his colleagues as the carbohydrate-binding site of crystalline Con A may in fact not be correct. Becker et al. (1971) reported the location of the saccharide binding site of Con A by soaking protein crystals in 1 mM o-iodophenyl β-D-glucopyranoside and interpreting the difference electron density maps so obtained. The sugar moiety was calculated to be approximately 20 Å from the Mn ion present in each protomer. Employing ^{13}C NMR, Brewer et al. (1973) and Villafranca and Biola (1974) place the sugar only 9-10 Å from the Mn ion. Furthermore, Hardman and Ainsworth (1973) recently reported that crystalline Con A contains a cavity which, in addition to binding a number of relatively simple nonpolar molecules, binds o-iodophenyl β-D-glucopyranoside and, very importantly, o-iodophenyl β-D-galactopyranoside. Whereas the glucoside is known to inhibit polysaccharide-Con A precipitation (Poretz and Goldstein, 1971), the galactoside does not (Hardman and Ainsworth, 1973). This nonpolar cavity studied by Hardman and Ainsworth coincides with the so-called saccharide-binding site of Edelman.

Our own studies relating to the Con A binding site involve the technique of UV difference spectroscopy (Fig. 6). Some years ago (Hassing and Goldstein, 1970) we showed that the binding of simple carbohydrate ligands to Con A is accompanied by small changes in the UV spectrum (Doyle et al., 1968). In the case of the chromogenic

1. BALANCE CELLS

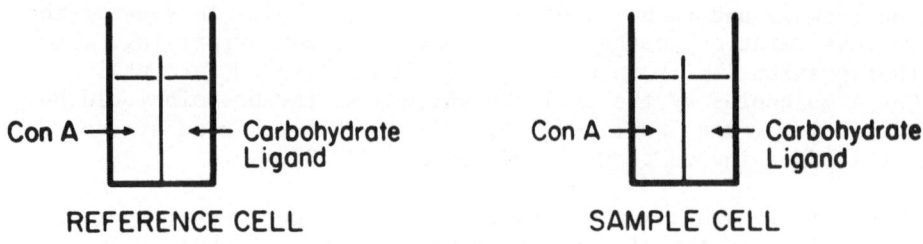

2. INVERT AND MIX SAMPLE CELL

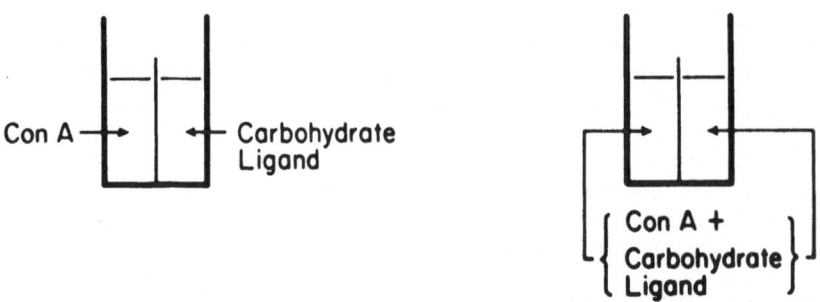

Fig. 6. Schematic representation of the principle in ultraviolet difference spectroscopy employing Yankeelov cuvettes.

p-nitrophenyl α-D-mannopyranoside, however, there is a rather pronounced spectral change. It is also possible to study the binding of a carbohydrate ligand to Con A by evaluating how the ligand displaces a chromogenic ligand such as p-nitrophenyl α-D-mannopyranoside from Con A. Such studies could also afford data on the strength of the interaction of the second ligand with the protein as well as information about the identity of the binding sites for the chromogenic and the competitor ligand. Studies involving displacement UV difference spectroscopy (Fig. 7) have recently been conducted in our laboratory (Bessler *et al.*, 1974) and I would like to summarize the results.

A number of ligands were examined for their ability to displace the chromogenic ligand p-nitrophenyl α-D-mannopyranoside. A displacement study is presented in Fig. 8, wherein it is shown that the addition of increasing concentrations of o-iodophenyl β-D-glucopyra-

1. BALANCE CELLS

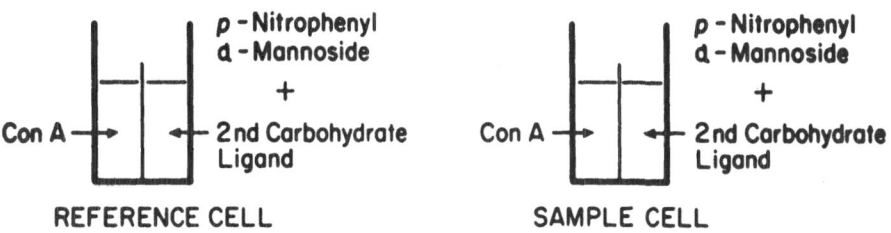

2. INVERT AND MIX SAMPLE CELL

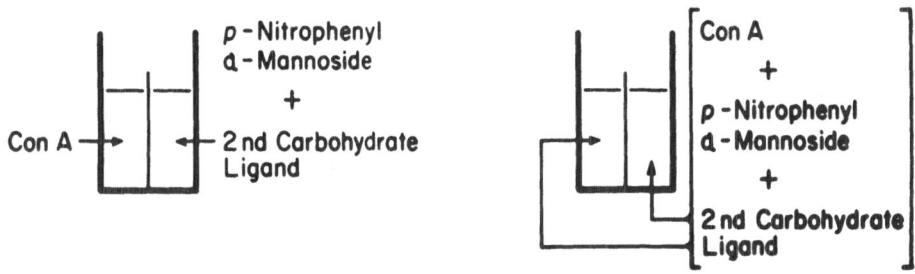

Fig. 7. Schematic representation of the principle involved in ultraviolet difference displacement spectroscopy employing Yankeelov cuvettes.

noside results in gradually abolishing the perturbation caused by p-nitrophenyl α-D-mannopyranoside. In this manner we were able to demonstrate that all carbohydrate ligands which inhibit Con A-polysaccharide interaction displaced p-nitrophenyl α-D-mannopyranoside from the protein binding site. On the other hand, no abolishment of the spectrum was observed upon the addition of o-iodophenyl β-D-galactopyranoside and myo-inositol, both shown to be noninhibitors of the Con A polysaccharide interaction.

Using this technique, we have also been able to determine the association constants of a number of ligands. For example a value of 3.4×10^4 M^{-1} was determined for p-nitrophenyl α-D-mannopyranoside at 4°; this is in good agreement with a value of 3.54×10^4 M^{-1} determined independently by Loontiens et al. (1973) by the technique of equilibrium dialysis. Similarly the binding constants of methyl

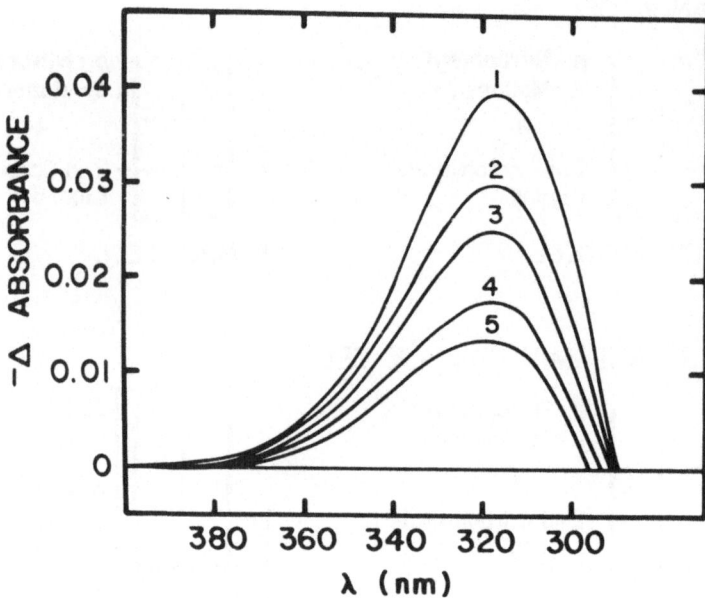

Fig. 8. "A Spectrophotometric Study of the Carbohydrate Binding Site of Con A." Reproduced by permission from *J. Biol. Chem.* 249, 2819, 1974.

α-D-mannopyranoside, maltose and a number of other sugars were determined and found to be in essential agreement with the values determined independently by other techniques (Yariv *et al.*, 1968; So and Goldstein, 1968b).

The data obtained from these studies can best be interpreted in terms of a single carbohydrate-binding site for the protein in solution. If a separate apolar binding site, distinct from the carbohydrate-binding site, exists when Con A is in the solution phase, then *p*-nitrophenyl α-D-mannoside is excluded from this site. There is only one *p*-nitrophenyl α-D-mannopyranoside binding site per monomeric unit of the protein and methyl α-D-mannopyranoside, a ligand which would not be expected to interact with an apolar binding site completely displaces *p*-nitrophenyl α-D-mannopyranoside from Con A whereas *o*-iodophenyl β-D-galactopyranoside fails to displace the *p*-nitrophenyl mannoside from the protein.

REFERENCES

1. Agrawal, B. B. L., Goldstein, I. J., Hassing, G. S., and So, L. L. (1968). "Protein-carbohydrate interaction. XVIII. The preparation and properties of acetylated concanavalin A, the hemagglutinin of the jack bean." *Biochemistry* 7, 4211.

2. Allen, A. K., Neuberger, A., and Sharon, N. (1973). "The purification, composition and specificity of wheat-germ agglutinin." *Biochem. J.* 131, 155.

3. Andersen, B. R. (1969). "Studies on the structure of the carbohydrate moiety of rabbit γG-globulin. I. Degradation with glycosidases." *Immunochem.* 6, 739.

4. Becker, J. W., Reeke, G. N., Jr., and Edelman, G. M. (1971). "Location of the saccharide binding site of concanavalin A." *J. Biol. Chem.* 246, 6123.

5. Bessler, W., Shafer, J. A., and Goldstein, I. J. (1974). "A spectrophotometric study of the carbohydrate binding site of concanavalin A." *J. Biol. Chem.* 249, 2819.

6. Brewer, C. F., Sternlicht, H., Marcus, D. M., and Grollman, A. P. (1973). "Interactions of saccharides with concanavalin A. Mechanism of binding of α- and β-methyl D-glucopyranoside to concanavalin A as determined by ^{13}C nuclear magnetic resonance." *Biochemistry* 12, 4448.

7. Brown, R. D., III, Brewer, C. F., and Koenig, S. H. (1975). "Evidence for conformational changes in concanavalin A upon binding of saccharides as determined from solvent water proton magnetic relaxation rate dispersion measurements." This book, p. 323.

8. Chase, P. S., and Miller, F. (1973). "Preliminary evidence for the structure of the concanavalin A binding site on human lymphocytes that induces mitogenesis." *Cell Immunol.* 6, 132.

9. Cook, G. M. W., and Stoddart, R. W. (1973). In: *Surface carbohydrates of the eukaryotic cell.* Academic Press, London and New York.

10. Doyle, R. J., Pittz, E. P., and Woodside, E. E. (1968). "Carbohydrate-protein complex formation. Some factors affecting the interaction of D-glucose and polysaccharide with concanavalin A." *Carbohyd. Res.* 8, 89.

11. Duke, J., Goldstein, I. J., and Misaki, A. (1972). "D-glucuronic acid: A noninhibitor of the concanavalin A system." *Biochim. Biophys. Acta* 271, 237.

12. Edelman, G. M., Cunningham, B. A., Reeke, G. N., Jr., Becker, J. W., Waxdal, M. J., and Wang, J. L. (1972). "The covalent and three-dimensional structure of concanavalin A." *Proc. Nat. Acad. Sci. U. S. A.* 69, 2580.

13. Edmundson, A. B., Ely, K. R., Sly, D. A., Westholm, F. A., Powers, D. A., and Liener, I. E. (1971). "Isolation and characterization of concanavalin A polypeptide chains." *Biochemistry* 10, 3554.

14. Goldstein, I. J., Hollerman, C. E., and Merrick, J. M. (1965). "Protein-carbohydrate interaction. I. The interaction of polysaccharides with concanavalin A." *Biochim. Biophys. Acta* 97, 68.

15. Goldstein, I. J., Holler, C. E., and Smith, E. E. (1965). "Protein-carbohydrate interaction. II. Inhibition studies on the interaction of concanavalin A with polysaccharides." *Biochemistry* 4, 876.

16. Goldstein, I. J., and So, L. L. (1965). "Protein-carbohydrate interaction. III. Agar gel-diffusion studies on the interaction of concanavalin A, a lectin isolated from jack bean with polysaccharides." *Arch. Biochem. Biophys.* 111, 407.

17. Goldstein, I. J., Poretz, R. D., So, L. L., and Yang, Y. (1968). "Protein-carbohydrate interaction. XVI. The interaction of concanavalin A with dextrans from *L. mesenteroides* B-512-F, *L. mesenteroides* (Birmingham), *Streptococcus bovis*, and a synthetic α-(1→6)-D-Glucan." *Arch. Biochem. Biophys.* 127, 787.

18. Goldstein, I. J., Reichert, C. M., Misaki, A., and Gorin, P. A. J. (1973). "An extension of the carbohydrate binding specificity of concanavalin A." *Biochim. Biophys. Acta* 317, 500.

19. Goldstein, I. J., Reichert, C. M., and Misaki, A. (1974). *Proc. N. Y. Acad. Sci.* In press.

20. Gottschalk, A. (1972). Personal communication.

21. Hardman, K. D., Wood, M. K., Schiffer, M., Edmundson, A. B., and Ainsworth, C. F. (1971). "Structure of concanavalin A at 4.25-Ångström resolution." *Proc. Nat. Acad. Sci. U. S. A.* 68, 1393.

22. Hardman, K. D., and Ainsworth, C. F. (1972). "Structure of concanavalin A at 2.4-Å resolution." *Biochemistry* 11, 4910.

23. Hardman, K. D., and Ainsworth, C. F. (1973). "Binding of nonpolar molecules by crystalline concanavalin A." *Biochemistry* 12, 4442.

24. Hassing, G. S., and Goldstein, I. J. (1970). "Ultraviolet difference spectral studies on concanavalin A-carbohydrate interaction." *Eur. J. Biochem.* 16, 549.

25. Hehre, E. J. (1960). "Contribution of classical immunology to the development of knowledge of dextran structures." *Bull. Soc. Chim. Biol.* 42, 1581.

26. Hellström, U. (1974). Private communication.

27. Iyer, R. N., and Goldstein, I. J. (1973). "Quantitative studies on the interaction of concanavalin A, the carbohydrate-binding protein of the jack bean, with model carbohydrate-protein conjugates." *Immunochem.* 10, 313.

28. Kabat, E. A. (1966). "The nature of an antigenic determinant." *J. Immunol.* 97, 1.

29. Kornfeld, R., Gregory, W. T., and Kornfeld, S. A. (1972). *Methods in enzymology.* XXVIII, p. 344.

30. Lloyd, K. O., Kabat, E. A., and Beychok, S. (1969). "Immunochemical studies on blood groups. XLIII. The interaction of blood group substances from various sources with a plant lectin, concanavalin A." *J. Immunol.* 102, 1354.

31. Loontiens, F. G., Van Wauwe, J. P., DeGussem, R., and DeBruyne, C. K. (1973). "Binding of para-substituted phenyl glycosides to concanavalin A." *Carbohyd. Res.* 30, 51.

32. Pauli, R. M., DeSalle, L., Higgins, P., Henderson, E., Norin, A., and Strauss, B. (1973). "Proliferation of stimulated human peripheral blood lymphocytes: preferential incorporation of concanavalin A by stimulated cells and mitogenic activity." *J. Immunol.* 111, 424.

33. Poretz, R. D., and Goldstein, I. J. (1970). "An examination of the topography of the saccharide binding sites of concanavalin A and of the forces involved in complexation." *Biochemistry* 9, 2890.

34. Poretz, R. D., and Goldstein, I. J. (1971). "Protein-carbohydrate interaction. On the mode of binding of aromatic moieties

to concanavalin A, the phytohemagglutinin of the jack bean." *Biochem. Pharm.* 20, 2727.

35. Powell, A. E., and Leon, M. A. (1970). "Reversible interaction of human lymphocytes with the mitogen concanavalin A." *Exp. Cell. Res.* 62, 315.

36. Quiocho, F. A., Reeke, G. N., Jr., Becker, J. W., Lipscomb, W. N., and Edelman, G. M. (1971). "Structure of concanavalin A at 4 Å resolution." *Proc. Nat. Acad. Sci. U. S. A.* 68, 1853.

37. Reeke, G. N., Jr. (1975). "Structure and function of concanavalin A." Personal communication.

38. Reichert, C. M., Goldstein, I. J., and Gorin, P. A. J. (1974). Unpublished results.

39. Reichert, C. F., Pan, P. M., Mathews, K. P., and Goldstein, I. J. (1973). "Lectin-induced blast transformation of human lymphocytes." *Nature New Biology* 242, 146.

40. Smith, E. E., and Goldstein, I. J. (1967). "Protein-carbohydrate interaction. V. Further inhibition studies directed toward defining the stereochemical requirements of the reactive sites of concanavalin A." *Arch. Biochem. Biophys.* 121, 88.

41. So, L. L., and Goldstein, I. J. (1967). "Protein-carbohydrate interaction. IX. Application of the quantitative hapten inhibition technique to polysaccharide-concanavalin A interaction. Some comments on the forces involved in concanavalin A-polysaccharide interaction." *J. Immunol.* 99, 158.

42. So, L. L., and Goldstein, I. J. (1968). "Protein-carbohydrate interaction. XIII. The interaction of concanavalin A with α-mannans from a variety of microorganisms." *J. Biol. Chem.* 243, 2003.

43. So, L. L., and Goldstein, I. J. (1969). "Protein-carbohydrate interactions. XXI. Interaction of concanavalin A with D-fructans." *Carbohyd. Res.* 10, 231.

44. So, L. L., and Goldstein, I. J. (1969). "Protein-carbohydrate interaction. XVII. The effect of polysaccharide molecular weight on the concanavalin A-polysaccharide precipitation reaction." *J. Immunol.* 102, 53.

45. So, L. L., and Goldstein, I. J. (1968). "Protein-carbohydrate interaction. XX. On the number of combining sites on concanavalin A, the phytohemagglutinin of the jack bean." *Biochim. Biophys. Acta* 165, 398.

46. Sumner, J. B., and Howell, S. F. (1935). "The non-identity of jack bean agglutinin with crystalline urease." *J. Immunol.* 29, 133.

47. Sumner, J. B., and Howell, S. F. (1936). "The identification of the hemagglutinin of the jack bean with concanavalin A." *J. Bacteriol.* 32, 227.

48. Toyoshima, S., Fukuda, M., and Osawa, T. (1972). "Chemical nature of the receptor site for various phytomitogens." *Biochemistry* 11, 4000.

49. Villafranca, J. J., and Viola, R. E. (1974). "The use of ^{13}C spin lattice relaxation times to study the interaction of α-methyl-D-glucopyranoside with concanavalin A." *Arch. Biochem. Biophys.* 160, 465.

50. Wang, J. L., Cunningham, B. A., and Edelman, G. M. (1971). "Unusual fragments in the subunit structure of concanavalin A." *Proc. Nat. Acad. Sci. U. S. A.* 68, 1130.

51. Waxdal, M. J., Wang, J. L., Pelumm, M. N., and Edelman, G. M. (1971). "Isolation and order of the cyanogen bromide fragments of concanavalin A." *Biochemistry* 10, 3343.

52. Yariv, J., Kalb, A. J., and Levitzki, A. (1968). "The interaction of concanavalin A with methyl α-D-glucopyranoside." *Biochim. Biophys. Acta* 165, 303.

4

^{13}C NMR STUDIES OF THE INTERACTION OF CONCANAVALIN A WITH SACCHARIDES*

C. F. Brewer, H. Sternlicht, D. M. Marcus, and
  A. P. Grollman
Albert Einstein College of Medicine
Bronx, New York, U. S. A., and
Bell Laboratories
Murray Hill, New Jersey, U. S. A.

ABSTRACT

The binding of α- and β-methyl-D-glucopyranoside and β-(o-iodopheryl)-D-glucopyranoside to concanavalin A has been studied by carbon-13 nuclear magnetic resonance techniques. The kinetics and binding orientations of these saccharides relative to the transition metal ion site in the protein have been determined.

I. INTRODUCTION

Concanavalin A (Con A) is a protein isolated from the jack bean (*Canavalia ensiformis*) and is a member of a class of proteins called lectins. Present interest in Con A derives from its unusual effects on animal cells (Sharon and Lis, 1972). These effects include selective agglutination of cells transformed by oncogenic viruses (Inbar and Sachs, 1969), and inhibition of growth of malignant cells in experimental animals (Shoham *et al.*, 1970).

The biological properties of Con A are related to its ability to selectively bind certain saccharides. The α-anomers of monosaccharides with the D-mannopyranosyl hydroxyl configuration at the 3, 4 and 6 positions (Goldstein *et al.*, 1965) selectively bind to Con A. The multiple binding valency of Con A and its ability to precipitate certain polysaccharides has also prompted use of the lectin as

---

*This work was supported by U. S. P. H. S. Grants No. GM-00065 and AI-05336-11 and the American Cancer Society Grant No. IC-61D.

a model for antibody-antigen reactions.

In the present paper, we will discuss our ^{13}C-NMR studies of the binding of α- and β-methyl-D-glucopyranoside (α- and β-MDG) to the zinc and manganese forms of Con A as well as our preliminary ^{13}C NMR data on the binding of β-(o-iodophenyl)-D-glucopyranoside (β-IPG). These data will be discussed in relation to the X-ray crystallographic studies of Con A and its crystalline complex with β-IPG.

II. METHODS

Synthesis of α- and β-methyl-D-glucopyranoside, uniformly enriched with 14% ^{13}C, and the preparation of zinc and manganese Con A (Zn- and Mn-Con A, respectively), have been previously reported (Brewer et al., 1973b). Synthesis of β-IPG, uniformly enriched with 14% ^{13}C, will be reported elsewhere.

A description of the materials and methods used in the NMR measurements has been published elsewhere (Brewer et al., 1973a).

III. RESULTS AND DISCUSSION

The proton-decoupled ^{13}C NMR spectrum of 14% ^{13}C-enriched α-MDG is shown in the upper part of Fig. 1. The carbon resonance assignments are listed above each peak. The ^{13}C spectrum of the α-anomer in the presence of Zn-Con A (Fig. 1A) or Mn-Con A shows line broadening which is eliminated if an excess of α-methyl-D-mannopyranoside (α-MDM), a competitive binding sugar, is added to solution (Fig. 1B). The line broadening in Fig. 1A is thus due to specific binding of α-MDG to Con A. Similar effects are observed for β-MDG in the presence of Con A.

The change in resonance line widths at half-peak height of the carbon resonances of the α-anomer showed a linear relation to the total saccharide concentration in the presence of a constant amount of protein. A similar plot for the β-anomer was non-linear, indicating incomplete binding of this saccharide to Con A. These data allow determination of the affinity constant for β-MDG which was calculated to be 70 ± 15 M^{-1} at 25°C. Fig. 2 shows a plot of the change in line width of the resonance of carbon-6 of the β-anomer as a function of f, the ratio of bound to unbound sugar. The linear relationship (Fig. 2) demonstrates that the observed line broadening of β-MDG also results from specific interactions with Con A.

The line widths of carbons 1 and 6 of both α- and β-MDG in the presence of native Con A is shown in Fig. 3 as a function of temperature between 4 and 45°C. The reciprocal absolute temperature is plotted against line broadening of the carbon resonances of the

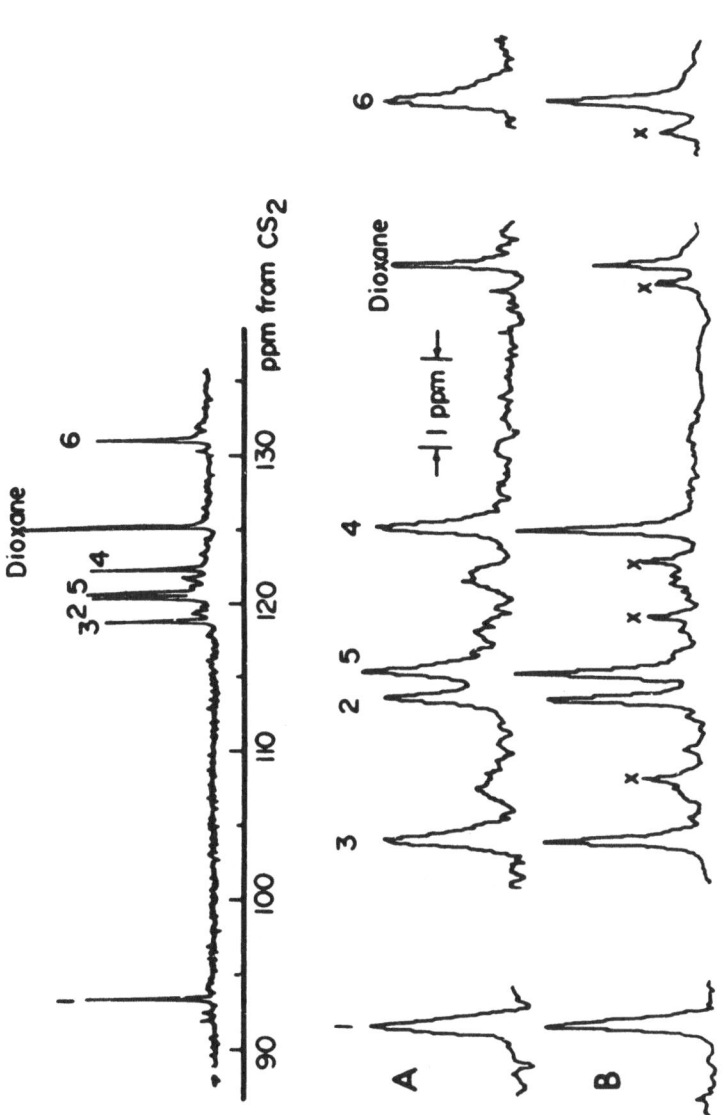

Fig. 1. Proton-decoupled ^{13}C-NMR spectra of ^{13}C-enriched α-methyl-D-glucopyranoside at pH 5.6 and 25°C and (A) in the presence of Zn-Con A; (B) in the presence of Zn-Con A and α-methyl-D-mannopyranoside. The concentrations used were: 1.25 x 10^{-2} M ^{13}C-enriched α-methyl-D-glucopyranoside. Spectra A and B are shown on an expanded scale: The small peaks (X) in spectrum B are those of α-methyl-D-mannopyranoside.

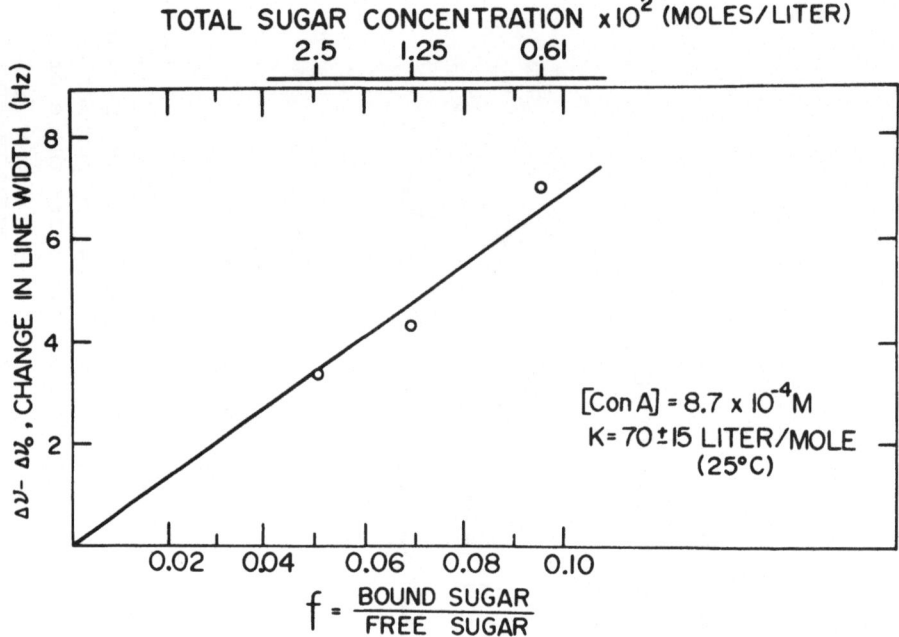

Fig. 2. Plot of the change in the C-6 line width of ^{13}C-enriched β-methyl-D-glucopyranoside in the presence of 8.7×10^{-4} M "native" Con A as a function of f, the ratio of bound to unbound sugar in solution. The total sugar concentration present ranged from 0.61×10^{-2} to 2.5×10^{-2} M. The affinity constant, K_a, is based on two independent sugar binding sites per Con A dimer (M.W. 54,000).

sugars, represented as $\log[f/\pi(\Delta\nu - \Delta\nu_0)]$, where $\Delta\nu$ and $\Delta\nu_0$ are the line widths at half peak height of the ^{13}C sugar resonances in the presence and absence of protein respectively. Between 4 and 25°C, the line widths of the carbon resonances of the α-anomer were nearly the same and broadened with increasing temperature. In this temperature range α-MDG exhanges slowly, on the NMR time scale, between free and bound environments, and the line broadening is dominated by the residence time of the sugar, τ_m, on the protein. At 25°C, τ_m α-MDG is 2.5×10^{-2} sec. Similar results were obtained for Zn- and Mn-Con A.

The results for the β-anomer (Fig. 3) are similar to those observed for the α-anomer, but differ in the magnitude of τ_m. τ_m for β-MDG is 2.5×10^{-3} sec at 25°C. These results suggest that the difference in affinity constants for α- and β-MDG binding to Con A is

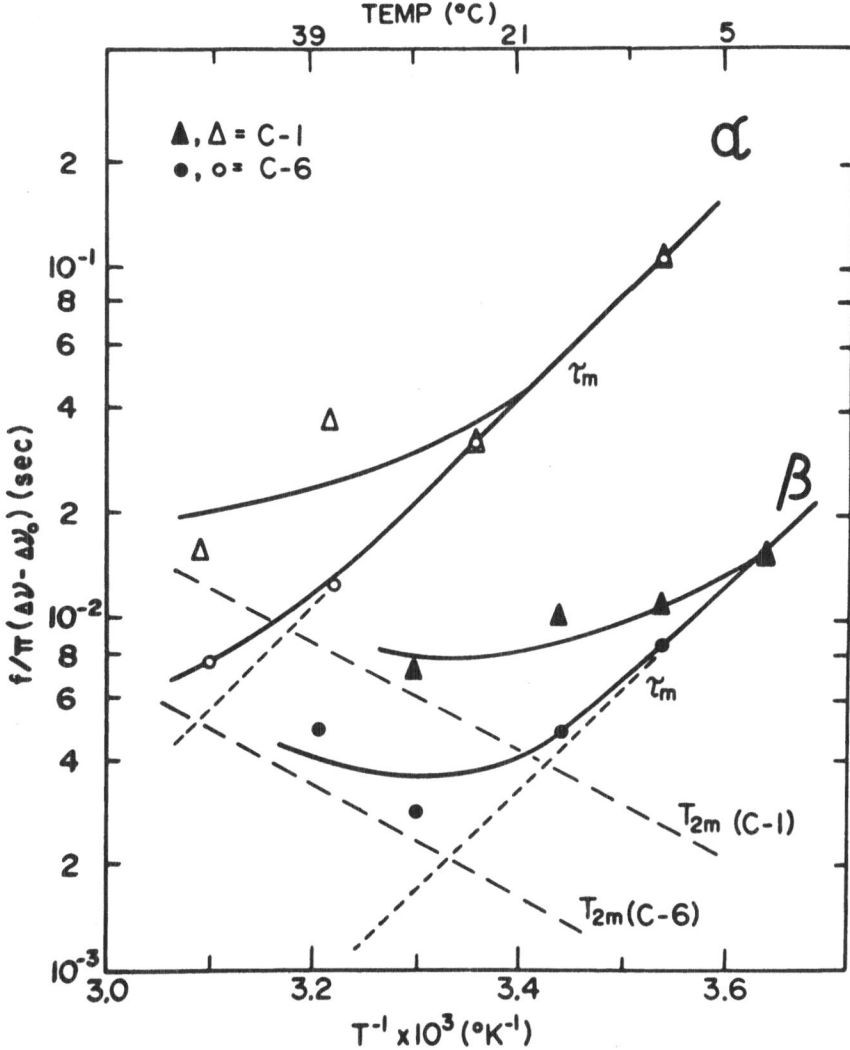

Fig. 3. Plot of the line broadening of the C-1 and C-6 resonances of ^{13}C-enriched α- and β-methyl-D-glucopyranoside at pH 5.60 in the presence of "native" Con A as a function of temperature. The line broadening effects are expressed as $\log[f/\pi(\Delta\nu-\Delta\nu_0)]$ where f denotes the ratio of bound to unbound sugar. The total sugar concentration varied from 0.61×10^{-2} to 5×10^{-2} M. The Con A concentration varied from 4.8×10^{-4} to 8.7×10^{-4} M. The measurements at 50°C were complicated by the instability of the protein at this temperature.

due primarily to the longer residence time of the former sugar when complexed to the protein. This indicates that the off-rate, k_{-1}, which is τ_m^{-1}, for α-MDG is slower than that for β-MDG by an order of magnitude.

The forward rate constants for binding of α- and β-MDG can also be calculated since their off-rates were determined and their affinity constants also measured by ^{13}C NMR techniques (Brewer et al., 1973a). The latter values are 1700 M^{-1} and 70 M^{-1} for α- and β-MDG at 25°C, respectively. The calculated forward rate constant at 25°C for α-MDG is 5 x 10^4 M^{-1} sec^{-1} while that for β-MDG is 2.8 x 10^4 M^{-1} sec^{-1}. Interestingly, the forward rate constant, k_1, for both sugars is essentially constant with increasing temperature and has a near zero activation energy (Brewer et al., 1972). The kinetics of α- and β-MDG binding to Con A is consistent with a multiple step binding mechanism in which isomerization of the protein-sugar complex occurs to give the final complex.

The spin-lattice (T_1) relaxation times of the ^{13}C carbons of α- and β-MDG were determined in the absence and presence of Zn- and Mn-Con A, and are shown in Table I and Table II, respectively. In the absence of protein, the T_1 values of the ring carbons of both sugars are approximately 1 sec. Addition of Zn-Con A results in a nearly uniform reduction in the observed T_1 values of the ring carbons of the sugars. This effect is consistent with the bound sugars having a much longer rotational correlation time corresponding to the tumbling time of Con A in solution. Equation 1 permits calculation of the T_1 values of the carbons of the sugars

$$T_1^{-1} = T_{10}^{-1} + f(T_{1m} + \tau_m)^{-1} \qquad \ldots (1)$$

when bound to Con A. T_1 and T_{10} are the spin-lattice relaxation times of the sugar in the presence and absence of Con A; f the fraction of bound to unbound sugar; T_{1m}, the spin-lattice relaxation time of the bound sugar; and τ_m, the residence time of the bound sugar on the protein. The T_{1m} values for α- and β-MDG bound to Zn-Con A are also shown in Tables I and II, respectively. Calculation of the rotational correlation of the sugars in the bound state (Abragam, 1961) demonstrate that they are firmly bound in the complex and tumble with the tumbling time of the protein.

In the presence of Mn-Con A, observed T_1 values of the ^{13}C carbons of both sugars were selectively shortened relative to their T_1 values in the presence of Zn-Con A (Tables I and II). These results suggest that the bound sugars experience additional relaxation due to the presence of the paramagnetic manganese ion in the protein. T_{1m} values for both sugars bound to Mn-Con A are given in Tables I and II.

The paramagnetic contribution of the manganese ion to the spin-

TABLE I

SPIN-LATTICE RELAXATION TIME OF THE ^{13}C CARBONS OF
α-METHYL-D-GLUCOPYRANOSIDE IN THE PRESENCE OF THE
ZINC AND MANGANESE TRANSITION METAL DERIVATIVES OF
CON A AT 25°C (τ_m = 0.025 ± 0.005 sec;
CONCENTRATION OF PROTEIN, 8.7 × 10^{-4} M)

Con A Transition Metal Derivative	Carbon	T_1 (sec)[a]			T_{1m} (sec)	T_{1p} (sec)[d,e]
		Free sugar	f = 0.071[b]	f = 0.14[c]		
Free sugar (no protein)	1	1.09				
	2	1.06				
	3	1.03				
	4	1.03				
	5	1.07				
	6	0.60				
Zn-Con A	1		0.86	0.71	0.26)	
	2		0.82	0.72	0.26)	
	3		0.82	0.72	0.29) 0.28[d]	
	4		0.84	0.72	0.30)	
	5		0.83	0.75	0.28)	
	6		0.45	0.40	0.12)	
Mn-Con A	1		0.56	0.45	0.069	0.091
	2		0.46	0.32	0.036	0.040
	3		0.40	0.29	0.026	0.028
	4		0.40	0.29	0.026	0.028
	5		0.49	0.38	0.048	0.057
	6		0.33	0.24	0.029	0.038

[a] T_1 times were determined accurately to ±0.03 sec.
[b] Total sugar concentration was 2.5 × 10^{-2} M.
[c] Total sugar concentration was 1.25 × 10^{-2} M.
[d] An average T_{1m} of ca. 0.28 sec for the ring carbons of sugars bound to Zn-Con A was used in calculating T_{1p}. The differences in the individual T_{1m} values reflect experimental errors in determining T_1 values.
[e] The estimated error in T_{1p} is ca. ±15% for the ring carbons and ca. ±25% for C-6.

TABLE II

SPIN-LATTICE RELAXATION TIME OF THE ^{13}C CARBONS OF β-METHYL-D-GLUCOPYRANOSIDE IN THE PRESENCE OF THE ZINC AND MANGANESE TRANSITION METAL DERIVATIVES OF CON A AT 25°C ($\tau_m = 2.5 \times 10^{-3}$ sec; CONCENTRATION OF PROTEIN, 8.7×10^{-4} M)

Con A Transition Metal Derivative	Carbon	$T_1{}^a$ (sec) Free sugar	$f = 0.049^b$	$f = 0.073^c$	T_{1m} (sec)	$T_{1p}{}^{d,e}$ (sec)
Free sugar (no protein)	1	1.08				
	2	1.05				
	4	1.06				
	3,5	1.06				
	6	0.56				
Zn-Con A	1		0.88		0.23)	
	2		0.85		0.22)–0.26^d	
	4		0.86		0.22)	
	3,5		0.81			
	6		0.43		0.09 0.11^d	
Mn-Con A	1		0.63	0.56	0.078	0.11
	2		0.55	0.48	0.058	0.075
	4		0.37	0.36	0.030	0.034
	3,5^f		0.50	0.42	0.045	0.054
	6		0.36	0.34	0.053	0.10

$^a T_1$ times were determined accurately to ±0.03 sec.
^{b}Total sugar concentration was 2.5×10^{-2} M.
^{c}Total sugar concentration was 1.25×10^{-2} M. f was calculated using an affinity constant K_1 equal to 70 ± 15.1 mol^{-1} at 25°C.
^{d}An average value for T_{1m} of the α-anomer bound to Zn-Con A is based on a $K_1 = 70 \pm 15.1$ mol^{-1}. An average of ca. 0.26 sec for the ring carbons and 0.11 sec for C-6 based on the combined data of α- and β-MDG was used in calculating T_{1p} (see Table I).
^{e}The estimated error in T_{1p} is slightly larger than that given for the α-anomer (Table I). The C-3 and C-5 resonances are unresolved. T_1 and the common T_{1m} and T_{1p} values given above represent apparent values and are subject to appreciable uncertainty.

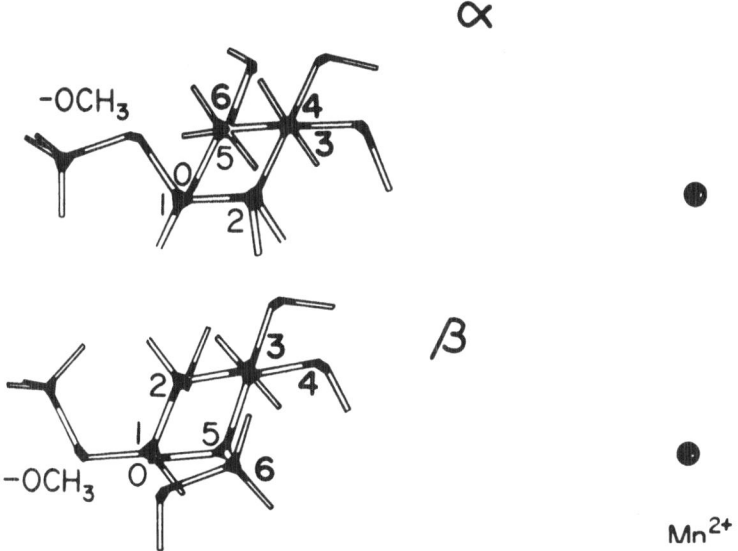

Fig. 4. Binding orientation of α- and β-methyl-D-glucopyranoside relative to the transition ion site in Con A illustrated as a side view with the Mn^{++} ion in the plane of the paper. In these projections using Fieser-Dreiding models, the ring carbons and oxygen (O) are labeled. The rotamer orientations of the methoxy and hydroxyl groups for both sugars are not known.

lattice relaxation time, T_{1p}, of the ^{13}C carbons of α- and β-MDG bound to Mn-Con A was calculated by subtracting the T_{1m}^{-1} values for Zn-Con A from the T_{1m}^{-1} values for Mn-Con A. The T_{1p} values are given in Tables I and II. The T_{1p} values of the carbons of both sugars differ from one another suggesting unique orientations of the bound sugars with respect to the manganese ion.

Equation 2 relates the experimentally determined T_{1p} values to the distance, r_{CM}, separating the ion from the carbons of the bound sugars.

TABLE III

EXPERIMENTALLY DETERMINED DISTANCES BETWEEN THE MANGANESE ION AND CARBONS OF α- AND β-MDG TO Mn-CON A AT 25°C

	Carbon	T_{1p} (sec)	Distance (Å) Exp[a]
α-MDG	1	0.091	11.7
	2	0.040	10.2
	3	0.028	9.6
	4	0.028	9.6
	5	0.057	10.8
	6	0.038	10.1
β-MDG	1	0.11	12.0
	2	0.075	11.2
	3[b]	0.054[b]	10.6
	4	0.034	9.9
	5[b]	0.054[b]	10.6
	6	0.10	11.8

[a] τ_r is 8×10^{-8} sec at 25°C. The estimated error in (Å) for this τ_r value is ± 0.3 Å.
[b] These resonances are unresolved at 23.4 kG. T_{1p} is an apparent average value for the relaxation time.

$$T_{1p}^{-1} = (2/5)g^2\beta^2\gamma_I^2 \, S(S+1) \, r_{CM}^{-6} \, \tau_c \, (1 + \omega_I^2 \tau_c^2)^{-1} \quad \ldots \quad (2)$$

The parameters given in Equation 2 can all be assigned experimentally determined values (Brewer et al., 1973b). The calculated carbon-Mn^{++} ion distances are given in Table III for α- and β-MDG. The results suggest that both sugars bind at a mean distance of 10 Å from the manganese ion in the protein. Furthermore, the data indicate that both sugars bind with their non-reducing ends facing the ion, and that they bind with different orientations. Figure 4 shows the binding orientations of the two sugars with respect to the ion that are consistent with the experimentally determined distance data in Table III. The orientation of the β-anomer is related to that of the α-anomer by inverting the former and rotating it relative to the latter. This places the 2-, 3-, and 4-hydroxyl groups of β-MDG at the positions in the bound complex found for the 6-, 4-, and 3-hydroxyl groups of α-MDG. Analysis of the binding data of Goldstein for derivatives of both anomers supports the assignment of two different binding orientations for α- and β-MDG. The two binding orientations found suggest that the difference in affinity constants between α- and β-MDG is primarily due to weak binding of the 2-hy-

TABLE IV

SPIN-LATTICE RELAXATION TIMES OF β-IPG IN THE PRESENCE OF Zn-CONCANAVALIN A AT 25°C (E_o = 1.70 x 10^{-3} M; S_o = 11.8 x 10^{-3} M)

Condition	Carbon	T_{10} (sec)	T_1 (sec)
Free β-IPG	1	0.85	
	2	0.78	
	3	0.76	
	4	0.78	
	5	0.79	
	6	0.41	
+ Zn-Con A	1		0.39
	2		0.36
	3		0.40
	4		0.35
	5		0.36
	6		0.24
+ Zn-Con A	1		---
+ α-MDM	2		0.44
	3		0.43
	4		0.45
	5		0.46
	6		0.27

droxyl of β-MDG at the position in the protein which binds the 6-hydroxyl-methyl group of α-MDG. A detailed analysis of binding orientations of α- and β-MDG is given elsewhere (Brewer et al., 1973b).

The mean value of 10 Å separating the Mn^{++} ion from bound α- or β-MDG in solution differs from the assignment made by Edelman and co-workers (Edelman et al., 1972) using X-ray diffraction techniques for the distance separating the carbohydrate binding site from the transition metal site in the protein. They have determined the location of the iodine atom of β-(o-iodophenyl)-D-glucopyranoside complexed to crystalline Con A to be approximately 20 Å from the transition metal site and located in the major cleft of each monomeric unit of the protein. Although the sugar ring was not observed in these studies, its position was inferred to be in the cleft above the iodine atom of the phenyl ring, with its non-reducing end facing the surface of the protein and directed away from the Mn^{++} ion. Thus, not only does the absolute distance assignment for the separation of the carbohydrate site from the transition metal ion site differ between the X-ray diffraction results and the NMR results, but also the orientations of the sugars with respect to the transition metal ion site.

TABLE V

COMPARISON OF $f_{app}(T_1^{-1}-T_{10}^{-1})$ FOR β-IPG AND α-MDG
IN THE PRESENCE OF Zn-CONCANAVALIN A AT 25°C

($S_o = 11.8 \times 10^{-3}$M; $E_o = 1.70 \times 10^{-3}$M)

Carbons	$T^{-1}(\text{sec})^{-1}$	$T_{10}^{-1}(\text{sec})^{-1}$	$f_{app}(T_1^{-1}-T_{10}^{-1})$ (β-IPG)	$3f_{app}(T_1^{-1}-T_{10}^{-1})$ (β-IPG)	$f_{app}(T_1^{-1}-T_{10}^{-1})$ (α-MDG)
1	2.56	1.17	0.104	0.312	0.26
2	2.77	1.28	0.096	0.288	0.26
3	2.50	1.32	0.122	0.366	0.29
4	2.85	1.28	0.092	0.276	0.30
5	2.78	1.27	0.095	0.285	0.28
6	4.16	2.43	0.083	0.250	0.12

$f_{app} = E_o/S_o$

TABLE VI

SPIN-LATTICE RELAXATION TIMES OF β-IPG IN THE
PRESENCE OF Mn-CONCANAVALIN A AT 25°C (E_o = 1.85
x 10^{-3} M; S_o = 11.8 x 10^{-3} M)

Con A transition Metal Derivative	Carbon	T_{10}(sec)	T_1(sec)	T_1(sec) + α-MDM
Free	1	0.85		
	2	0.78		
	3	0.76		
	4	0.78		
	5	0.79		
	6	0.41		
Zn-Con A (f_{app} = 0.144)	1		0.39	---
	2		0.36	0.44
	3		0.40	0.43
	4		0.35	0.45
	5		0.36	0.46
	6		0.24	0.27
Mn-Con A (f_{app} = 0.156)	1		0.21	0.37
	2		0.19	0.38
	3		0.16	0.32
	4		0.12	0.38
	5		0.18	0.36
	6		0.10	0.19

In order to help resolve this apparent discrepancy between the NMR solution results and the X-ray diffraction results for the assignment of the carbohydrate site in the protein, we synthesized β-IPG uniformly labelled with 14% ^{13}C in the glucose moiety, and examined its binding to Zn- and Mn-Con A in solution using the ^{13}C NMR techniques described in this paper. Data obtained for β-IPG binding to Zn-Con A are shown in Table IV. In the presence of Zn-Con A, the observed T_1 values of the aryl glycoside decrease as were observed for α- and β-MDG. However, addition of an excess of unenriched α-MDM to the solution did not return the observed T_1 values of β-IPG to its free sugar values as were observed when either α- or β-MDG in the presence of Zn-Con A were similarly treated. This suggests that a fraction of α-IPG binds to Zn-Con A by a mechanism not involving its carbohydrate moiety. Data suggesting more than 1 equivalent of β-IPG bound per protein monomer are shown in Table V. This table compares the relative decrease in the T_1 values of α-MDG and β-IPG in the presence of Zn-Con A under the same conditions. Comparison of the $f_{app}(T_1^{-1} - T_{10}^{-1})$ data for β-IPG with the same data for α-MDG suggests that there is nearly three times as much of the former bound to Zn-Con A as there is the latter under the same condi-

TABLE VII

T_{1p} VALUES FOR β-IPG AND α-MDG
BOUND TO Mn-CONCANAVALIN A AT 25°C

Carbon	T_{1p} (sec) β-IPG	T_{1p} (sec) α-MDG
1	0.066	0.091
2	0.048	0.040
3	0.038	0.028
4	0.021	0.028
5	0.043	0.057
6	0.029	0.038

tions. The data in Tables IV and V suggest that β-IPG binds to Con A at a site that also binds α-MDM, and at one or two other sites which do not bind the latter sugar.

Evidence that β-IPG binds to the "10 Å site" found for α- and β-MDG is provided by the ^{13}C NMR data of β-IPG in the presence of Mn-Con A, shown in Table VI. In the presence of Mn-Con A, the T_1 values for β-IPG are selectively shortened relative to its T_1 values in the presence of Zn-Con A. Addition of excess β-MDM to the β-IPG-Mn-Con A solution results in only a partial increase in the T_1 values of β-IPG relative to its free sugar values, as was also observed in the case of Zn-Con A. The increased selective shortening of the T_1 values of β-IPG in the presence of Mn-Con A relative to Zn-Con A suggest that the sugar moiety of β-IPG also binds close enough to the manganese ion to experience relaxation due to the paramagnetic ion. Table VII shows the calculated T_{1p} values of β-IPG and those of α-MDG. It is apparent that β-IPG also binds at a mean distance of 10 Å from the manganese ion with its non-reducing end facing the ion.

Thus, the ^{13}C NMR data for β-IPG in the presence of Zn- and Mn-Con A suggest that β-IPG binds to two and possibly three different sites in the protein. One of these sites is 10 Å from the transition metal ion and also binds α- and β-MDG and α-MDM, while the other sites appear to bind by hydrophobic interactions with the aryl glycoside.

Evidence that the site observed by Edelman and co-workers for β-IPG complexed to crystalline Con A is due to the binding of the o-iodophenyl ring of the molecule was obtained from further X-ray crystallographic studies. Collaborating with Karl Hardman (Hardman and Ainsworth, 1973), we have demonstrated that β-(o-iodophenyl)-D-galactopyranoside (β-IPGal), an analogue of β-IPG that does not inhibit hemagglutination or dextran precipitation by Con A, forms a crystalline complex with Con A in which the iodine atom of β-IPGal occupies the same site in the protein as that of β-IPG. Inspection of the crystallographic data for the protein crystal complexes with β-IPG and β-IPGal indicates that phenylalanine 192 of Con A in the major cleft forms aromatic "stacking" complexes with the phenyl rings of both aryl glycosides. In addition, the occupancy factor, which measures the fractional degree of binding of molecules to crystalline Con A, was nearly 1.0 (100%) for β-IPGal compared with approximately 0.50 for β-IPG. This indicates that β-IPGal has a higher affinity for this site. Though β-IPGal binds with nearly 100% occupancy to crystalline Con A, the galactose moiety is not observed in the diffraction data, just as the glucose moiety of β-IPG is not observed in its complex. This strongly suggests that the carbohydrate groups of β-IPG and β-IPGal are free to move above their iodophenyl rings in the cleft, which effectively smears out their diffraction patterns. This observation is inconsistent with the carbohydrate moieties of these compounds providing specific binding interactions with the protein. In fact, Hardman and Ainsworth (1973) have shown that o-iodobenzoic acid as well as a variety of other aromatic molecules bind to this site. It thus appears that the observed complexes formed between β-IPG or β-IPGal with crystalline Con A are due to their iodophenyl rings binding to the protein, and not their carbohydrate moieties.

It is interesting to note that other investigators have observed the binding of aromatic molecules to Con A. Specifically, Gray and Glew (1973) report that dyes such as rose bengal apparently bind to Con A. The binding of these aromatic molecules may be related to the type of binding observed for β-IPGal with crystalline Con A. The relation between the "aromatic" binding site and the carbohydrate binding site in Con A remains to be determined.

REFERENCES

1. Abragam, A. (1961). *Principles of Nuclear Magnetism*, New York, N. Y., Oxford University Press, Chapt. 8.

2. Brewer, C. F., Sternlicht, H., Marcus, D. M., and Grollman, A. P. (1973). In *Lysozyme*. Osserman, E., Beychok, S., and Canfield, R., Eds., New York, N. Y., Academic Press. In press.

3. Brewer, C. F., Sternlicht, H., Marcus, D. M., and Grollman, A. P. (1973). "Binding of ^{13}C-enriched α-methyl-D-glucopyranoside to concanavalin A as studied by carbon magnetic resonance." *Proc. Nat. Acad. Sci. U. S. A.* 70, 1007.

4. Brewer, C. F., Sternlicht, H., Marcus, D. M., and Grollman, A. P. (1973b) "Interactions of sacharides with concanavalin A.

Mechanism of binding of α- and β-methyl-D-glucopyrannoside to concanavalin A as determined by ^{13}C nuclear magnetic resonance." *Biochemistry* 12, 4448.

5. Edelman, G. M., Cunningham, B. A., Reeke, G. M., Jr., Becker, J. W., Waxdal, M. J., and Wang, J. L. (1972). "The covalent and three-dimensional structure of concanavalin A." *Proc. Nat. Acad. Sci. U. S. A.* 69, 2580.

6. Goldstein, I. J., Hollerman, C. E., and Smith, E. E. (1965). "Protein-carbohydrate interaction. II. Inhibition studies on the interaction of concanavalin A with polysaccharides." *Biochemistry* 4, 876,

7. Gray, R. D., and Glew, R. H. (1973). "The kinetics of carbohydrate binding to concanavalin A." *J. Biol. Chem.* 248, 7547.

8. Hardman, K. D., and Ainsworth, C. F. (1973). "Binding of nonpolar molecules by crystalline concanavalin A." *Biochemistry* 12, 4442.

9. Inbar, M., and Sachs, L. (1969). "Interaction of the carbohydrate-binding protein concanavalin A with normal and transformed cells." *Proc. Nat. Acad. Sci. U. S. A.* 63, 1418.

10. Sharon, N., and Lis, H. (1972). "Lectins: Cell-agglutinating and sugar-specific proteins." *Science* 177, 949.

11. Shoham, J., Inbar, M., and Sachs, L. (1970). "Differential toxicity on normal and transformed cells *in vitro* and inhibition of tumor development *in vivo* by concanavalin A." *Nature* 227, 1244.

5

SELF-ASSOCIATION, CONFORMATION AND BINDING EQUILIBRIA
OF CONCANAVALIN A

        William H. Sawyer
        University of Melbourne
        Victoria, Australia

ABSTRACT

    The discovery of distinct intact and fragmented forms of Con A, together with the observation that Con A self-associates near neutrality raises questions that may be important when interpreting experiments concerned with the biological actions of the protein. Do intact and fragmented units have the same affinity for carbohydrate? Do intact and fragmented units differ in conformation? Are all dimeric units of a homologous type or do hybrid dimers consisting of one intact and one fragmented unit also exist? Can all dimeric types self-associate to the tetramer form? Do dimer and tetramer species differ in their affinity for carbohydrate?

    These questions have been made amenable to investigation by the development of a method which separates intact and fragmented species under conditions which do not cause time-dependent or irreversible changes in protein conformation. It is found that intact dimeric units preferentially associate to the tetramer form. Under appropriate conditions of pH and ionic strength, dimer and tetramer species, and therefore fragmented and intact forms, can be separated by chromatography on Bio Gel P-100. Hybrid dimers are not present in appreciable amounts. Both types of homologous dimers (intact and fragmented) have similar affinity for carbohydrate, but dimer and tetramer species show significant differences. The results of near UV circular dichroism studies indicate that fragmented units possess slightly different conformation than intact units. An ionization-linked conformational transition in Con A does not appear to be linked directly with the self-association of the protein between pH 5 and 7.

Ligand-induced changes in the conformation of Con A are now being examined in detail. Pflumm et al. (1971) have shown that occupation of the sugar binding site of Con A results in a perturbation of conformation as revealed by near UV circular dichroism measurements. The perturbation is relatively small and does not result in more than 1-2% increase in the rotational relaxation time (Shinitzky et al., 1973). On the other hand, removal of metal ions causes a hydrodynamic change sufficient to increase the frictional coefficient and to decrease the sedimentation coefficient ($S_{20,w}$) from 3.98 S to 3.78 S. Differences between the native and the apoprotein conformation are now being examined using fluorescence polarization and the hydrophobic fluorescent probe 1-anilinonaphthalene-8-sulfonate.

I. INTRODUCTION

The central problem that has exercised the minds of protein chemists over recent decades has been the relationship between the primary sequence of amino acids, the solution properties, and the biological actions of proteins. In this regard, concanavalin A (Con A) not only presents intriguing and unique problems, but the consequences of solving these problems are far reaching as is evidenced by surveying the contributions to this book. The discovery of distinct intact and fragmented forms of Con A (Olson and Liener, 1967; Wang et al., (1971), together with the observation that the protein self-associates near neutrality raises questions that may be important when interpreting experiments concerned with the biological actions of the protein. The wide variety of these applications and biological actions necessitates close scrutiny of the conformation, the state of association-dissociation and the binding properties of the protein existing under any given set of environmental conditions. While X-ray diffraction studies provide us with a valuable yet static picture of the three dimensional architecture of the molecule, the behavior of the protein in solution is certainly of a more dynamic nature. In the extreme, neglect of these factors may result in misinterpretation of important experimental data.

The basic property of Con A which is responsible for its biological actions is the binding of carbohydrate. In this chapter, we examine the solution behavior of the protein with the aim of relating the binding property to the conformation and state of association of the molecule. Indeed, detailed knowledge of the physical state of the acceptor is essential for the accurate interpretation of binding data. The microheterogeneity of Con A in the form of fragmented and intact units presents additional complications but the development of a mild method which separates these forms has made the following questions amenable to investigation. Do intact and fragmented units have the same affinity for carbohydrate? Do intact and fragmented units differ in conformation? Are all dimeric units of a homologous type or do hybrid dimers consisting of one intact and one

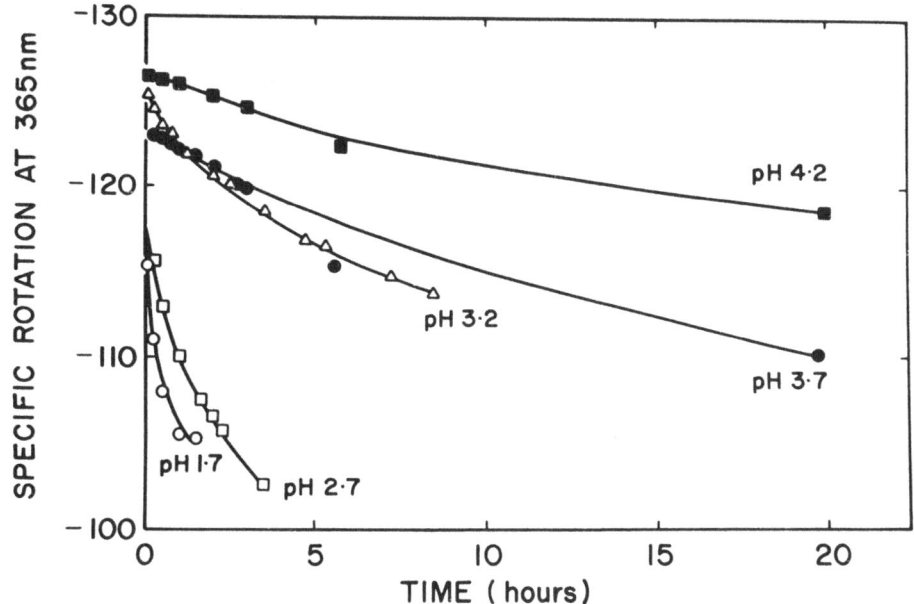

Fig. 1. Time dependent changes in the levorotation of concanavalin A solutions adjusted from pH 5.0 to acid pH values. Measurements were made at 25°C and all buffers were of ionic strength 0.1; HCl/KCl buffers, pH 1.7 and 2.7; formate buffer, pH 3.2; acetate buffers, pH 3.7 and 4.2 (from McKenzie et al., 1972).

fragmented unit also exist? Can all dimeric species self-associate to the tetramer form? Do dimer and tetramer species differ in their affinity for carbohydrate? Are all dimeric units of a homologous type or do hybrid dimers consisting of one intact and one fragmented unit also exist? Can all dimeric species self-associate to the tetramer form? Do dimer and tetramer species differ in their affinity for carbohydrates?

II. METHODS, RESULTS, AND DISCUSSION

A. The Transition from Native to Apoprotein

The early experiments of Sumner and Howell (1936a,b) showed that the agglutinating power of Con A was lost on exposure of the protein to acid pH, but that the activity could be restored by the addition

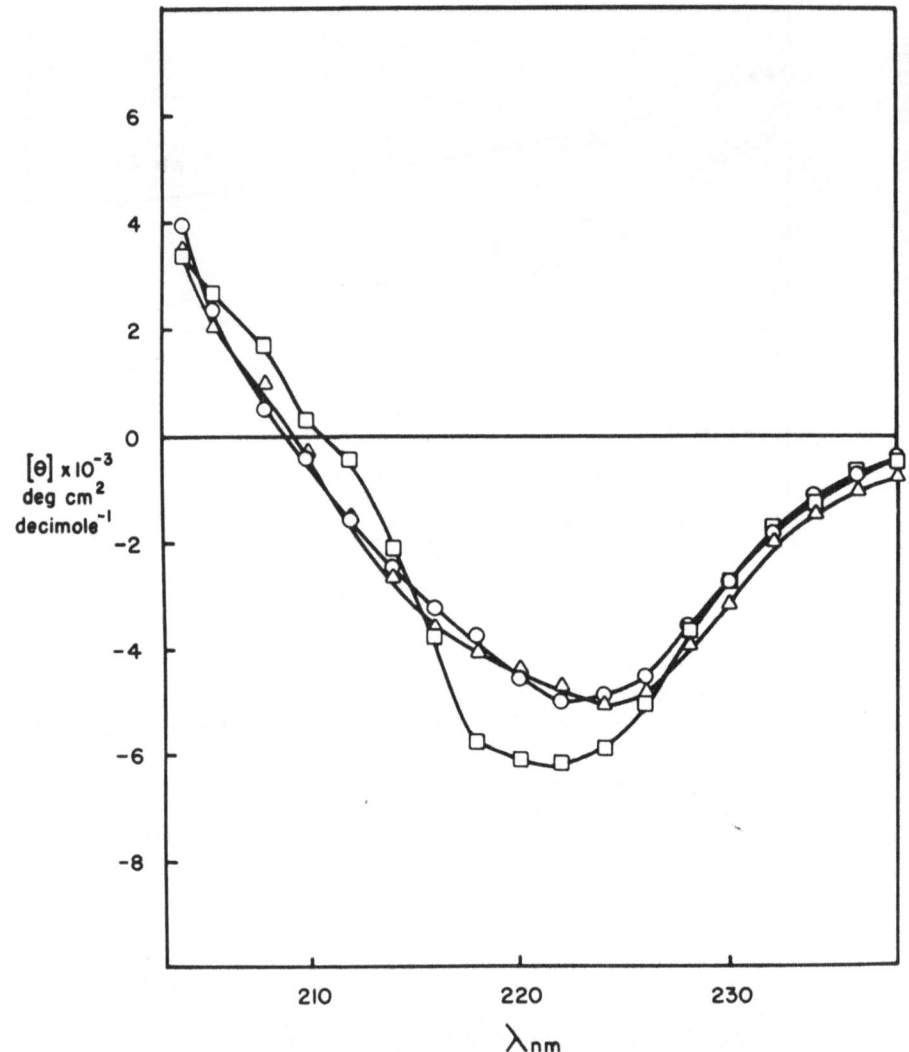

Fig. 2. Far ultraviolet circular dichroic spectra of concanavalin A. □, pH 5.0 (acetate buffer, I = 0.1); ⊙, pH 2.1 (NaCl/HCl buffer, I = 0.1); ▲, pH 7.0 (phosphate buffer, I = 0.3). (Adapted from Sawyer et al., 1974).

of Ca^{++} and Mn^{++} at neutral pH. The metal ion requirement of Con A has since been studied in detail (Yariv et al., 1968; Kalb and Levitski, 1968; Brewer et al., 1973; Sherry and Cottam, 1973). The dissociation of metal ions at acid pH involves a concomitant change in protein conformation and may therefore be monitored by a number of spectroscopic methods. Fig. 1 shows that the reaction involves a decrease in levorotation and is time dependent, becoming faster as the pH is lowered. The change in optical rotation arises from a decrease in the magnitude of the ellipticity of the protein at 200 nm (Fig. 2), an observation which suggests a slight decrease in the proportion of β-structure in the molecule. Near ultraviolet circular dichroism (CD) transitions are also perturbed at acid pH (Fig. 3) indicating a change in the environment of the aromatic residues in the demetallized protein (McCubbin et al., 1971; Sawyer et al., 1974). The general increase in ellipticity at all wavelengths between 255 nm and 295 nm indicates that phenylalanine, tyrosine and tryptophan are all perturbed although the largest change is seen at a tyrosine transition (283 nm). However, these near ultraviolet perturbations are not large enough to affect optical rotatory dispersion in the visible region of the spectrum.

The acid induced conformational transition is sufficiently large to cause a change in the hydrodynamic behaviour of the molecule. There is no change in the molecular weight of Con A between pH 5.0 and pH 1.5, yet there is a small decrease in the sedimentation coefficient as depicted in Fig. 4. The decrease in $S_{20,w}$ from 3.98 S at pH 5.0 to 3.77 S at pH 1.7 corresponds to an increase in the frictional coefficient from 5.97×10^{-8} to 6.28×10^{-8}. Although significant, this increase is less than one fifth of that observed for the acid transition of bovine serum albumin (Harrington et al., 1956). The conformational change is also expressed as a lack of isomorphism in crystals of native and demetallized protein (Jack et al., 1971).

It is appropriate at this point to comment on differences in the near ultraviolet CD found for different preparations of Con A. Commercially available material prepared by the Sumner-Howell crystallization procedure (Sumner and Howell, 1936a) frequently shows near ultraviolet CD spectra which are more characteristic of the demetallized protein (Fig. 3). The data of Yariv et al. (1968) show that the addition of metal ions to commercially prepared Con A sometimes increases the number of binding sites determined at pH 5.0. However, no such effect has been found for Con A prepared by the affinity chromatography method of Agrawal and Goldstein (1967), and McKenzie and Sawyer (1973). Similarly, Uchida and Matsumoto (1972) observed that a proportion of commercially prepared Con A did not bind to Sephadex G-50 but that the addition of metal ions restored the full binding capacity. In the original Sumner-Howell method, recrystallizations are accomplished by dissolving crystals in dilute hydrochloric acid, a procedure that cannot be recommended in the light of the above discussion. Recently, Karlstam (1973) has confirmed the close dependence of the binding property on the metal ion content of the protein.

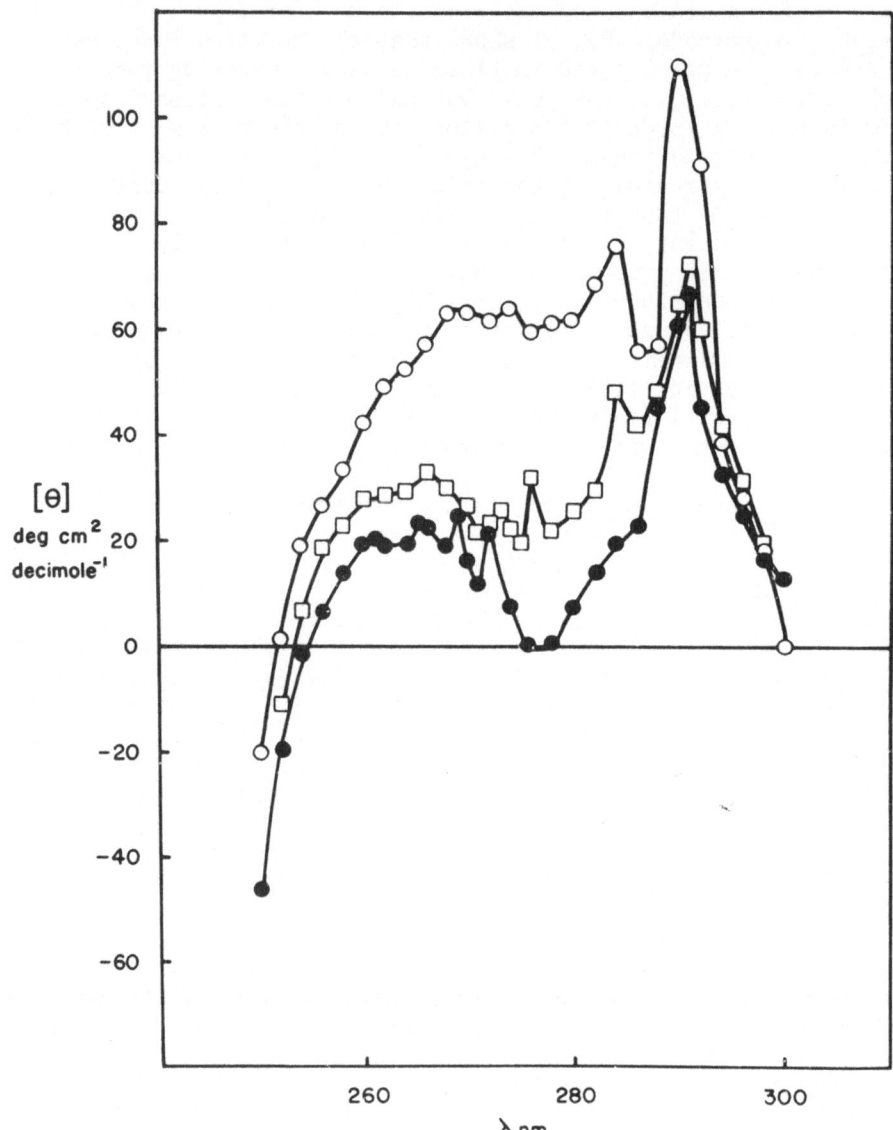

Fig. 3. Near ultraviolet circular dichroic spectra of concanavalin A. ●, Protein prepared by the method of Agrawal and Goldstein (1968), (pH 5.0, acetate buffer, I = 0.1); ◻ , Miles-Yeda concanavalin A prepared by the procedure of Sumner and Howell (1936a), (pH 5.0, acetate buffer, I = 0.1); ⊙, Agrawal and Goldstein concanavalin A at pH 2.1 (NaCl/HCl buffer, I = 0.1).

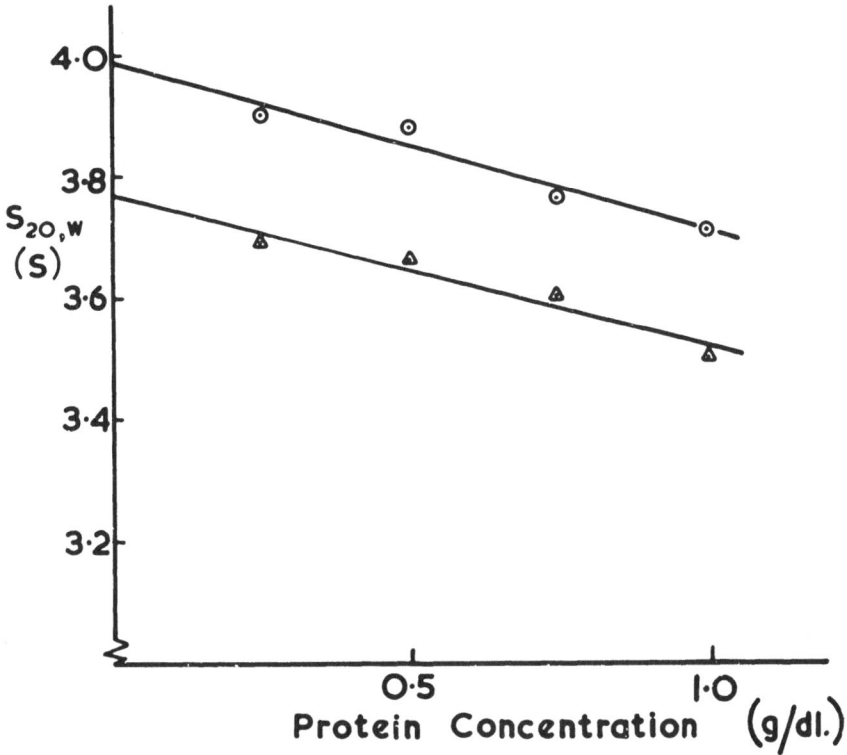

Fig. 4. Concentration dependence of the sedimentation coefficient of concanavalin A at pH 1.7 (△, HCl/NaCl buffer, I = 0.1) and pH 5.0 (⊙, acetate buffer, I = 0.1). Solutions of the same concentration at the two pH values were compared in a single double cell run, thus ensuring comparisons at the same speed and temperature.

B. Stability, Self-Association and the Separation of Intact and Fragmented Species

No time dependent changes occur in the conformation of Con A between pH 5.0 and 6.6, but at higher pH values a time dependent increase in the levorotation occurs and eventually solutions become opalescent on standing at room temperature. However, at pH 7.0 this instability can be controlled either by increasing the ionic strength

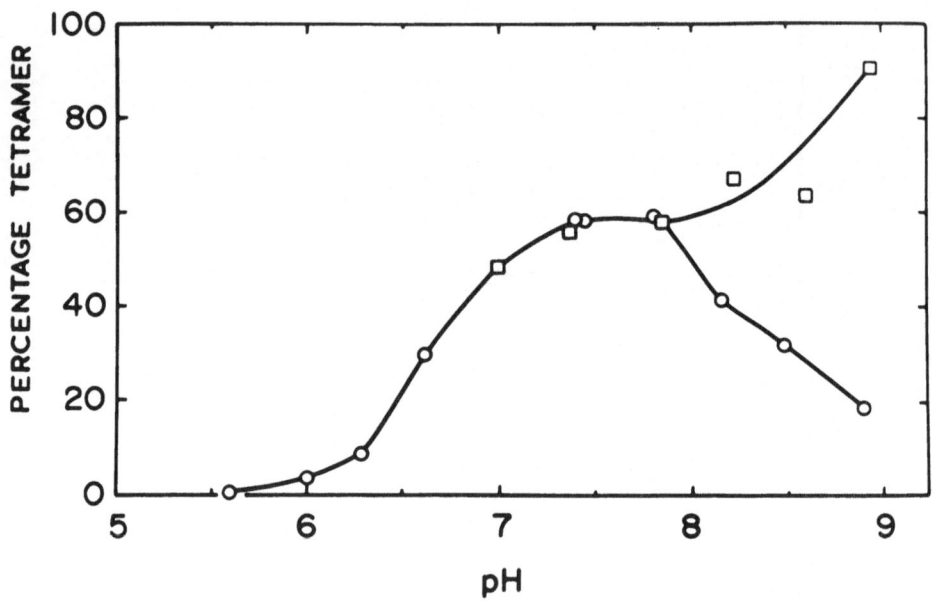

Fig. 5. The pH dependence of tetramer formation measured at 1 hr (◉) and 24 h (□) after adjustment of the pH from 5.0. The temperature of solutions during velocity sedimentation and during the 24-h storage period was 20 ± 1°C. All buffers were of ionic strength 0.3. (From McKenzie et al., 1972).

of the solvent to 0.3 or above, or by adding a carbohydrate ligand (e.g., 10 mM glucose). This observation is important since it allows us to examine the self-association and binding properties of the protein at neutral pH without damaging protein structure during analysis.

At pH 5.0, Con A exists as a dimeric molecule of molecular weight 54,000. Above pH 5.5, the protein rapidly associates to a tetramer until at pH 7.8 (I = 0.1) approximately 60% of the protein is in the tetrameric form (Fig. 5). Above pH 7.8, the association becomes time dependent and most of the protein which has not precipitated exists as tetramer. The self-association from dimer to tetramer appears irreversible and is not of the classical Gilbert type (Gilbert, 1959). Certainly, the weight average molecular weight shows no sign of concentration dependence within the range 0.42 - 1.20 g/dl. Thus, the two species may be separated by transport techniques, such as velocity

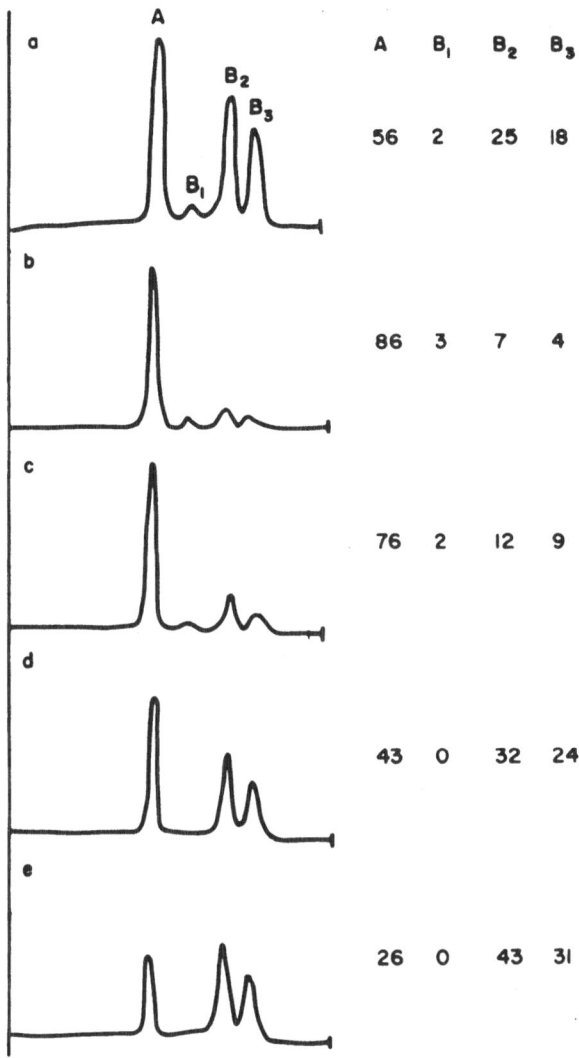

Fig. 6. Electrophoresis of fractionated concanavalin A on SDS-polyacrylamide gel. a, Unfractionated concanavalin A; b, c, d and e, cuts from the tetramer to dimer side, respectively, of a gel filtration profile (BioGel P-100). Fraction b is almost pure tetramer and consists of intact units; pure dimer consisting of fragmented units is more difficult to obtain (Fraction e). Densitometric tracings are shown on the left, and the percentage of each band on the right; A represents the intact monomer unit and B_2 and B_3 are the two major fragments. Gels were stained with Coomassie brilliant blue and scanned densitometrically at 500 nm. (From McKenzie and Sawyer, 1973).

sedimentation or gel filtration chromatography. Nevertheless, the extent of association is controlled within limits by the temperature and ionic strength of the solvent in a way that suggests that intermolecular hydrophobic bonds predominate at the dimer-dimer plane of interaction (McKenzie et al., 1972; Sawyer and Puckridge, 1973). Con A has an isoelectric point of approximately 7.0 (Agrawal and Goldstein, 1968). Thus, reduction of net positive charge as neutrality is approached from the acid side would assist such hydrophobic association.

The failure to achieve complete association to tetramer at pH 7.0 suggests the presence of material which is reluctant or incapable of entering the association reaction. Indeed, electrophoretic analyses of the dimer and tetramer fractions across a gel filtration profile (BioGel p-100) reveal that the tetramer consists entirely of intact polypeptide chains, whereas the material which fails to associate, namely dimer, consists predominently of fragmented forms (Fig. 6). Thus, the preferential association of intact units provides a useful and mild method for separating intact and fragmented forms. However, it should be emphasized that such resolution by gel filtration chromatography must be conducted under specified conditions of pH (7.0), ionic strength (I = 0.3) and temperature 20°C). Careful resolution and area analysis of sedimentation velocity patterns and gel chromatography profiles obtained under these conditions has confirmed that the amount of tetramer formed agrees with quantitative gel electrophoresis measurements of the amount of intact units present in the original solution (56% intact, 44% fragmented).

Interestingly, above pH 7.8, the fragmented units (i.e., dimer) undergo infinite association finally precipitating and leaving stable tetramer units in solution (Cunningham et al., 1972; McKenzie et al., 1972). This reaction is pH and temperature dependent and is enhanced by certain salts (HCO_3^-, SO_4^-). However, the use of this reaction to prepare intact units is unwise because of the irreversible conformational changes which occur at alkaline pH (Pflumm et al., 1971; Zand et al., 1971). Cunningham et al., (1972) have shown that the reaction affects both native Con A and the isolated intact subunits, and suggest that the CD spectral shifts result from an increase in intermolecular β-structure. The rate of change in ellipticity was dependent on solvent conditions and could be slowed by the addition of α-methyl-D-mannoside. The presence of this ligand also inhibits the development of turbidity at alkaline pH by stabilizing the dimeric form of the protein (McKenzie et al., 1972).

The preferential incorporation of intact units into the tetramer structure now raises the question of whether crystals of Con A are enriched in intact species, since the unit cell of the crystal is made up of two tetramer units (Reeke, 1975). For example, we might expect the dimer-tetramer association to be a prerequisite for crystallization. We have carried out electrophoretic analysis of crystals during the course of crystallization at pH 6.8 following the conditions of Quiocho et al. (1971). The mother liquor is depleted of intact units whilst the crystals are enriched with intact units as the crystallization progresses. However, the incorporation of intact units into the crystal lattice is not exclusive of some fragmented units. The result is consistent with observations of Abe et al. (1971) who observed a higher proportion of intact units in protein prepared by Sumner-Howell crystallization procedure (Sumner and Howell, 1936a) than in protein prepared by the method of Agrawal and Goldstein (1967). Variation in the subunit composition of different batches of commercially available protein prepared by the Sumner-Howell method was also observed.

C. An Ionization-Linked Transition Between pH 5.0 and 7.0

Although Con A is stable between pH 5 and 7 in the sense that no time dependent changes in conformation occur, the specific rotation does vary within this pH range (Sawyer et al., 1974). No change is observed in the Moffitt-Yang parameter b_0. However, the a_0 parameter varies from -204 at pH 5.0 to -155 at pH 7.0 as depicted in Fig. 7 (Sawyer et al., 1974). Analysis of the transition in terms of Tanford's theory of ionization-linked changes in conformation (Tanford, 1961; Tanford and Taggart, 1961), cannot match the experimental data with either a one or a two proton ionization, deviations predominating at extremes of the transition. Such deviation could be due to the occurrence of two ionizations of differing pK. Alternatively, a_0 may be influenced by the self-association of the protein as well as by its ionization; such instances are well documented in the literature (Tominatsu et al., 1966; McKenzie et al., 1967; Dessen and Pantaloni, 1969). Certainly, the transition in a_0 does not match the self-association transition which starts at a slightly higher pH.

The changes in visible optical rotatory dispersion do not arise from perturbation of near ultraviolet aromatic transitions. The similarity of the near ultraviolet CD spectra at pH 5 and 7 indicates that the environment of aromatic residues is not perturbed as a result of either the association reaction or the ionization-linked conformational transition. However, it is noted that the ellipticity at 220 nm is significantly lower at pH 7 than at pH 5 (Fig. 2). Use of the Kronig-Kramers transforms (Moscowitz, 1960; Schellman and Schellman, 1964) has shown that this change in the far ultraviolet CD is of sufficient magnitude and is of the correct sign to cause the observed changes in the visible optical rotatory dispersion. Moreover, the decrease in ellipticity at 220 nm indicates a slight reduction in the proportion of β-structure in the molecule. Thus, the transition requires consideration when interpreting X-ray crystallographic data for Con A crystallized at various pH values between 5 and 7 (Edelman et al., 1972; Hardman and Ainsworth, 1972).

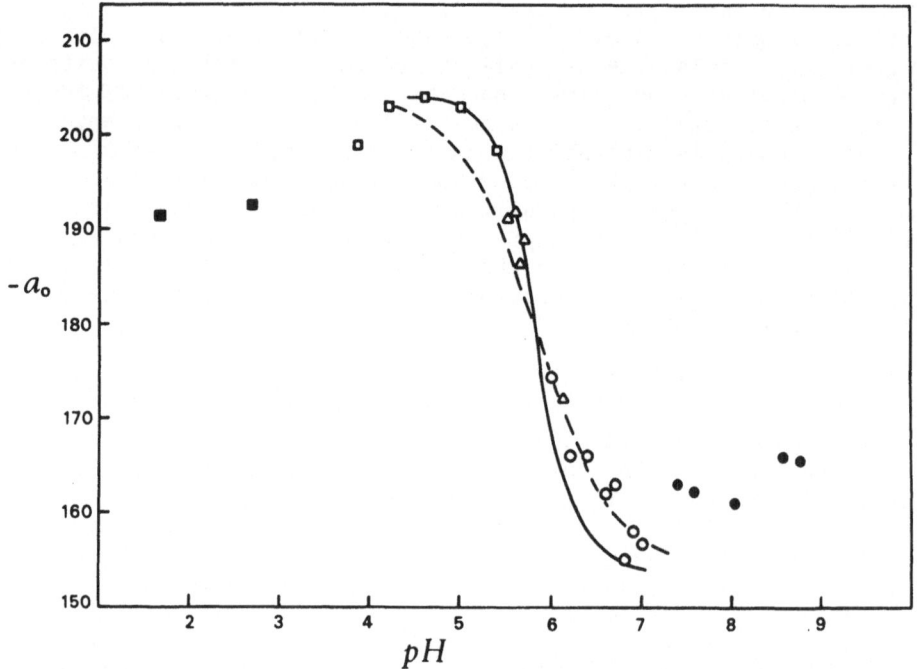

Fig. 7. The ionization-linked conformational transition in concanavalin A as revealed by the pH dependence of the Moffitt-Yang parameter, a_0. The symbols represent experimental points whilst the broken and solid lines represent the curves simulated for one and two proton transitions, respectively, according to equations of Tanford (1961). Solid symbols represent pH regions where a_0 is time dependence and values have been extrapolated to zero time. Buffers (I = 0.1): ■, HCl/NaCl; ▫, acetate; △, cacodylate; ⊙, phosphate; ●, veronal.

D. Conformation of Intact and Fragmented Units

The single break in the polypeptide chain in the fragmented monomer unit occurs between residues asx 119 and ser 120 (Edelman, et al., 1972). In terms of the three dimensional structure of the molecule, these residues are situated on a lobe of β-structure which, besides being topographically close to the dimer-dimer plane of interaction, separates the carbohydrate binding site from the contact site between monomer units in the dimer. Thus if the single peptide cleavage causes a perturbation in structure, the effect may alter both the carbohydrate binding and self-association characteristics of the molecule. The separation of fragmented and intact forms as

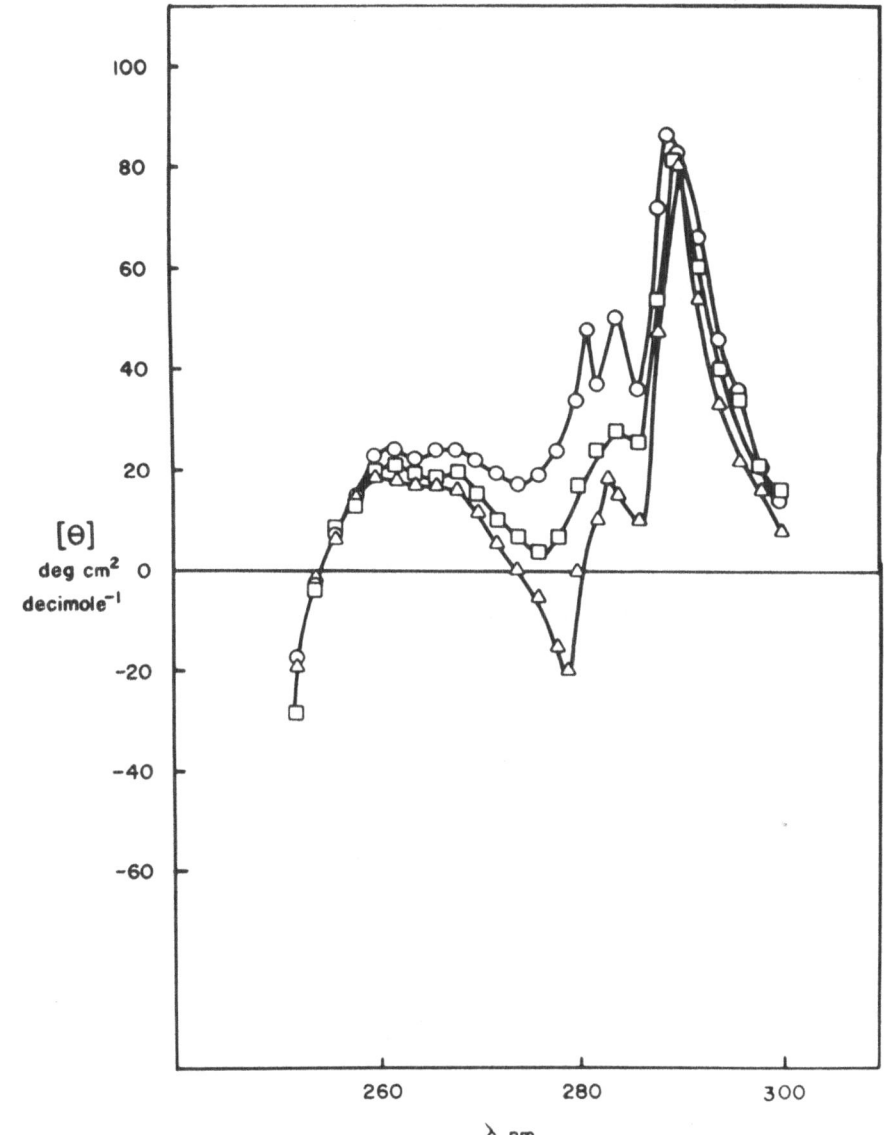

Fig. 8. Near ultraviolet circular dichroic spectra of concanavalin A. ◻, pH 5.0 (acetate buffer, I = 0.1); ⊙, dimer and ▲, tetramer fractions at pH 7.0 (phosphate buffer, I = 0.3). The spectrum for unfractionated concanavalin A at pH 7.0 (at both I = 0.1 and 0.3) was similar to that obtained at pH 5.0. Dimer and tetramer fractions were prepared by chromatography on BioGel P-100 (From Sawyer et al., 1974).

dimer and tetramer, respectively, permits examination of their conformational differences.

The near ultraviolet CD spectra of dimer and tetramer and a mixture of the two forms is shown in Fig. 8 (Sawyer *et al.*, 1974). The spectrum of the tetramer is more negative than that of the dimer whilst the unfractionated mixture of the two forms assumes intermediate ellipticities. These differences do not represent characteristics of dimer and tetramer since the spectra at pH 5.0 (dimer) and pH 7.0 (dimer-tetramer mixture) are the same to within experimental error. Rather they reflect characteristic differences between intact and fragmented conformations. The detection of these differences together with the proximity of the site of cleavage to the monomer-monomer and dimer-dimer interaction sites, provides a molecular basis for the preferential association of intact dimer units elaborated earlier.

E. Homologous and Hybrid Dimers

The occurrence of intact and fragmented monomer units means that three types of dimer are possible at pH 5.0; homologous dimers of either two intact or two fragmented units, and a hybrid molecule of one intact and one fragmented form. The spearation of fragmented dimers from intact tetramers completely accounts for the proportion of intact and fragmented material in the original solution. Thus hybrid dimers are essentially absent, a result which is also confirmed by the analysis of binding data. However, we have not been able to obtain a dimer fraction which is completely free of intact material, and thus small amounts of hybrid material might be present although incomplete chromatographic resolution of dimer from tetramer is the more likely cause of this behavior (McKenzie and Sawyer, 1973).

F. Covalent Modification of Quaternary Structure

The multivalency of Con A may play an important role in determining its effects on cellular systems. Certainly, Con A may cross-link receptors of two different cells so causing their agglutination. Moreover, inhibition of immunoglobulin receptor mobility and of phagocytosis by Con A may result from the cross-linking of adjacent receptor sites on the cell surface (Yahara and Edelman, 1972; Berlin, 1972). Investigation of these aspects using dimeric and tetramic Con A is unrealistic because of the possibility of dissociation of the tetrameric molecule once it has bound to the cell surface. The problem has been approached by maleylation and succinylation of Con A in the expectation of dissociating the molecule without disruption of the carbohydrate binding site. Both these treatments lead to dissociation of tetramer to dimer (but not to monomer) and have little effect on the affinity of the molecule for carbohydrate (Gunther *et al.*, 1973; Young, 1974). However, the addition of negative charge does appear to affect those properties dependent on intermolecular association. For example, maleylation improves the stability of the protein at pH 7 and above but destroys its ability to precipitate with dextran (Young, 1974). Similarly, succinylation decreases the capacity to agglutinate sheep erythrocytes, and inhibits cap formation by glycoprotein receptors on lymphocyte cell surfaces: the mitogenic property of the succinylated derivative is unaffected (Gunther *et al.*, 1973). Alternative methods of fixing dimer and tetramer forms, for example, by crosslinking with glutaraldehyde, are deserving of investigation.

G. <u>Heterogeneous Binding Equilibria</u>

The binding of small molecular weight ligands to polymerized or isomerized forms of macromolecular acceptors which are in dynamic equilibrium has been treated by Nichol, Jackson and Winzor (1967). Such systems provide an important means of metabolic control (Frieden, 1971). However, in Con A at pH 5.0 and at pH 7.0 we have present a mixture of acceptors which do not interact, and it has been shown that the binding function r (grams of ligand bound per gram of acceptor) is given by a sum of expressions, each individually describing a rectangular hyperbola (McKenzie and Sawyer, 1973).

$$r = \frac{M_s}{C_T} \cdot \sum_{i=1}^{x} \frac{\bar{C}_i \tau_i K_{i,s} [S]}{M_i (1 + K_{i,s} [S])} \qquad \ldots (1)$$

The intrinsic association constant, $K_{i,s}$ (M^{-1}) pertains to each acceptor species i, M_s and M_i are the molecular weights of S (ligand) and i, respectively, C_T is the total concentration of acceptor, and τ_i is the number of equivalent and independent binding sites on i. In the case where the noninteracting species present are monomer (i = 1, molecular weight M_1) and higher polymers (molecular weight iM_1) of an acceptor, it is convenient to define a molar binding function r_m as the number of moles of ligand bound per mole of monomer. Modifying Equation (1) we may write:

$$r_m = \frac{M_1 r}{M_s} = \frac{M_1}{C_T} \cdot \sum_{i=1}^{x} \frac{\bar{c}_i \tau_i K_{i,s} [S]}{M_i (1 + K_{i,s} [S])} \qquad \ldots (2)$$

With respect to a double reciprocal plot, the ordinate intercept may be written:

$$\frac{1}{S} \xrightarrow{\lim} 0 \left(\frac{1}{r_m}\right) = \frac{C_T}{M_1} \left[\sum_{i=1}^{x} \frac{\bar{C}_i \tau_i}{iM_1}\right]^{-1} \qquad \ldots (3)$$

These general equations have been used to examine binding models and to curve fit binding data at pH 5.0 and 7.0 (McKenzie and Sawyer, 1973). The binding data were obtained using the steady-state dialysis technique of Colowick and Womack (1969). The technique employs a single dialysis cell and allows the accumulation of data sufficient for a complete binding curve within a period of approximately 30 min. Thus, problems of long term instability which interfere with equilibrium dialysis experiments are overcome.

H. Binding Equilibria of Intact and Fragmented Dimer Species

Three types of Con A dimer are possible at pH 5; homologous dimers of two intact or two fragmented units, or a hybrid dimer. Each of these species might potentially possess 0, 1 or 2 binding sites for carbohydrate depending on whether binding sites are preserved or destroyed during dimerization. The binding data at pH 5.0 are shown in double reciprocal form in Fig. 9, r being expressed on a monomer molecular weight basis. With reference to Equation (2), it can be shown that a mixture of dimeric forms with different values of τ must exist since the ordinate intercept of the reciprocal plot does not give an integral number of binding sites. Indeed, forms possessing less than two binding sites must be present since the ordinate intercept (Fig. 9) is greater than unity. Furthermore, the dimeric types present at pH 5 must have similar affinity for carbohydrate since the double reciprocal plot is linear, providing an apparent association constant of 1.65 ± 0.03 M^{-1} at 20°C. At this point we should note that nonintegral values for the number of binding sites have been obtained by other workers, even in the presence of added metal ions (So and Goldstein, 1968; Yariv et al., 1968; Young, 1974).

The value of the ordinate intercept (Equation 3) is particularly important since it permits discrimination between various binding models. For example, if fragmented monomer units do not bind carbohydrate, the ordinate intercept equals $C_T/\bar{C}_1$, and thus provides a measure of the concentration of the binding species present ($\bar{C}_1$). Such estimates disagree with results of electrophoresis experiments which show that 44% of the protein is in the fragmented form, and therefore the model can be discounted. Of twelve models considered (McKenzie and Sawyer, 1973) one agrees well with experimental data. In this model, dimers of two intact units possess two binding sites, whilst dimers of two fragmented units possess one binding site; hybrid dimers are absent. The possession of only one binding site on dimers of two fragmented units suggests some type of asymmetric association between monomer units such that one binding site is rendered inaccessible or is destroyed. An extreme example of this is the "head-to-tail" mechanism which has been proposed for the dimerization of lysozyme (Sophianopoulos, 1969; Howlett and Nichol, 1972). The detail of such association may well be provided by X-ray crystallographic studies.

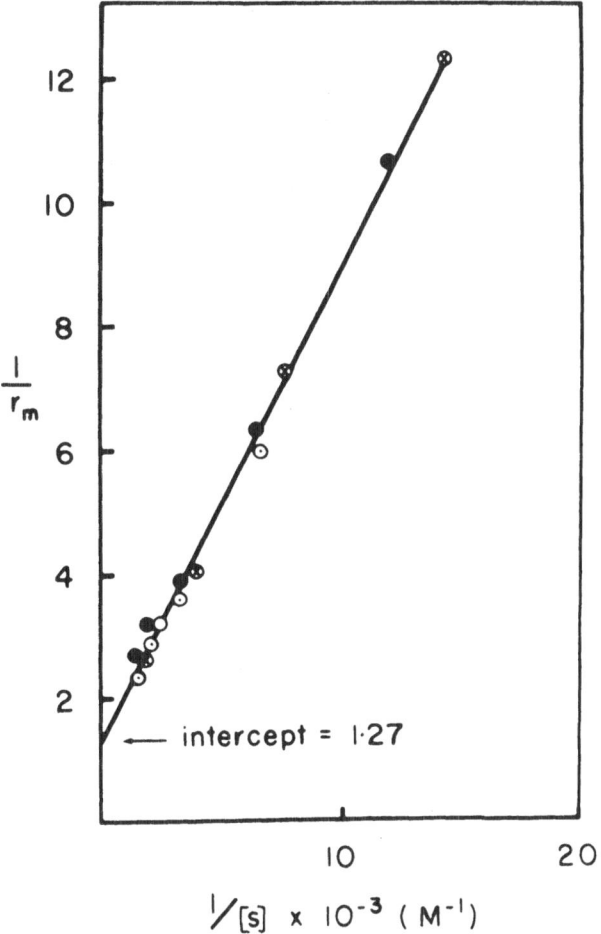

Fig. 9. Double reciprocal plot for the binding of α-D-methyl-glucopyranoside to concanavalin A at pH 5.0. The symbols represent separate experiments one of which was carried out in the presence of added metal ions (Ca^{++}, Mg^{++}, Mn^{++}). r is expressed on the basis of the monomer molecular weight. (From McKenzie and Sawyer, 1973).

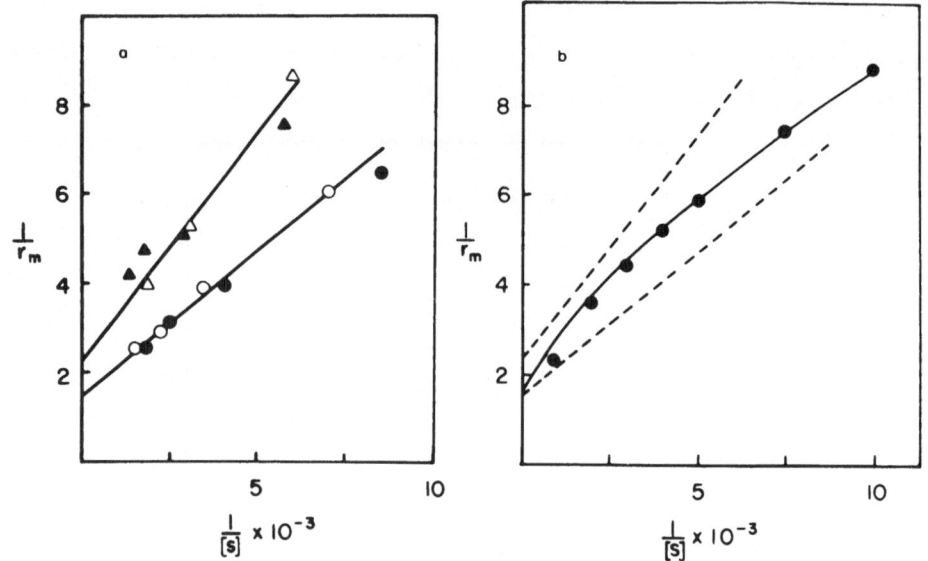

Fig. 10. Double reciprocal plots of binding data obtained at pH 7.0 (phosphate buffer, I = 0.3, 25.0°C). r_m is calculated on the basis of the monomer molecular weight. Closed and open symbols refer to separate experiments. a, Dimer (△, ▲) and tetramer (○, ●) fractions. b, Unfractionated concanavalin A containing approximately equal proportions of dimer and tetramer (●). The solid line is an attempt to fit the data according to Equation 2 using the following parameters: $M_1 = 2M_2$, $\tau_1 = 4, \tau_2 = 1$, $K_{1,s} = 0.3 \times 10^3 \text{ M}^{-1}$, $K_{2,s} = 6.5 \times 10^3 \text{ M}^{-1}$. The broken lines are for tetramer and dimer fractions as in a. (From McKenzie and Sawyer, 1973).

I. Binding Equilibria of Dimer and Tetramer

The binding data for dimeric and tetrameric Con A are presented in double reciprocal form in Fig. 10 (McKenzie and Sawyer, 1973). The slopes and intercepts for these fractions are significantly different. Analysis reveals that the data for the dimer fraction are consistent with dimers possessing one binding site for carbohydrate. Since electrophoretic analysis shows the dimer fraction to consist predominantly of fragmented units (Fig. 6) this conclusion substantiates that reached at pH 5.0. Several models are possible for the tetramer fraction but the value for the ordinate intercept suggests that the tetramer possesses four binding sites and is rich in intact units as indeed electrophoretic results show (Fig. 6).

The differing slopes and intercepts observed in Fig. 10a suggest that dimer and tetramer fractions possess different affinities for carbohydrate and thus that the double reciprocal plot of a mixture of these forms would be nonlinear. Fig. 10b indicates that such is the case, the solid line being an attempt to fit the data to Equation (2), assuming that the tetramer possesses four binding sites and the dimer one ligand binding site. In general, the binding data at pH 7.0 are "noisier" than those at pH 5.0, probably reflecting the greater instability of the protein. This, together with the fact that the dimer fraction contained approximately 20% tetramer prevents any rigorous quantitative analysis of the results at pH 7.0.

III. CONCLUSION

The molecular state and stability of Con A merits attention when investigating the biological actions of the protein. Interpretations may be clearest at pH 5.0 where the protein is dimeric and hybrid dimers are absent. Although homologous dimers of intact and fragmented forms exist and may have different valency, they have similar affinity for carbohydrate.

However, many biological experiments are carried out near neutrality where Con A exists as a mixture of dimer and tetramer forms, the proportion depending on conditions of pH, ionic strength and temperature. At present it is not known if these different polymer forms (and therefore the valency of the molecule) influence the biological actions of the protein. However, the question is open to experimental investigation since dimer and tetramer forms can be separated chromatographically. Because of preferential association of intact dimers, such separation also provides a mild method for isolating intact and fragmented species. The slight conformational differences noted between intact and fragmented species provides a molecular basis for the preferential association of intact dimers as well as for the differing binding affinities of dimer and tetramer expressed at pH 7.0.

REFERENCES

1. Abe, Y., Iwabuchi, M., and Ishii, S-I. (1971). "Multiple forms in the subunit structure of concanavalin A." *Biochem. Biophys. Res. Commun.* 45, 1271.

2. Agrawal, B. B. L., and Goldstein, I. J. (1967). "Protein-carbohydrate interaction. VI. Isolation of concanavalin A by specific adsorption on cross-linked dextran gels." *Biochim. Biophys. Acta* 147, 262.

3. Agrawal, B. B. L., and Goldstein, I. J. (1968). "Protein-carbohydrate interaction. VII. Physical and chemical studies on concanavalin A, the hemagglutinin of the jack bean." *Arch. Biochem. Biophys.* 124, 218.

4. Berlin, R. D. (1972). "Effect of concanavalin A on phagocytosis." *Nature New Biol.* 235, 44.

5. Brewer, C. F., Sternlicht, H., Marcus, D. M., and Grollman, A. P. (1973). "Binding of ^{13}C-enriched α-methyl-D-glucopyranoside to concanavalin A as studied by Carbon Magnetic Resonance." *Proc. Natl. Acad. Sci., U. S. A.* 70, 1007.

6. Colowick, S. P., and Womack, F. M. (1969). "Binding of diffusible molecules by macromolecules: rapid measurement by rate of dialysis." *J. Biol. Chem.* 244, 774.

7. Cunningham, B. A., Wang, J. L., Pflumm, M. N., and Edelman, G. M. (1972). "Isolation and proteolytic cleavage of the intact subunit of concanavalin A." *Biochemistry* 11, 3233.

8. Dessen, P., and Pantaloni, D. (1969). "Glutamate déshydrogénase. Structure quaternaire et proporiétés rotatoires." *Eur. J. Biochem.* 8, 292.

9. Edelman, G. M., Cunningham, B. A., Reeke, G. N., Becker, J. W., Waxdal, M. J., and Wang, J. L. (1972). "The covalent and three-dimensional structure of concanavalin A." *Proc. Natl. Acad. Sci., U. S. A.* 69, 2580.

10. Frieden, C. (1971). "Protein-protein interaction and enzymatic activity." *Ann. Rev. Biochem.* 40, 653.

11. Gilbert, G. A. (1959). "Sedimentation and electrophoresis of interacting substances. I. Idealized boundry shape for a single substance aggregating reversibly." *Proc. R. Soc.* A250, 377.

12. Gunther, G. R., Wang, J. L., Yahara, I., Cunningham, B. A., and Edelman, G. M. (1973). "Concanavalin A derivatives with altered biological activities." *Proc. Natl. Acad. Sci. U. S. A.* 70, 1012.

13. Hardman, K. D., and Ainsworth, C. F. (1972). "Structure of concanavalin A at 2.4 Å resolution." *Biochemistry* 11, 4910.

14. Harrington, W. F., Johnson, P., and Ottewill, R. H. (1956). "Bovine serum albumin and its behaviour in acid solution." *Biochem. J.* 62, 569.

15. Howlett, G. J., and Nichol, L. W. (1972). "Computer stimulation of sedimentation equilibrium distributions for systems involving heterogenous associations. The interaction of N-acetylglucosamine with lysozyme." *J. Biol. Chem.* 247, 5681.

16. Jack, A., Weinzierl, J., and Kalb, A. J. (1971). "An X-ray crystallographic study of demetallized concanavalin A." *J. Mol. Biol.* 58, 389.

17. Kalb, A. J., and Levitzki, A. (1968). "Metal-binding sites of concanavalin A and their role in binding of α-methyl-D-glucopyranoside." *Biochem. J.* 109, 669.

18. Karlstam, B. (1973). "Evidence of the requirement of a full complement of manganese and calcium ions for optimal binding of carbohydrates to electrophoretically homogeneous concanavalin A." *Biochim. Biophys. Acta* 329, 295.

19. McCubbin, W. D., Oikawa, K., and Kay, C. M. (1971). "Circular dichroism studies on concanavalin A." *Biochem. Biophys. Res. Commun.* 43, 666.

20. McKenzie, H. A., Sawyer, W. H., and Smith, M. B. (1967). "Optical rotatory dispersion and sedimentation in the study of association-dissociation: Bovine β-lactoglobulins near pH 5." *Biochim. Biophys. Acta* 147, 73.

21. McKenzie, G. H., Sawyer, W. H., and Nichol, L. W. (1972). "The molecular weight and stability of concanavalin A." *Biochim. Biophys. Acta* 263, 283.

22. McKenzie, G. H., and Sawyer, W. H. (1973). "The binding properties of dimeric and tetrameric concanavalin A: Binding of ligands to noninteracting macromolecular acceptors." *J. Biol. Chem.* 248, 549.

23. Moscowitz, A. (1960). In *Optical Rotatory Dispersion*. Djerassi, C., Ed., p. 150, McGraw-Hill, New York.

24. Nichol, L. W., Jackson, W. J. H., and Winzor, D. J. (1967). "A theoretical study of binding of small molecules to a polymerizing protein system. A model for allosteric effects." *Biochemistry* 6, 2449.

25. Olson, M. O. J., and Liener, I. E. (1967). "The association and dissociation of concanavalin A, the phytohemagglutinin of the jack bean." *Biochemistry* 6, 3801.

26. Pflumm, M. N., Wang, J. L., and Edelman, G. M. (1971). "Conformational changes in concanavalin A." *J. Biol. Chem.* 246, 4369.

27. Quiocho, F. A., Reeke, G. N., Becker, J. W., Lipscomb, W. N., and Edelman, G. M. (1971). "Structure of concanavalin A at 4 Å resolution." *Proc. Natl. Acad. Sci. U. S. A.* 68, 1853.

28. Reeke, G. (1975). "Structure and function of concanavalin A." Personal communication.

29. Sawyer, W. H., and Puckeridge, J. (1973). "The dissociation of proteins by chaotropic salts." *J. Biol. Chem.* 248, 8429.

30. Sawyer, W. H., McKenzie, G. H., and Nichol, L. W. (1974). "Conformational transition in concanavalin A. Conformational difference between intact and fragmented units." *Aust. J. Biol. Sci.* 27, 1.

31. Schellman, J. H., and Schellman, C. (1964). In *The Proteins*. Neurath, H., Ed., 2nd edn., vol. 2, p. 1, Academic Press, New York.

32. Sherry, A. D., and Cottam, G. L. (1973). "Proton relaxation rate and fluorometric studies of manganese and hare earth binding to concanavalin A." *Arch. Biochem. Biophys.* 156, 665.

33. Sumner, J. B., and Howell, S. F. (1936a). "The identification of the hemagglutinin of the jack bean with concanavalin A." *J. Bact.* 32, 227.

34. Sumner, J. B., and Howell, S. F. (1936b). "The role of divalent metals in the reversible inactivation of jack bean hemagglutinin." *J. Biol. Chem.* 115, 583.

35. So, L. L., and Goldstein, I. J. (1968). "Protein-carbohydrate interaction. XX. On the number of combining sites on concanavalin A, the phytohemagglutinin of the jack bean." *Biochim. Biophys. Acta* 165, 398.

36. Sophianopoulas, A. J. (1969). "Association sites of lysozyme in solution. I. The active site." *J. Biol. Chem.* 244, 3188.

37. Tanford, C. (1961). "Ionization-linked changes in protein conformation. I. Theory." *J. Amer. Chem. Soc.* 83, 1628.

38. Tanford, C., and Taggart, V. G. (1961). "Ionization-linked changes in protein conformation. II. The N→R transition in β-lactoglobulins." *J. Amer. Chem. Soc.* 83, 1634.

39. Tominatsu, Y., Vitello, L., and Gaffield, W. (1966). "Effect of aggregation on the optical rotatory dispersion of poly(α,L-glutamic acid)." *Biopolymers* 4, 653.

40. Uchida, T., and Matsumoto, T. (1972). "Heterogeneity of commercially available concanavalin A with respect to carbohydrate-binding ability." *Biochim. Biophys. Acta* 257, 230.

41. Wang, J. L., Cunningham, B. A., and Edelman, G. M. (1971). "Unusual fragments in the subunit structure of concanavalin A." *Proc. Natl. Acad. Sci., U. S. A.* 68, 1130.

42. Yahara, I., and Edelman, G. M. (1972). "Restriction of the mobility of the mobility of lymphocyte immunoglobulin receptors by concanavalin A." *Proc. Natl. Acad. Sci., U. S. A.* 69, 608.

43. Yariv, J., Kalb, A. J., and Levitzki, A. (1968). "The interaction of concanavalin A with methyl α-D-glucopyranoside." *Biochim. Biophys. Acta* 165, 303.

44. Young, N. M. (1974). "The effects of maleylation on the properties of concanavalin A." *Biochim. Biophys. Acta* 336, 46.

45. Zand, R., Agrawal, B. B. L., and Goldstein, I. J. (1971). "pH-dependent conformational changes of concanavalin A." *Proc. Natl. Acad. Sci. U. S. A.* 68, 2173.

6

STUDIES ON THE INTERACTION OF CONCANAVALIN A WITH GLYCOPROTEINS*

Avadhesha Surolia, S. Bishayee, Ateeq Ahmad,
  K. A. Balasubramanian, D. Thambi-Dorai, S. K. Podder,
  and B. K. Bachhawat
Christian Medical College Hospital
Vellore, India, and
Indian Institute of Science
Bangalore, India

ABSTRACT

Lectins (phytohaemagglutinin) are known to have the unique property of binding with certain specific sugars, polysaccharides and glycoproteins. Although the kinetics of interaction between lectins and sugar have been extensively studied, the binding characteristics of the lectins with various glycoproteins are not well understood. In this laboratory a systematic study has been initiated in relation to the interaction of lectins with glycoproteins.

Concanavalin A is known to bind α-glucosides, mannosides and biopolymers having these sugar configurations. A galactose binding protein from castor bean has been purified to homogeneity and was found to contain mannose. This lectin was used as the source of glycoprotein for studying its interaction with concanavalin A. This study showed that the interaction is temperature dependent and the dissociation is time and α-methyl glucoside concentration dependent. This has led to speculate a model for cell-lectin interaction. Using concanavalin A it has been shown that all the lysosomal enzymes from brain studied were glycoprotein in nature. Moreover, using Sepharose-bound concanavalin A it has been possible to devise a method by which these lysosomal enzymes could be purified considerably. With the knowledge that the interaction between lectin and glyco-

---

*The work was supported by a grant from the Council of Scientific and Industrial Research, India.

protein is not only dependent on the specific sugar present in the glycoprotein, but also on the nature of the glycoprotein it was possible to develop a novel method for immobilizing various glycoprotein enzymes, such as arylsulphatase A, hyaluronidase and glucose oxidase.

I. INTRODUCTION

Concanavalin A (Con A), a lectin derived from *Canavalia ensiformis*, was purified by Sumner as early as in 1919 (Sumner, 1919). However, the work of Goldstein et al. (1965) has established that Con A forms precipitates with biopolymers, such as α-mannopyranoside, α-glucopyranoside and α-*N*-acetyl-D-glucosamine, which contain the correct determinant sugars towards which Con A is specific. Considering the fact that Con A has the property of binding specific sugars, a number of interesting applications of Con A have emerged utilizing this unique property of this protein (Inbar and Sachs, 1969; Donnelly and Goldstein, 1970; Lloyd, 1970; Novogrodski and Katchalski, 1971; Dufau et al., 1972; Bessler and Goldstein, 1973; Nicolson and Singer, 1972; Mackler, 1972; Noonan et al., 1973; Allen et al., 1972; Cuatrecasas and Tell, 1973; Cuatrecasas, 1973). In brief, we would like to discuss some of the aspects mentioned below in which we have been interested for the last few years.

1. The use of lectins bound to insoluble matrix for affinity chromatography and purification of glycoproteins.

2. Investigation of the chemical nature of lysosomal acid hydrolases.

3. A new method of immobilization of glycoprotein enzymes using Con A.

4. The nature of specificity involved in the interaction between the glycoproteins and lectins.

II. METHODS

Purification of lysosomal enzymes by the use of Con A-Sepharose has been achieved (Bishayee and Bachhawat, 1973a and 1973b). This method can be used for the purification of other glycoprotein enzymes also. As shown in Fig. 1 when crude lysosomal fraction was passed through the Con A-Sepharose column, a number of enzymes were retained and these enzymes could be eluted with α-methyl-D-glucopyranoside (MG) at different pHs (Bishayee and Bachhawat, 1974a). It is apparent that pH has a profound effect even in the presence of MG for dissociating the Sepharose-Con A-enzyme complex. This has led us to conclude that binding specificity between lectin and glycoprotein is not only dependent on the nature of the sugar present in the

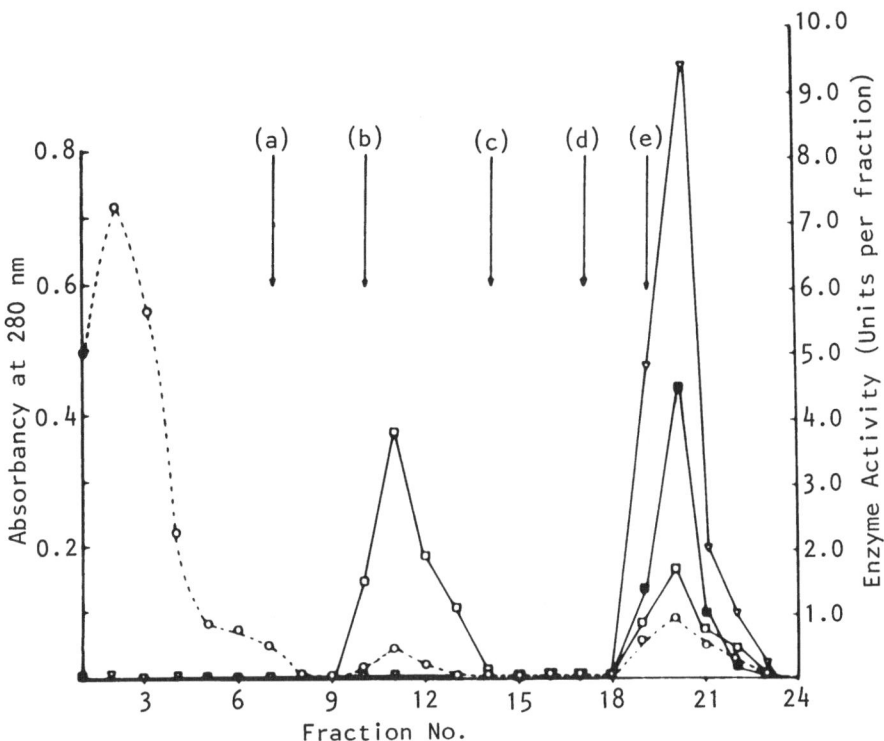

Fig. 1. Elution profile of lysosomal acid hydrolases from Con A-Sepharose column. Soluble crude lysosomal acid hydrolases from sheep brain were prepared according to the method described earlier by Bishayee and Bachhawat (1974b). The enzymes were assayed also according to this method. The arrows indicate (a) washing the column with 0.05M Na-acetate buffer, pH 5.0; (b) elution with 0.5M MG in 0.05M Na-acetate buffer; (c) washing with 0.05M Na-acetate buffer, pH 5.0; (d) equilibration with 0.05M tris-HCl buffer, pH 8.0; and (e) elution with 0.5M MG in 0.05M tris-HCl, pH 8.0. Absorbency at 280 nm (-O-), acid phosphatase (-□-), β-N-acetyl hexosaminidase (-■-), arylsulphatase A (-Δ-).

glycoprotein, but also on the nature of the glycoprotein (Surolia et al., 1973; Bachhawat and Bishayee, 1973; Bishayee and Bachhawat, 1974a), such as its isoelectric point. In this way arylsulphatase A, acid phosphatase, β-N-acetyl hexosaminidase and β-galactosidase have been purified 20-50 fold (Bishayee and Bachhawat, 1974a; Bishayee and Bachhawat, 1974b).

The glycoprotein nature of the enzymes eluted from the Con A-Sepharose column is shown by (i) the presence of carbohydrates in pH 5.0 and 8.0 eluates after extraction with chloroform-methanol (Table

TABLE I

CARBOHYDRATE COMPOSITION OF GLYCOPROTEINS FROM DIFFERENT FRACTIONS

Fractions	Neutral sugar (µg)/mg protein	Amino sugar (µg)/mg protein	NANA (µg)/mg protein
pH 5.0 supernatant fraction	13.0 ± 3.0	—	8.0 ± 2.0
Unretarded	5.0 ± 2.0	—	4.0 ± 1.0
pH 5.0 eluate	121.0 ± 10.0	13.5 ± 2.0	36.0 ± 6.0
pH 8.0 eluate	80.0 ± 7.0	7.2 ± 2.0	30.0 ± 5.0

Various pH fractions were obtained from Sepharose-Con A column as described in Fig. 1. N-acetyl neuraminic acid, neutral sugar and protein were determined according to the method of Bishayee and Bachhawat (1974b).

INTERACTION WITH GLYCOPROTEINS

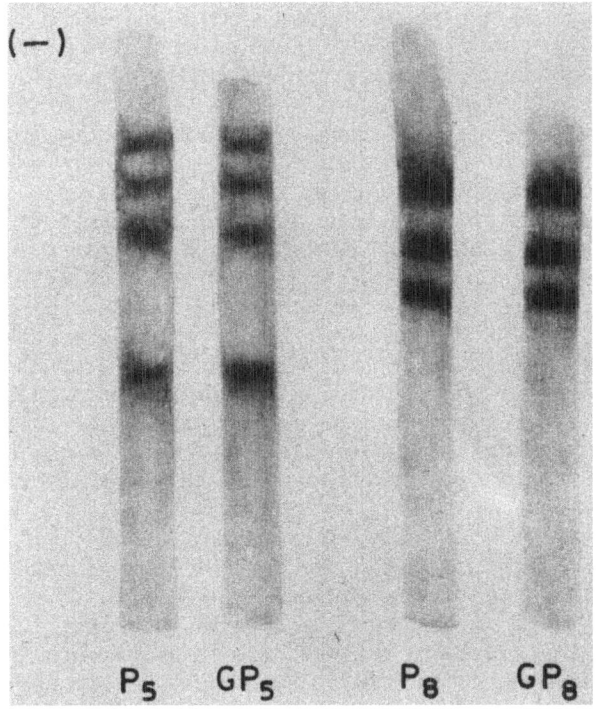

Fig. 2. Gel electrophoresis pattern of eluted materials at pH 5.0 and 8.0. Electrophoresis and staining for proteins and glycoproteins were carried out according to the method described by Bishayee and Bachhawat (1974b). P_5 and GP_5 indicate the protein and glycoprotein stains of the fraction eluted at pH 5.0 from Con A-sepharose respectively. The protein and glycoprotein staining of the fraction eluted from Con A-sepharose at pH 8.0 are indicated by P_8 and GP_8 respectively.

I), (ii) the presence of increased amount of sialic acid in the pH 5.0 and pH 8.0 eluate (Table I), and (iii) the glycoprotein stain given by all the protein bands on electrophoresis of the column eluates (Fig. 2) (Bishayee and Bachhawat, 1974b).

The specificity of binding of glycoproteins to Con A-Sepharose is indicated by the fact that the eluted fractions are enriched and the unretarded fraction is markedly depleted in neutral sugar relative to the original sample. The presence of glucose and mannose as detected by paper chromatography in the eluted fractions is in accordance with the binding specificity of Con A (Bishayee and Bachhawat, 1974b). Glucose which is not frequently present in glycoproteins has also been reported by us earlier in the highly purified sheep brain alkaline phosphatase (Saraswathi and Bachhawat, 1971). Van Nieuw Amerongen *et al.* (1972) have isolated an electrophoretically homogeneous sialoglycoprotein (GP-350) from calf brain which contains 4% glucose; Goldstone and Koenig (1973) have also detected glucose in purified lysosomal fractions of rat kidney and liver.

The elution profiles of enzyme activities indicate the presence of isoenzymes. Only a fraction of acid phosphatase and β-N-acetylhexosaminidase was released at pH 5.0 and the rest was eluted at pH 8.0. This shows the presence of at least two types of enzymes of both acid phosphatase and β-*N*-acetylhexosaminidase-differing in their MG-induced release at pH 5.0. Thus one type binds to Con A more strongly than the other at pH 5.0. Either the difference in the carbohydrate moieties or the protein-chains may lead to this differential elution of the enzymes. Since the carbohydrate compositions of the pH 5.0 and 3.0 eluates do not differ markedly except in amino sugar, the protein portion of these enzymes may play an important role in the elution.

Recently, we have developed a simple and rapid procedure for the purification of chicken brain arylsulfatase A. In this method, a Con A-Sepharose affinity column was used and 30 fold purification was achieved in a single step. The chicken brain arylsulfatase was partially purified up to the ammonium sulfate fractionation step, according to the method of Farooqui and Bachhawat (1972), and directly applied to a Con A-Sepharose column at room temperature. The binding of the enzyme to Con A attached to Sepharose occurred rapidly. However, it was observed that the enzyme is not tightly bound to Con A and it is leached out during repeated washing. Thus it was necessary to slightly modify the method of Bishayee and Bachhawat (1974b) so as to prevent the loss of enzyme. This is done by rechromatography of the washings containing the enzyme on a second Con A-Sepharose column. Thus the chicken brain arylsulfatase A interacts with covalently bound Con A differently than when compared to sheep brain arylsulfatase A and B at pH 7.4 (Bishayee and Bachhawat, 1974b). The difference in binding of the enzyme to Con A-Sepharose is not surprising since chicken brain arylsulfatase A differs from other arylsulfatase A in physico-chemical properties (Farooqui and Bachhawat, 1972). Studies are in progress to elucidate the carbohydrate structure of arylsulfatase A from different sources.

The elucidation of the carbohydrate composition of the lysosomal enzymes will throw some light on the origin of lysosomes. Milsom and Wynn (1973) have reported the presence of galactose, mannose and glucose in order of decreasing concentration in lysosomal membrane from rat liver. The work from this laboratory (Bishayee and Bachhawat, 1974b) shows the maximum concentration of mannose and the minimum of galactose are present in the lysosomal enzymes. The difference in the carbohydrate composition between lysosomal membrane and lysosomal enzymes might indicate the difference in their origin. However, these studies have not been done in the same tissue and as such it is premature to postulate any hypothesis about their origin. Using different lectins, Hennig and Uhlenbruck (1973) have shown the similarity of agglutination patterns of rat liver plasma membrane and lysosomal membrane. These indicate the possibility of a lysosomal membrane being formed from plasma membrane.

It became apparent during our studies on interaction between Con A and lysosomal acid hydrolases that lysosomal acid hydrolases form an insoluble precipitate with Con A and this was found to be stable and it is not easily dissociated. The only way one can dissociate it is to prepare the Con A specific sugar at certain conditions of pH and ionic strength. This observation led us to study the immobilization of three different glycoprotein enzymes. Thus we have employed (1) arylsulphatase A, (2) hyaluronidase, and (3) glucose oxidase. All three enzymes are known to be glycoproteins (Nichol and Roy, 1965; Borders and Raftery, 1968; Pazur *et al.*, 1965). As has been pointed out earlier, Con A specific sugars are of alpha-configuration and the use of hyaluronidase for this purpose is of special interest. Hyaluronidase degrades hyaluronic acid to form tetra-saccharide containing N-acetyl glucosamine and glucuronic acid (Aaranson and Davidson, 1967). Although N-acetyl glucosamine is a Con A specific sugar, this sugar in both the hyaluronic acid and in the enzymatically degraded product are in the β-anomeric form. Thus, it is of interest to observe the stability of the Con A-hyaluronidase complex during the enzymic reaction. Glucose oxidase employs the Con A-specific sugar, glucose, as the substrate. It was of interest to see whether we could immobilize glucose oxidase with Con A when using proper ionic strength and pH.

The immobilization procedure is quite simple. One can simply add Con A to a glycoprotein in a definite proportion in the presence of salt to obtain the Con A-enzyme complex. The glycoprotein enzyme interacts with Con A through the carbohydrate moiety of the glycoprotein enzyme and forms a precipitate. The enzyme is quite active in this complex form. This way we have been able to immobilize arylsulphatase A (Ahmad *et al.*, 1973), hyaluronidase (Balasubramanian and Bachhawat, 1974), and glucose oxidase (Thambi-Dorai and Bachhawat, 1974).

III. RESULTS

The kinetic studies indicate that the immobilized arylsulphatase A and soluble arylsulphatase A are almost identical (Ahmad *et al.*, 1973). However, the thermal stability studies indicated that this enzyme in immobilized form is extremely stable when compared to soluble enzyme (Fig. 3). It may be also mentioned that this complex was found to be stable and the enzyme could be reused 5-6 times without any appreciable loss of activity.

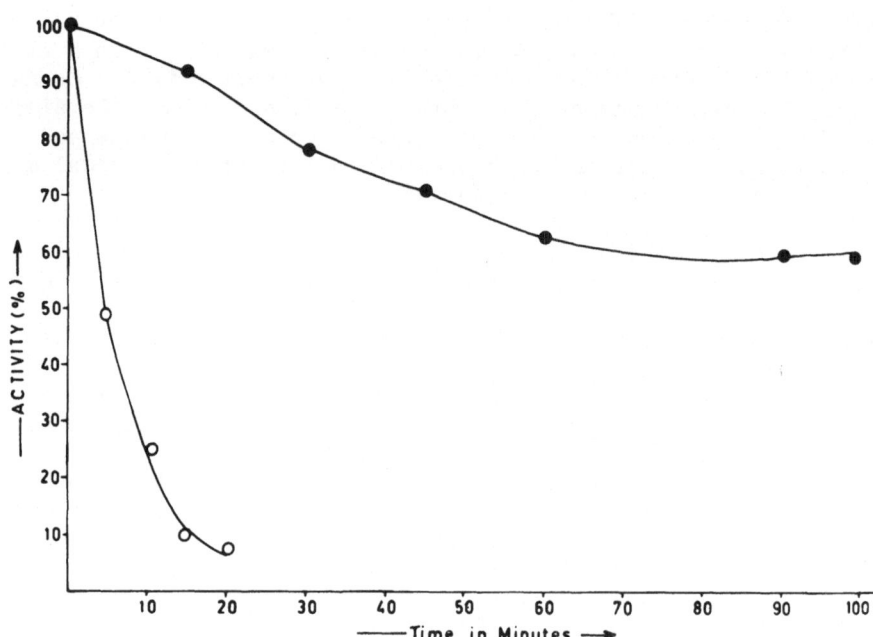

Fig. 3. Thermal inactivation of free and Con A bound arylsulphatase A at 55°C in 0.05M sodium acetate buffer, pH 5.0. Both free and Con A bound arylsulphatase A were prepared according to the method of Ahmad et al. (1973). Protein concentration used for both enzymes was 1.1 mg/ml. At different intervals of time, an aliquot of the enzyme solution was taken out and quickly chilled in ice. Arylsulphatase A, both free and immobilized, was assayed according to the method of Farooqui and Bachhawat (1972). The percentage of enzyme activity remaining after heat treatment was plotted against time.

-●- immobilized arylsulphatase A
-O- free arylsulphatase A

In the case of hyaluronidase we found that when hyaluronidase was immobilized with Con A this enzyme could be used repeatedly. However, the immobilization leads to an apparent loss of the total activity of the enzyme, and this activity could be restored if the Con A enzyme was dissociated by alpha-methyl glucoside (Table II). The kinetic studies indicate that although both free and bound hyaluronidase had the same K_m there was a marked difference in the V_{max} (Table III). These studies indicate that the diffusion of the macromolecular substrate into the immobilized matrix may limit the enzyme activity, depending on the substrate employed and also on the molecular weight of the substrate as determined by gel filtration.

TABLE II

EFFECT OF DEPOLYMERIZATION OF HYALURONIC ACID ON ENZYMATIC ACTIVITY

Substrate	Ratio of Activity of Dissociated Enzyme/Bound Enzyme
Polymerized hyaluronic acid	3.0
Depolymerized hyaluronic acid	2.1
Chondroitin sulphate	1.46

Polymerized and depolymerized hyaluronic acid was prepared according to the method of Matsumura and Pigman (1965). Assay of hyaluronidase and the immobilization of hyaluronidase were made according to the method of Aronson and Davidson (1967).

TABLE III

COMPARATIVE KINETIC DATA OF SOLUBLE AND CONCANAVALIN A-BOUND HYALURONIDASE USING VARIOUS SUBSTRATES

Enzyme	Substrate	$K_m \times 10^{-1}$ (mg)	$V_{max} \times 10^2$
Soluble	Polymerized hyaluronic acid	1.3	0.75
Bound	Polymerized hyaluronic acid	1.2	0.225
Soluble	Depolymerized hyaluronic acid	1.8	0.895
Bound	Depolymerized hyaluronic acid	1.3	0.396
Soluble	Chondroitin sulphate	3.6	1.86
Bound	Chondroitin sulphate	3.1	0.546

Conditions of the experiments were the same as described in Table II.

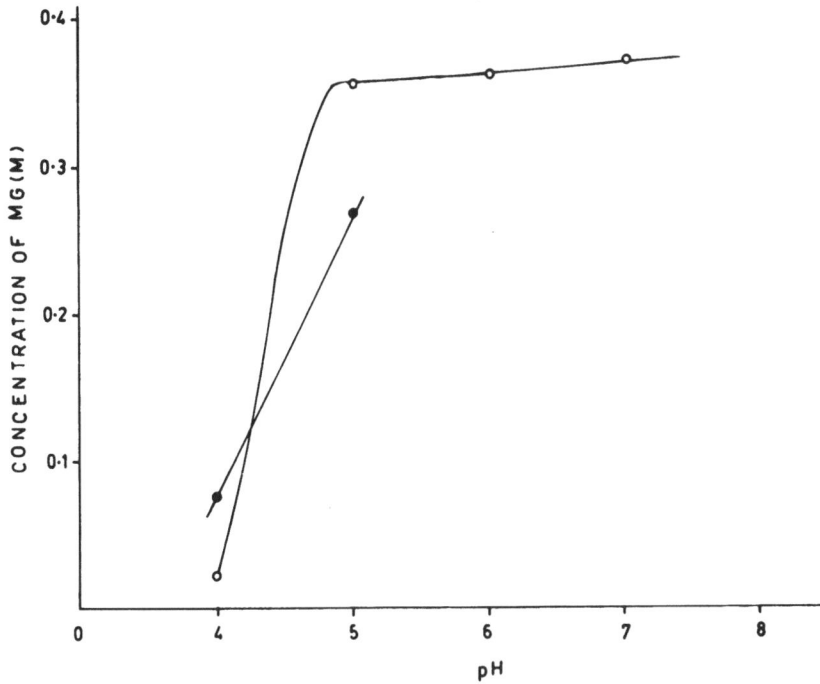

Fig. 4. Effect of pH and ionic strength on the dissociation of Con A-glucose oxidase complex by α-methyl-D-glucopyranoside. Concentration (in moles per litre, M) of α-methyl-D-glucopyranoside required for 50% dissociation of Con A-glucose oxidase complex at different pH values in the presence (-●-) and absence (-O-) of 0.2M NaCl. Procedures for assay of glucose oxidase and dissociation studies are according to the method of Thambi-Dorai and Bachhawat (1974).

Studies with Con A-glucose oxidase complex indicate that this complex could be dissociated at a lower pH with high concentration of glucose. However, the concentration of α-methyl-D-glucopyranoside required for the 50% dissociation of this Con A-enzyme complex at various pHs is dependent on pH and ionic strength (Fig. 4). Using pH 6.0 for the assay of the enzyme, it was possible to show that the glucose oxidase could be immobilized and could be used repeatedly without appreciable loss due to the dissociation of the Con A-glucose oxidase complex.

Studies on the immobilization of arylsulphatase A and B, acid phosphatase, β-N-acetyl hexosaminidase, β-galactosidase, β-glucuronidase, hyaluronidase and glucose oxidase, using Con A, showed that

Con A is a good tool for the immobilization of enzymes belonging to the glycoprotein group, provided they have the constituent sugars which are Con A specific. These Con A immobilized enzymes can be stored for a long time, and free enzymes can be obtained by the dissociation of the complex with either α-methyl-D-glucopyranoside or α-methyl-D-mannopyranoside (α-MM). A limitation of this immobilization technique is that it cannot be used with α-glucosidases and α-mannosidases, since the substrates for these enzymes will dissociate the complex. However, the studies on the Con A-glucose oxidase complex revealed that α-D-glucose, which is also a Con A-specific sugar, can be used, without hindering the stability of the complex, if the environmental conditions like pH, ionic strength, etc., are properly controlled. The fact that these immobilized enzymes can be re-used several times without appreciable loss of activity indicates the economical use of the immobilized enzymes for analytical purposes.

In recent years a number of reports have appeared regarding the binding of lectins with cell membrane glycoproteins (Kubanck et al., 1973; Janson and Burger, 1973; Allen et al., 1972). However, there had been no report regarding the nature of interaction between lectins and glycoproteins. In our studies on lectins we were able to isolate the lectins from castor bean (RC_1) in homogeneous form. This lectin was a glycoprotein and was found to contain the Con A specific sugars, mannose and glucose (Podder et al., 1974). This led us to investigate the interaction of Con A and RC_1.

The interaction of the glycoproteins-RC_1 with Con A was studied with regard to the following points:

1. whether specificity is solely determined by the sugar moiety;

2. whether the protein moiety participates in stabilizing the lectin-glycoprotein complex; and

3. if so, what the nature of this specificity is, whether it is electrostatic or hydrophobic.

The stoichiometry of the insoluble complex Con A-RC_1, as determined by analysis of the neutral sugar in the supernatant (from RC_1), revealed that a molecule of tetrameric Con A binds per molecule of RC_1, (Fig. 5). However, it is not clear how many of the four subunits of Con A are directly involved with the specific sugar residues of RC_1.

The studies of the rates of formation and dissociation of Con A-RC_1 complex were done by monitoring the change in absorbance at 320 nm upon mixing the components. When Con A was mixed with RC_1 the complex formation resulted in a time-dependent appearance of turbidity. The pseudofirst order rate constants k_f', calculated according

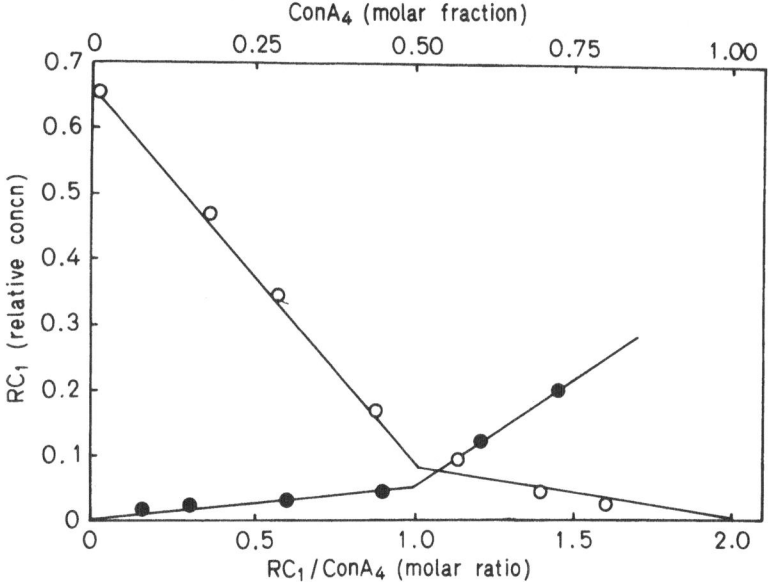

Fig. 5. Mixing experiment with Con A and RC_1 at room temperature (29°C) in 0.05 M phosphate buffer, 0.2 M NaCl, pH 6.8. In one case RC_1 was added gradually to a solution of Con A (1 μM) and the neutral sugar content of RC_1 remaining in supernatant after centrifugation was estimated from absorbance at 485nm. In the other case, the total amount of protein (2.3 μM) was kept constant but its composition varied during the filtration and neutral sugar in the supernatant was estimated similarly. The relative concentration of RC_1 remaining in the supernatant expressed in terms of concentration of neutral sugar was plotted against the molar ratio of $RC_1/Con\ A_4$ (●) and the molar fraction of Con A (○).

to the Guggenheim method, are summarized in Table IV. In these experiments only the last phase of the reaction is considered, the evaluated forward rate (k_f) constant refers to the slowest step in the consecutive reaction of Con A with RC_1, leading to the precipitation of complex (Podder et al., 1974). The rates of complex formation are found to be dependent on ionic strength as shown in Table IV. It is apparent from the table that the calculated forward rate constants (k_f) are about 1000-fold smaller than those of the N-acetyl glucosamine to lysozyme reaction (Chapman and Schimmel, 1968). This is not surprising in view of the fact that large steric factors are involved in the proper recognition of a specific sugar moiety in the glycoprotein.

TABLE IV

FORMATION RATE CONSTANT OF THE COMPLEX RC_1-Con A_4 AT 29°C IN PHOSPHATE BUFFER, pH 6.8

Medium	Conc. Con A_4	Con A_4/RC_1	k_f'	$k_f' = k_f'$ (Con A_4)
	µM		ks^{-1}	µM^{-1}ks^{-1}
0.05M Phosphate buffer + 0.2M NaCl, pH 6.8, ionic strength 0.31	0.83 1.62 3.26	3.55 7.4 14.7	2.84 4.8 8.4	3.4 2.95 2.56
0.05M Phosphate buffer pH 6.8, ionic strength 0.11	3.26	14.7	2.84	0.87

For the determination of the formation rate constant the kinetic measurements were performed under pseudofirst order condition, i.e., when one component was present in large excess. The data were analyzed according to Guggenheim (1926), as mentioned by Podder et al. (1974). The pseudo-first order rate constant thus obtained has been denoted as k_f'.

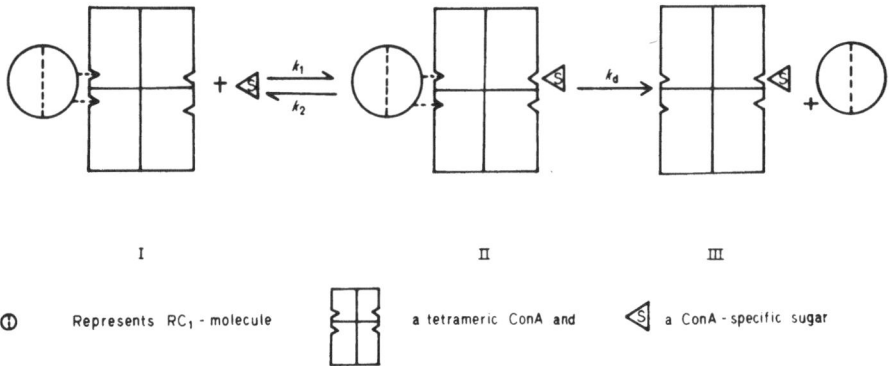

Fig. 6. Schematic representation of RC_1-Con A_4 complex: mechanism of dissociation of complex in the presence of Con A-specific sugars. K_1 and K_2 are the forward and backward rate constants respectively and K_d is the rate constant of dissociation of the complex RC_1-Con A_4. The involvement of the RC_1-binding site in complex I is not indicated. I, II, and III indicate RC_1-Con A_4, RC_1-Con A_4-sugar and Con A_4-sugar complexes respectively.

The rate constant of dissociation of Con A_4-RC_1 calculated for different concentrations of Con A specific sugars, namely α-methyl-D-glucopyranoside and α-methyl-D-mannopyranoside, are given in Table V. The rate of dissociation is found to be concentration dependent. The concentration dependence of the calculated first order rate constant from the semilog plot of turbidity vs. time can be taken to indicate that the carbohydrate recognizing sites of Con A are not completely blocked by the specific sugar residues of the glycoprotein. The ratio of first order rate constant to glycoside concentration is constant (Table V), thereby suggesting that dissociation may take place as depicted schematically in Fig. 6. The step I→II is quite fast. Therefore the step II→III is rate limiting. The proposed mechanism is consistent with the observed kinetic behavior of the α-methyl-D-glucopyranoside-induced dissociation of the lysosomal enzyme-Con A complexes at different pHs (Bishayee et al., 1974).

The concentration dependence of methyl mannoside of the Con A_4-RC_1 dissociation suggests that some of the carbohydrate binding sites of Con A bound to glycoprotein are still available to interact simultaneously with the other glycoprotein present at the cell surface (Podder et al., 1973). Consequently, the cell surface would appear to be multivalent with regard to Con A, leading to extensive agglutination. The ratio of average forward rate constant ($3mM^{-1}s^{-1}$) and dissociation rate constant ($0.07 Ks^{-1}$) gives a value of 43 μM^{-1} for the association constant of the lectin-glycoprotein complex. The

TABLE V

RATE CONSTANT OF DISSOCIATION OF THE COMPLEX RC_1-Con A_4 AT 29°C IN PRESENCE OF CONCANAVALIN A-SPECIFIC SUGAR IN 0.05M PHOSPHATE CONTAINING 0.5M NaCl, pH 6.8

Sugar	Sugar conc. c_s	k_d^1	k_d^1/c_s ($= k_dK$)	k_d^a
	mM	ks^{-1}	$mM^{-1} \times ks^{-1}$	ks^{-1}
α-Methyl-D-glucopyranoside	47	9.4	0.200	0.087
	83	16.7	0.201	0.0875
	141	27.4	0.0195	0.0846
α-Methyl-D-mannopyranoside	47.5	25.5	0.536	0.056
	74	55.0	0.730	0.076
	83	38.4	0.463	0.048
	123	64	0.52	0.054

For the study of dissociation rate constant the Con A_4-RC_1 complex was centrifuged and suspended in 0.05M Na-phosphate buffer, pH 6.8, containing 0.2M NaCl. To this suspension Con A specific sugars at different concentration were added. The time dependent disappearance of turbidity caused by Con A-specific sugars, e.g., α-MG and α-MM was followed at 320 nm and the first order rate k_d' was calculated according to Guggenheim (1926) as explained elsewhere (Podder et al., 1974).

kinetically determined K values are 1000-fold larger than the K_a of binding of specific sugar to Con A (So and Goldstein, 1968). This indicates that even though the initial recognition may take place via a carbohydrate moiety in the glycoprotein, once this has taken place further stabilization due to protein-protein interaction occurs, which amounts to 3-4 KCal. This conclusion is further supported when one compares the rate of formation and dissociation of insoluble complexes that are formed when glycogen and Con A are mixed together. The calculated formation rate constant has a value of 24 $mM^{-1}s^{-1}$ based on the fact that one Con A molecule is involved with every 300 sugar residues (Olson and Liener, 1967). The rate of dissociation of Con A glycogen by α-methyl-D-mannopyranoside is too fast to be measured by the method described here. This is to be expected if only one sugar residue is involved in the binding with Con A. This conclusion is further justifiable when one compares the affinity constants of mono-, di-, and oligosaccharides of varying lengths, suggesting that the mono- and oligosaccharides interact with Con A by a similar mechanism, most likely involving a single saccharide residue as suggested by Brown *et al.* (1975). However, the present work is not sufficient to make any quantitative statement about the nature of the stabilizing force. The fact that the dissociation of Con A-lysosomal enzyme complex by α-methyl-D-glucopyranoside is pH dependent and the forward rate constant increases with an increase in the ionic strength, seems to suggest that the reactions between Con A and RC_1 and Con A and lysosomal enzymes are ionic in character. At the same time it is pertinent to mention that the lower enthalpy value (ΔH) of -10 KCal for binding of Con A with acid phosphatase, as compared with saccharides, indicates the hydrophobic nature of the interactions (Bishayee *et al.*, 1974).

REFERENCES

1. Ahmad, A., Bishayee, S., and Bachhawat, B. K. (1973). "A novel method for immobilization of chicken brain arylsulphatase A using concanavalin A." *Biochem. Biophys. Res. Commun.* **53**, 730.

2. Allan, D., Auger, J., and Crumpton, M. J. (1972). "Glycoprotein receptors for concanavalin A isolated from pig lymphocyte membrane by affinity chromatography in sodium deoxycholate." *Nature New Biol.* **236**, 23.

3. Aaranson, N. W., Jr., and Davidson, E. A. (1967). "Lysosomal hyaluronidase from rat liver." *J. Biol. Chem.* **242**, 437.

4. Bachhawat, B. K., and Bishayee, S. (1973). "Interaction between concanavalin A and brain lysosomal acid hydrolases." *IX. Int. Cong. Biochem.*, Stockholm. Abs., p. 399.

5. Balasubramanian, K. A., and Bachhawat, B. K. (1974). Unpublished

data.

6. Bessler, W., and Goldstein, I. J. (1973). "Phytohemagglutinin purification: A general method involving affinity and gel chromatography." *FEBS Lett.* 34, 58.

7. Bishayee, S., Farooqui, A. A., and Bachhawat, B. K. (1973a). "Purification of brain lysosomal arylsulfatases by concanavalin A-Sepharose column chromatography." *Ind. J. Biochem. Biophys.* 10, 1.

8. Bishayee, S., and Bachhawat, B. K. (1973b). "Use of concanavalin A in the purification of brain lysosomal acid hydrolases." *4th Meet. Int. Soc. Neurochem.*, Tokyo, Abs., p. 399.

9. Bishayee, S., and Bachhawat, B. K. (1974a). "Interaction between concanavalin A and brain acid hydrolases." *Biochim. Biophys. Acta* 334, 378.

10. Bishayee, S., and Bachhawat, B. K. (1947b). "A study on the glycoprotein nature of brain acid hydrolases." *Neurobiol.* 4, 48.

11. Bishayee, S., Bachhawat, B. K., and Podder, S. K. (1974). "Specificity of concanavalin A towards lysosomal acid hydrolases: A kinetic study of the interaction of concanavalin A with arylsulfatase A and acid phosphatase from sheep brain." Personal Communication.

12. Borders, C. L., and Raftery, M. A. (1968). "Purification and partial characterization of testicular hyaluronidase." *J. Biol. Chem.* 243, 3756.

13. Brown, R. D., Brewer, C. F., and Koenig, S. H. (1975). "Evidence for conformational changes in concanavalin A upon binding of saccharides as determined from solvent water proton magnetic relaxation rate dispersion measurements." This book, Abstract Section, p. 323.

14. Chapman, D. M., and Schimmel, P. R. (1968). "Dynamics of lysozyme saccharide interactions." *J. Biol. Chem.* 243, 3771.

15. Cuatrecasas, P., and Tell, G. P. E. (1973). "Insulin like activity of concanavalin A and wheatgerm agglutinin - direct interaction with insulin receptors." *Proc. Nat. Acad. Sci., U. S. A.* 70, 485.

16. Cuatrecasas, P. (1973). "Interaction of concanavalin A and wheatgerm agglutinin with the insulin receptors on fat cells and liver." *J. Biol. Chem.* 248, 3228.

17. Donnelly, E. H., and Goldstein, I. J. (1970). "Glutaraldehyde insolubilized concanavalin A. An adsorbant for the specific isolation of polysaccharides and glycoproteins." *Biochem. J.* 118, 679.

18. Dufau, M. L., Tsuruhara, T., and Catt, J. (1972). "Interaction of glycoprotein hormones with agarose concanavalin A." *Biochim. Biophys. Acta* 278, 281.

19. Farooqui, A. A., and Bachhawat, B. K. (1972). "Purification and properties of arylsulphatase A from chicken brain." *Biochem. J.* 126, 1025.

20. Goldstein, I. J., Hollerman, C. F., and Smith, E. E. (1965). "Protein carbohydrate interaction. II. Inhibition studies on the interaction of concanavalin A with polysaccharide." *Biochemistry* 4, 876.

21. Goldstone, A., and Koenig, H. (1973). "Physicochemical modifications of lysosomal hydrolases during intracellular transport." *Biochem. J.* 132, 267.

22. Hennig, R., and Uhlenbruck, G. (1973). "Detection of carbohydrate structure on isolated subcellular organelles of rat liver by heterophile agglutinins." *Nature New Biol.* 242, 120.

23. Inbar, M., and Sachs, L. (1969). "Interaction of carbohydrate binding protein concanavalin A with normal and transformed cells." *Proc. Nat. Acad. Sci., U. S. A.* 63, 1418.

24. Janson, V. K., and Burger, M. M. (1973). "Isolation and characterization of agglutinin sites. II. Isolation and partial characterization of a surface membrane receptor for wheatgerm agglutinin." *Biochim. Biophys. Acta* 291, 127.

25. Kubanck, J., Entlicher, G., and Kocourek, J. (1973). "Studies on phytohemagglutinins. XIII. A phytohemagglutinin receptor from human erythrocytes." *Biochim. Biophys. Acta* 304, 93.

26. Lloyd, K. O. (1970). "Preparation of two insoluble forms of the phytohemagglutinins concanavalin A and their interaction with polysaccharides and glycoproteins." *Arch. Biochem. Biophys.* 137, 460.

27. Mackler, B. F. (1972). "Effect of concanavalin A on human lymphoid cell lines and normal peripheral lymphocytes." *Jour. Natl. Cancer. Inst., U. S. A.* 49, 935.

28. Matsumura, G. and Pigman, W. (1965). "Catalytic role of copper and iron ions in the depolymerization of hyaluronic acid by as-

corbic acid." *Arch. Biochem. Biophys.* 110, 526.

29. Milsom, D. W., and Wynn, C. H. (1973). "Protein and carbohydrate composition of lysosomal membranes." I. *Biochem. Soc. Trans.*, 1, 426.

30. Nichol, L. W., and Roy, A. B. (1965). "The sulfatase of ox liver. IX. The polymerization of sulfatase A." *Biochemistry* 4, 386.

31. Nicolson, G. L., and Singer, S. J. (1972). "Electron microscopic localization of macromolecules on membrane surfaces." *Ann. N. Y. Acad. Sci.*, N. Y. 195, 368.

32. Noonan, K. D., Ranger, H. C., Basilico, C., and Burger, M. M. (1973). "Surface changes in temperature sensitive simian virus-40 transformed cells." *Proc. Natl. Acad. Sci., U. S. A.* 70, 347.

33. Novogrodsky, A., and Katchalski, E. (1971). "Lymphocyte transformation by concanavalin A and its reversion by α-methyl-D-mannopyranoside." *Biochim. Biophys. Acta* 228, 579.

34. Olson, M. O. J., and Liener, I. E. (1967). "Some physical and chemical properties of concanavalin A, the phytohemagglutinin from jack bean." *Biochemistry* 6, 105.

35. Pazur, J. H., Kleppe, K., and Cepure, A. (1965). "A glycoprotein structure for glucose oxidase from *Aspergillus niger*." *Arch. Biochem. Biophys.* 111, 351.

36. Podder, S. K., Surolia, A., and Bachhawat, B. K. (1974). "On the specificity of carbohydrate lectin recognition. The interaction of a lectin from *Ricinus communis* with simple saccharides and concanavalin A." *Eur. J. Biochem.* 44, 151.

37. Saraswathi, S., and Bachhawat, B. K. (1971). "Glycoprotein nature of alkaline phosphatase from sheep brain." *Third International Meeting of I. S. N.*, Budapest, Abst., p. 359.

38. So, L. L., and Goldstein, I. J. (1968). "Protein-carbohydrate interaction. XX. On the number of combining sites on concanavalin A, the phytohemagglutinin of the jack bean." *Biochim. Biophys. Acta* 165, 398.

39. Sumner, J. B. (1919). "The globulins of the jack bean *Canavalia ensiformis*." *J. Biol. Chem.* 38, 137.

40. Surolia, A., Prakash, N., Bishayee, S., and Bachhawat, B. K. (1973). "Isolation and comparative physico-chemical studies of

concanavalin A from *Canavalia ensiformis* and *Canavalia gladiata*." *Ind. J. Biochem. Biophys.* 10, 145.

41. Thambi-Dorai, D., and Bachhawat, B. K. (1974). Unpublished data.

42. Van Nieuw Amerongen, A., Van Den Eijnden, D. H., Heijlman, J., and Roukema, P. A. (1972). "Isolation and characterization of soluble glucose containing sialoglycoprotein from the cortical grey matter of calf brain." *J. Neurochem.* 19, 2195.

7

INTERACTION OF CONCANAVALIN A WITH THE SURFACE OF

VIRUS - INFECTED CELLS*

George Poste
Roswell Park Memorial Institute
Buffalo, N. Y., U. S. A.

ABSTRACT

Infection of untransformed cells with a wide-range of non-oncogenic enveloped viruses causes a significant increase in their susceptibility to agglutination by concanavalin A (Con A). The increased Con A agglutinability of these cells is not caused by an increase in the number of Con A sites on the cell surface but involves alteration in the surface properties of infected cells to allow redistribution of Con A receptors to form "patches" following binding of Con A to the cell surface. Similarities between Con A-mediated agglutination of normal cells infected with non-oncogenic viruses and the agglutination response of cells transformed by oncogenic viruses will be reviewed. Finally, the use of Con A as an experimental tool to modify the replication and cytopathogenicity of non-oncogenic viruses grown in mammalian cells will be presented.

I. INTRODUCTION

The interaction of concanavalin A (Con A) with carbohydrate-containing receptors on the surface of mammalian cells has emerged as a potent experimental technique for studying the organization of the cell periphery and for determining alterations in cell surface pro-

---

*This work was supported by grants from the Medical and Agricultural Research Council of Great Britain and the Cancer Research Campaign, and by Grant No. CA-13393 from the National Institutes of Health.

perties. Considerable attention has been devoted to the use of Con A and various other plant lectins as probes to monitor differences in the surface organization of normal cells and cells transformed by oncogenic viruses or carcinogenic chemicals. As reviewed elsewhere in this book (Nicolson, 1975; Sachs, 1974), numerous studies have shown that cells transformed by RNA- or DNA-containing tumor viruses are agglutinated by Con A at doses which fail to affect untransformed cells, but the latter can be made susceptible to agglutination by low doses of Con A by brief trypsinization.

Susceptibility to agglutination by low doses of Con A is not, however, unique to the transformed cell state. Several studies have reported examples of transformed or tumor cells which fail to show an increased susceptibility to Con A agglutinability compared with their normal counterparts, and a number of normal cell types, notably cells obtained from embryonic tissues, are agglutinated by low doses of Con A without prior trypsinization (Nicolson, 1974). It has also been demonstrated recently that normal cells become susceptible to agglutination by low doses of Con A when infected with a wide variety of non-oncogenic enveloped viruses (Table I). In this paper, the mechanism of Con A-induced agglutination of cells infected with non-oncogenic viruses will be reviewed and possible similarities with the agglutination response in cells transformed by oncogenic viruses outlined. Finally, the use of Con A as an experimental tool to modify the replication and cytopathogenicity of non-oncogenic viruses growing in mammalian cells will be described.

II. METHODS, RESULTS, AND DISCUSSIONS

A. Non-Oncogenic Viruses as Experimental Tools in Cell Biology:

The diverse properties shown by the major virus groups in their mode of entry to the cell, the mechanism and site of their intracellular replication and the way in which they are assembled and released from the cell creates a complicated pattern of host cell response to virus infection. This diversity reflects the great potential of viruses in the study of cellular organization, since different viruses or individual strains of the same virus can be selected to induce any one of a wide range of changes affecting specific cellular functions.

Most of the experimental work to be described in this chapter has been done with Newcastle disease virus (NDV), a large RNA-containing virus of the paramyxovirus group, which provides an excellent experimental model for studying virus-induced alterations in the cell periphery. There are a large number of stable and well-characterized NDV strains available which differ markedly in their ability to alter the lectin agglutinability of infected cells. By using NDV strains of appropriate virulence a complete graded scale of cellular response to infection, including alterations in susceptibility to Con A aggluti-

TABLE I

CONCANAVALIN A AGGLUTINATION OF NORMAL CELLS INFECTED WITH NON-ONCOGENIC VIRUSES

Myxoviruses

Influenza A_2 Becht et al. (1972)
Fowl plague

Paramyxoviruses

Newcastle disease virus (NDV)   Becht et al. (1972)
                                         Poste and Reeve (1972)

Sendai (parainfluenza 1)   Poste and Newhouse (1974)
Parainfluenza 3
Mumps
Measles
Canine distemper
Respiratory syncytial

SV5   Becht et al. (1972)

Rhabdoviruses

Vesicular stomatitis virus   Becht et al. (1972)

Alfaviruses

Sindbis virus   Becht et al. (1972)
                                         Birdwell and Strauss (1973)

Semliki forest virus   Becht et al. (1972)

Poxviruses

Vaccinia   Zarling and Tevethia (1972)

Herpesviruses

Herpes simplex   Poste (1972a)
                                         Tevethia et al. (1972)

nation, can be achieved (Table II). This broad spectrum of graded cellular response to infection by NDV strains of differing virulence is also interesting from a virological standpoint since similar amounts of new infective virus particles are released from cells infected with virulent, mesogenic and avirulent strains, indicating that differences in the cytopathogenicity of the various virus strains are not due simply to more rapid multiplication of the more virulent strains (Reeve et al., 1974).

The origin and properties of the cloned NDV strains described in this paper and the methods for infection of cell cultures, measurement of viral infectivity, hemagglutination and virus RNA synthesis are described in detail elsewhere (Alexander et al., 1970; 1973a,b; Reeve and Poste, 1971a,b; Poste et al., 1972a,b; 1974). Con A used in experiments was purified by affinity-chromatography as described (Poste and Reeve, 1974) and succinyl-Con A and ^{3}H-Con A were prepared from affinity-chromatography purified Con A using the methods given in full elsewhere (Reeve et al., 1974).

B. The Mechanism of Con A-Induced Agglutination of Normal Cells Infected with Non-Oncogenic Viruses: Increased Mobility and Redistribution of Con A-Binding Sites on NDV-Infected Cells:

NDV and other enveloped viruses are released from infected cells by budding from the plasma membrane. During this process, virus envelope proteins are incorporated into discrete areas of the plasma membrane followed by alignment of the virus nucleocapsid beneath the modified areas of plasma membrane. New virus particles (virions) are then released by a budding process involving envelopment of the nucleocapsid by a portion of plasma membrane containing the viral envelope proteins. Since NDV envelope proteins bind Con A molecules (Becht et al., 1972; Poste et al., 1972; Reeve et al., 1974), their insertion into the plasma membrane of infected cells would be expected to increase the number of available Con A sites which would account for the increased susceptibility of infected cells to agglutination by Con A. However, measurements of the binding of ^{3}H-Con A to uninfected cells and cells infected with different NDV strains has shown similar binding to both infected and uninfected cells (Poste and Reeve, 1974) (Fig. 1). Also, cells infected with avirulent NDV strains (Queensland; Ulster) that fail to increase cellular susceptibility to Con A agglutination bind similar amounts of ^{3}H-Con A as cells infected with virulent strains (Herts; Texas; Warwick and Beaudette C) causing a marked increase in cell agglutinability (Fig. 1). These results agree with observations on cells infected with Sindbis virus (Birdwell and Strauss, 1972) and Influenza virus (Nicolson, 1973a) in which significant changes in ^{3}H-Con A binding to infected cells were not found despite an increased susceptibility of the infected cells to agglutination by Con A.

TABLE II

SUSCEPTIBILITY OF BHK AND CHICK EMBRYO (CE) CELLS INFECTED WITH DIFFERENT NEWCASTLE DISEASE VIRUS STRAINS TO AGGLUTINATION BY CONCANAVALIN A (500 µg/ml[a])

Virus Strain	Virulence	Detection of agglutination response (hours after infection)		Maximum agglutination response[b]	
		BHK	CE	BHK	CE
uninfected	control	N.F.	N.F.	0	0
Herts	high	5	5	++++	++++
Texas	high	6	6	++++	++++
Field pheasant	high	6	6	++++	++++
Warwick	high	8	7	++++	++++
Lamb	high	9	9	+++	+++
Beaudette C	medium	10	10	+++	+++
F	vaccine strain	12	12	+	+
Queensland	avirulent	N.F.	N.F.	0	0
Ulster	avirulent	N.F.	N.F.	0	0

[a]data compiled from Poste (1972), Poste and Reeve (1972), Poste et al. (1972, 1974).

[b]agglutination scored as 0, +, ++, +++ or ++++ for 0, 25, 50, 75% or > 90% of cells agglutinated, respectively, using the methods described by Poste (1972a).

N.F. - not found

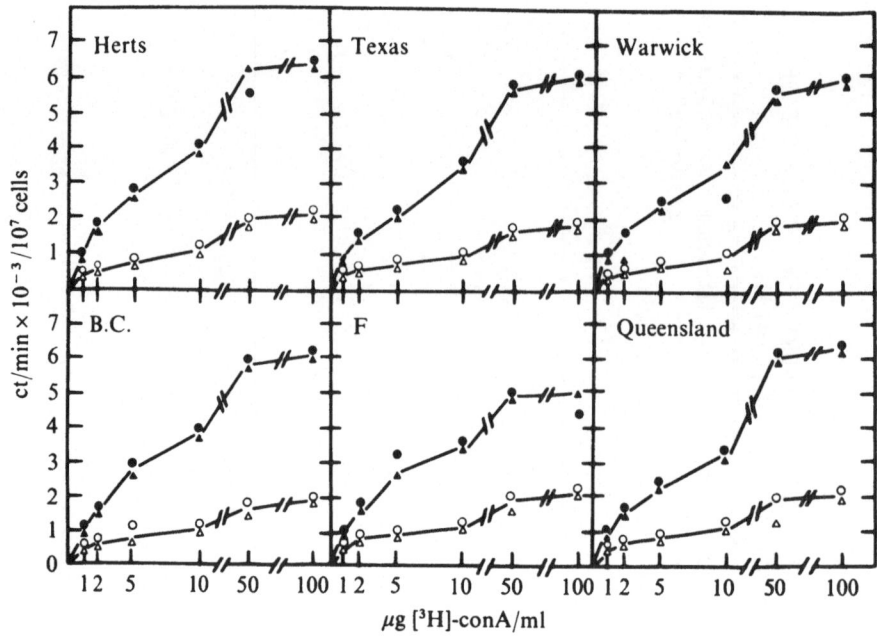

Fig. 1. Specific binding of ^{3}H-concanavalin A at 21°C and 37°C (closed symbols) or 0°C (open symbols) to BHK cells and BHK cells infected with Newcastle disease virus strains Herts, Texas, Warwick, Beaudette C (B.C.), F and Queensland (●, ○) = infected cells; (▲, △) = uninfected cells. The results for ^{3}H-Con A binding at 21°C and 37° were identical and for convenience are shown together. Measurements were made 7, 8, 8 and 10 hours after infection in cells infected with strains Herts, Texas, Warwick and Beaudette C, respectively, at which time the cells had become susceptible to agglutination by 500 μg/ml Con A which failed to agglutinate uninfected control cells. Binding of ^{3}H-Con A to cells infected with strains F and Queensland was measured 16 hours after infection. Specific binding of ^{3}H-Con A was calculated by subtracting the amount bound in the presence of 0.1 M α-methylmannoside from the amount bound in the absence of this compound. ^{3}H-Con A was prepared from affinity-chromatography purified Con A by the methods described by Poste and Reeve (1974).

The possibility that the failure to detect differences in ^{3}H-Con A binding in infected and uninfected cells might reflect changes in the size and surface area of infected cells has been excluded (Reeve et al., 1974). However, there are a number of other possible explanations for the similarity of Con A binding to infected and uninfected cells. An increase in the number of Con A binding sites on infected cells might not be detected if the number of new sites provided by insertion of viral envelope proteins into the plasma membrane was very small compared to the number of existing Con A sites. This is considered unlikely, however, in view of the large amounts of NDV envelope glycoproteins synthesized in infected cells (Alexander and Reeve, 1972) and also because the assembly and release of new virions normally occurs over a very large proportion of the cell periphery (Yunis and Connelly, 1969; Reeve et al., 1974). It is possible, however, that the failure to detect an increase in ^{3}H-Con A-binding to infected cells might be due to steric hindrance between Con A molecules. For example, the Con A-binding sites, including any new sites created by insertion of viral envelope proteins into the plasma membrane, might be able to pack more closely than the ^{3}H-Con A molecules used to detect these sites. This problem applies equally, however, to the numerous other reports in the literature where significant differences in Con A-binding have not been detected in cells with different susceptibilities to agglutination (Lis and Sharon, 1973).

The failure to detect an increase in Con A-binding sites on NDV-infected cells showing increased susceptibility to agglutination by Con A resembles previous findings on Con A agglutination of virus-transformed cells and protease-treated normal cells. With the exception of the results published from Burger's laboratory (Noonan and Burger, 1973), numerous studies have shown that agglutination of transformed cells does not result from an increase in the number of Con A-binding sites exposed at the cell periphery but involves topographical redistribution of Con A-binding sites on the cell surface to form "clusters" or "patches" (Lis and Sharon, 1973; Nicolson, 1973b). Differences in cellular susceptibility to agglutination by Con A are presumed therefore to reflect differences in the organization of the cell periphery which affect the ease with which Con A-binding sites can be redistributed to form patches.

Evidence has been obtained recently in this laboratory (Poste and Reeve, 1974) that the increased agglutinability of normal cells infected with NDV involves topographical re-arrangement of Con A-binding sites on the cell surface to form "patches", similar to the clustering of Con A-binding sites accompanying Con A agglutination of cells transformed by oncogenic viruses (Nicolson, 1971, 1972, 1973c; Rosenblith et al., 1973; Guerin et al., 1974).

Immunofluorescence measurements on the distribution of Con A-binding sites on uninfected cells indicate that the sites are distributed in a random dispersed pattern at both 4°C and 37°C (Fig. 2a) but are redistributed to form patches when the cells are trypsinized and incubated at 37°C (Fig. 2b). The change in the distribution of Con A-binding sites induced by trypsinization is also accompanied by an increase in the susceptibility of the cells to agglutination by Con A (500 µg/ml), but non-trypsinized cells remain insusceptible to

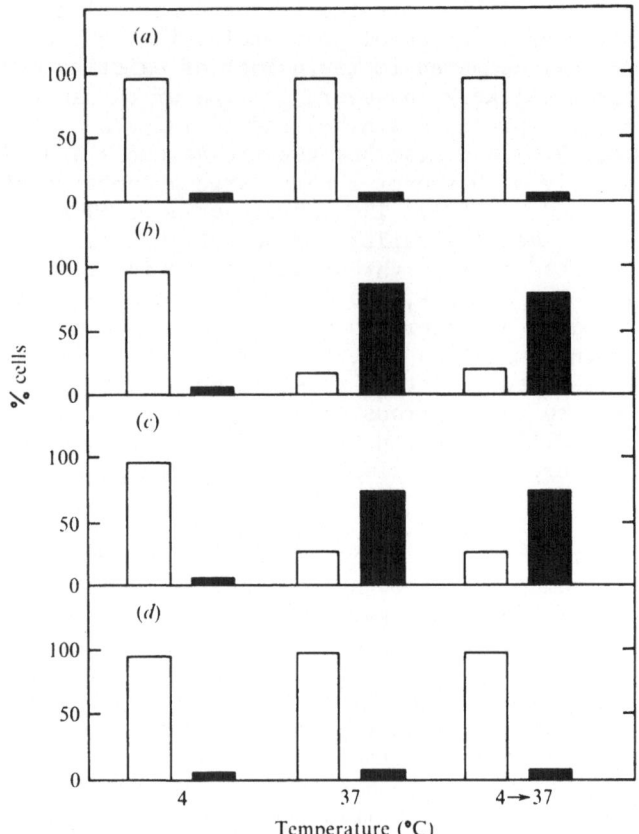

Fig. 2. The distribution of Con A-binding sites on BHK cells detected by indirect immunofluorescence using fluorescein-conjugated rabbit anti-Con A antibodies (F-anti-Con A). Cells were examined under ultraviolet illumination using a UG-1 excitation and K-430 barrier filter. Cells were exposed to Con A at 4°C, 37°C or exposed at 4°C followed by warming and incubation at 37°C for 30 minutes (4 →37°C) before addition of F-anti-Con A. The results are expressed as the percentage of cells showing uniform ring fluorescence (▭) characteristic of randomly dispersed Con A-binding sites or showing "patchy" fluorescence (■) characteristic of clustered Con A-binding sites. The results are derived from three replicate experiments and a minimum of at least 50 cells were examined in each experiment. a; distribution of Con A-binding sites at different temperatures on uninfected BHK cells exposed to Con A and F-anti-Con A before final fixation with 2.5% glutaraldehyde. b; distribution of Con A-binding sites at different temperatures on BHK cells 1 hour after treatment with 0.01% twice crystallized trypsin for 10 minutes at 37°C and treated with Con A and F-anti-Con A before fixation. c; distribution

agglutination by this concentration of Con A. Observations on BHK cells infected with the Beaudette C strain of NDV which shown an increased susceptibility to agglutination by Con A reveal a dispersed distribution of Con A-binding sites on cells exposed to Con A at 4°C, but a majority of infected cells show redistribution and clustering of Con A sites when incubated with Con A at 37°C (Fig. 2c). Similarly, treatment of infected cells with Con A at 4°C followed by warming to 37°C is accompanied by an increase in Con A agglutinability and redistribution of the previously randomly distributed Con A sites to produce patches (Fig. 2c). The change in the distribution of Con A-binding sites to form patches on the surface of infected cells occurs as cells become increasingly susceptible to agglutination by Con A (Fig. 3a). In contrast, cells infected with the avirulent DNV strain Queensland, which do not show an increased susceptibility to Con A agglutination, retain a consistent pattern of randomly dispersed Con A sites throughout infection (Fig. 3b). Further observations on Beaudette C - infected cells that were fixed with glutaraldehyde before addition of Con A revealed that Con A-binding sites are randomly distributed at all temperatures (Fig. 2d). This last observation indicates that the inherent distribution of Con A-binding sites is similar on both infected and uninfected cells. This suggests that the increased agglutinability of cells infected with virulent or mesogenic NDV strains does not result directly from virus-induced changes in the topography of Con A sites on the cell surface. Instead, the redistribution of Con A sites to form patches on infected cells appears to result from the greater mobility of the sites within the membrane compared with uninfected cells, and that redistribution of Con A sites is a secondary phenomenon induced by binding of Con A to the cell surface.

These results reinforce the similarity between Con A or agglutination of normal cells infected with non-oncogenic viruses and the agglutination response in cells transformed by tumor viruses. Three independent studies have shown recently that neither transformation nor proteolytic modification of the cell surface lead directly to clustering of Con A receptors (Inbar and Sachs, 1973; Nicolson, 1973b, c; Rosenblith et al., 1973). The inherent topography of Con A-binding sites on these cells is no different from normal cells and clustering of sites occurs only after binding of Con A to the cell surface at room temperature or above.

---

of Con A-binding sites on BHK cells 10 hours after infection with NDV strain Beaudette C exposed to Con A and F-anti-Con A before fixation. The cells were susceptible to agglutination by 500 μg/ml Con A at 21°C or 37°C but not at 4°C. d; distribution of Con A-binding sites at different temperatures on BHK cells 10 hours after infection with NDV strain Beaudette C and fixed with 2.5% glutaraldehyde before exposure to Con A and F-anti-Con A (reproduced with permission, Poste and Reeve, 1974).

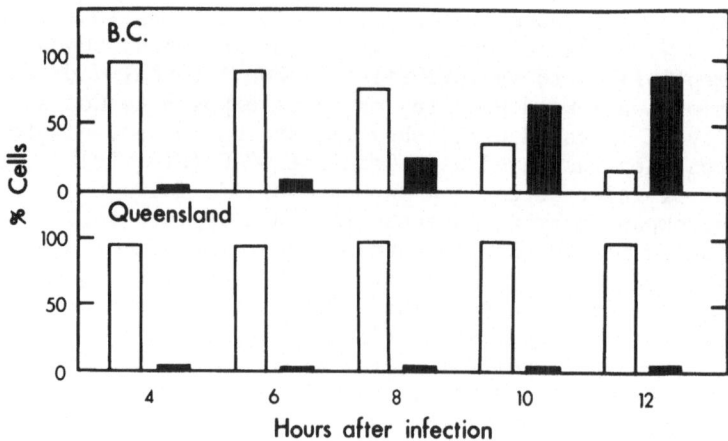

Fig. 3. The distribution of Con A-binding sites on BHK cells at intervals after infection with NDV strains Beaudette C (B.C.) and Queensland, using the methods described in the legend to Fig. 2. Cells were exposed to Con A and F-anti-Con A at 37°C before final fixation. (▭) = uniform fluorescence characteristic of randomly dispersed binding sites and (▮) = patchy fluorescence characteristic of clustered binding sites. Cells infected with strain B.C. showed increased susceptibility to agglutination by 500 μg/ml Con A from 8 hours after infection and by 12 hours after infection more than 90% of the cell population was agglutinated. Cells infected with strain Queensland were not susceptible to agglutination by this dose of Con A at any time during the experimental period up to 72 hours after infection (reproduced with permission, Poste and Reeve 1974).

The precise mechanism by which infection with NDV and a number of other non-oncogenic enveloped viruses alters the organization of the cell surface to permit increased mobility and clustering of the Con A-binding sites is unknown. Theoretically, several mechanisms might be involved. Firstly, the intrinsic fluidity of the lipid phase of the plasma membrane might be altered after infection by those NDV strains that enhance cellular agglutinability. Secondly, the Con A-binding sites themselves might be structurally modified as a result of infection and become more mobile within the membrane. A third possibility is that the movement of the plasma membrane glycoproteins that carry Con A-binding sites is controlled by peripheral membrane-associated contractile proteins attached to the inner face of the plasma membrane and the function of such proteins might be altered by infection. Currently, no information is available to distinguish between these possibilities. Nicolson (1974) has reviewed the evidence for the contribution of similar mechanisms to the increased mobility and clustering of Con A binding sites on virus-transformed and other neoplastic cells. There appears to be little available data for the involvement of either of the first two mechanisms but circumstantial evidence is available which suggests that intracellular contractile elements, probably microtubules, associated with the inner face of the plasma membrane may control the distribution of Con A-binding sites and other surface molecules on the outer face of the plasma membrane (Edelman et al., 1973; Nicolson and Painter, 1973; Ji and Nicolson, 1974). Even if a system of trans-membrane control of the distribution of Con A-binding sites by intracellular contractile proteins is eventually proven it remains to be established how not only malignant transformation but also infection with non-oncogenic viruses and a variety of enzymic alterations in the cell periphery could all influence the activity of these proteins.

C. <u>Enzymic Modification of the Cell Periphery as a Common Factor in Con A Agglutinability of Virus-Transformed Cells and Normal Cells Infected with Non-Oncogenic Viruses</u>:

The release of enzymes from lysosomes is a common cellular response to a wide range of injurious stimuli. The intracellular release of lysosomal enzymes under these conditions often leads to cell injury and death. Similarly, the massive extracellular release of enzymes from damaged lysosomes is considered to contribute to cell death and tissue damage in a wide range of allergic and other inflammatory conditions. However, there is a growing appreciation that the release of lysosomal enzymes is not always lethal to the cell. Evidence is accumulating to support the theory of sub-lethal autolysis (Weiss, 1967; Poste, 1971) which proposes that controlled extracellular release and activity of lysosomal enzymes may be an important mechanism in the modification of the cell periphery.

The concept that enzymic modification of the cell periphery might contribute to the altered surface properties of transformed and other tumor cells has gained strength in the last few years from observations made in a number of laboratories. As discussed earlier, brief proteolytic treatment of the surface of normal untransformed cells increases their susceptibility to agglutination by Con A (Burger, 1969; Inbar and Sachs, 1969; Ozanne and Sambrook, 1971; Nicolson, 1972; Poste, 1972a). This observation prompts the obvious question of whether naturally occurring enzymic modification of the cell periphery by lysosomal proteases and/or glycosidases might contribute to the increased Con A agglutinability and other altered surface properties found in transformed cells.

Some support for this concept is provided by the finding that brief proteolytic treatment of normal cell not only increases Con A agglutinability but also stimulates cell division in stationary cultures (Burger, 1970; Sefton and Rubin, 1970), reduces intracellular cyclic AMP levels to those found in rapidly-growing transformed cells (Sheppard, 1972) and "exposes" antigenic determinants characteristically found only on transformed or tumor cells (Burger, 1971; Tarro, 1973). These similarities between the surface properties of protease-treated normal interphase cells and transformed cells may be coincidental, but they invite the speculation that transformation is accompanied by enhanced enzymic modification of the cell surface.

Another related observation has been made by Hynes (1973) and Wickus and Robbins (1973) who have detected a major cell surface protein which is expressed in normal cells but is detectable only in small amounts or not at all in virus transformed cells. Importantly, mild trypsinization of normal cells results in a reduction or disappearance of this protein (at least in its original form). Hynes has interpreted this finding as consistent with the idea that the small amounts or absence of this protein on the surface of transformed cells is due to its removal by proteolytic digestion. The finding that several virus-transformed cell lines possess higher levels of lysosomal proteases and glycosidases compared with their untransformed counterparts (Bosmann *et al.*, 1974) and that transformed cells release proteases into their external environment (Rubin, 1971; Taylor *et al.*, 1972; Ossowski *et al.*, 1973; Unkeless *et al.*, 1973) lend support to this suggestion. Additional circumstantial support for the view that the surface of transformed cells is subject to continuous proteolytic modification is also offered by recent experiments which have demonstrated that treatment of virus-transformed or other tumor cells with protease-inhibitors modified their growth behavior so that they more closely resembled normal untransformed cells (Schnebli and Burger, 1972; Borek *et al.*, 1973; Latner *et al.*, 1973; Ossowski *et al.*, 1973; Taylor and Lambach, 1973; Whur *et al.*, 1973).

Modification of the cell periphery by lysosomal enzymes might also provide the mechanism by which the surface properties of normal cells change during mitosis, including the increased susceptibility of normal mitotic cells to agglutination by Con A (Burger, 1973). Non-lethal extracellular release of lysosomal enzymes is a well documented feature of mitosis (Allison, 1969) and it is tempting to speculate that modification of the cell periphery by these enzymes is responsible for the alteration in cellular agglutinability. Also, direct treatment of normal cells with isolated lysosomal enzymes has been shown to increase the susceptibility of normal cells to agglutination by Con A (Mosser *et al.*, 1973) and treatment of normal cells with agents such as vitamin A that induce extracellular release of lysosomal enzymes increases their agglutinability by Con A without impairing cell viability (Poste, 1972a; Borek *et al.*, 1973).

The evidence summarized above provides considerable circumstantial support for the view that lysosomal enzymes may represent an important mechanism in modifying the surface properties of both normal and transformed cells. It is suggested that modification of the cell periphery by lysosomal enzymes may be responsible for the increased susceptibility of mitotic normal cells and virus-transformed cells to agglutination by Con A.

There is also evidence to suggest that modification of the cell periphery by lysosomal enzymes may be responsible for the enhanced Con A agglutinability of cells infected with NDV and other non-oncogenic enveloped viruses.

Infection of normal BHK or CE cells by NDV is accompanied by significant loss of radio labelled cell surface macromolecules (Reeve et al., 1974). This loss of ^{14}C-labelled surface macromolecules from infected cells labelled with ^{14}C-glucosamine before infection is detected before new virions are released from the cell surface. The time of onset and the extent of the loss of surface macromolecules from infected cells varies, however, depending on the virus strain involved (Fig.4). Infection with virulent (Herts; Texas; Warwick) or mesogenic strains (Beaudette C) causes significant loss of ^{14}C-labelled macromolecules but infection with attenuated vaccine strains (F) or avirulent strains (Queensland) produces little detectable loss over and above that occurring in uninfected control cells (Fig. 4). There is also a close temporal correlation between the onset of loss of labelled surface macromolecules and the first detectable increase in the susceptibility of the infected cells to agglutination by Con A (Fig. 4). In addition, both these processes show a temporal correlation with the onset of detectable extracellular release of lysosomal enzymes from cells infected with various NDV strains (Reeve et al., 1972, 1974). A causal relationship between the extracellular release of lysosomal enzymes and removal of cell surface molecules in infected cells is further suggested by the finding that the loss of surface molecules is significantly reduced by treatment of cells before infection with lysosomal stabilizing agents (Reeve et al., 1974).

Evidence of enzymic modification of the surface of NDV-infected cells is also provided by recent results on the protein composition of the envelope of different NDV strains. Electrophoretic separation of detergent-treated NDV particles reveals four major structural proteins (Table III). Two of these proteins are glycosylated and have been identified as components of the virus envelope (Scheid and Choppin, 1973). The larger envelope glycoprotein, designated HN, appears to be similar in different strains of NDV. However, the second smaller envelope glycoprotein is present as Fo or F depending on the strain of NDV involved and the host cell type in which the virus was grown. The Fo glycoprotein appears to be a precursor for the F glycoprotein since brief trypsinization of virions containing

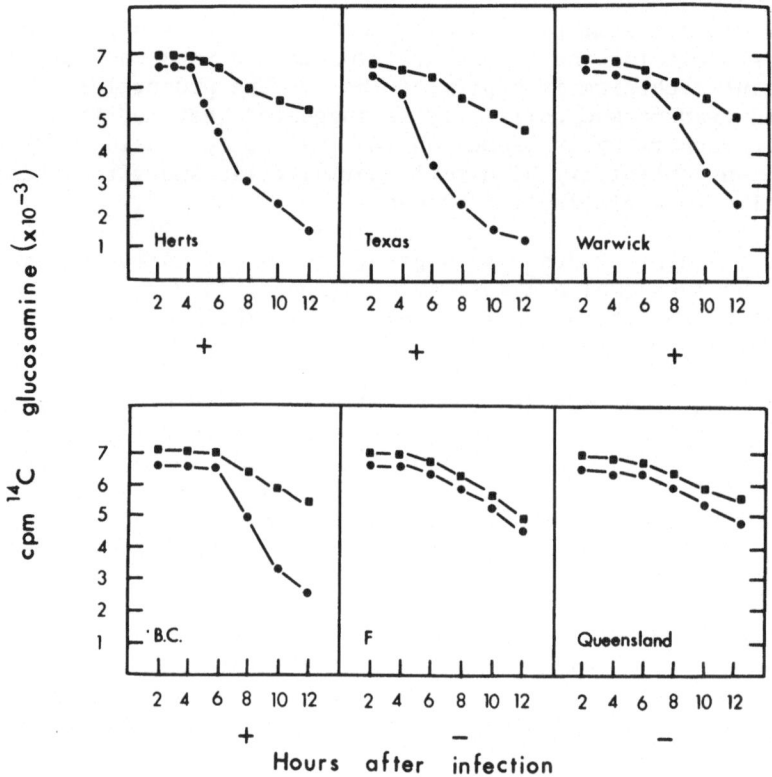

Fig. 4. Fate of steady-state ^{14}C-glucosamine label in BHK cells infected with different strains of Newcastle disease virus (●——●) and uninfected control cells (■——■). Cells were grown in medium containing 0.5 µCi/ml of ^{14}C-glucosamine (specific activity 50.8 µCi/mmole) for 24 hours, washed twice with PBS and then infected with the various NDV strains. Samples were taken at intervals after infection and the amount of cell-associated ^{14}C-glucosamine measured as described by Reeve et al. (1974). The susceptibility of the infected cells to agglutination by Con A (500 µg/ml) was also tested at intervals; + = onset of increased agglutinability; - = cells were not agglutinated by Con A at any time over the 12 hour experimental period.

Fo results in the appearance of the F glycoprotein (Fig. 5). Observations on the presence of Fo or F envelope glycoproteins in NDV virions of differing virulence grown in BHK or CE cells reveals that the Fo precursor is found only in virions of the avirulent strains Queensland and Ulster (Table IV). However, the host cell in which NDV is grown

TABLE III

PROPERTIES OF NEWCASTLE DISEASE VIRUS PROTEINS[a]

Designation	Molecular[b] Weight (daltons)	Glycosylated	Function
HN	76000	+	hemagglutinin, neuraminidase
(Fo)	71000	+	precursor of F protein (limited number of strains and host-cell induced variation)
F	62000	+	cell fusion, hemolysis
NP	62000	–	nucleocapsid subunit
M	41000	–	membrane protein (linking envelope and nucleocapsid?)

[a] data compiled from Mountcastle et al. (1971), Shapiro and Bratt (1971), Samson and Fox (1973), Scheid and Choppin (1973), and Poste and Newhouse (1974).

[b] small amounts of two additional proteins have been identified whose functions have yet to be established.

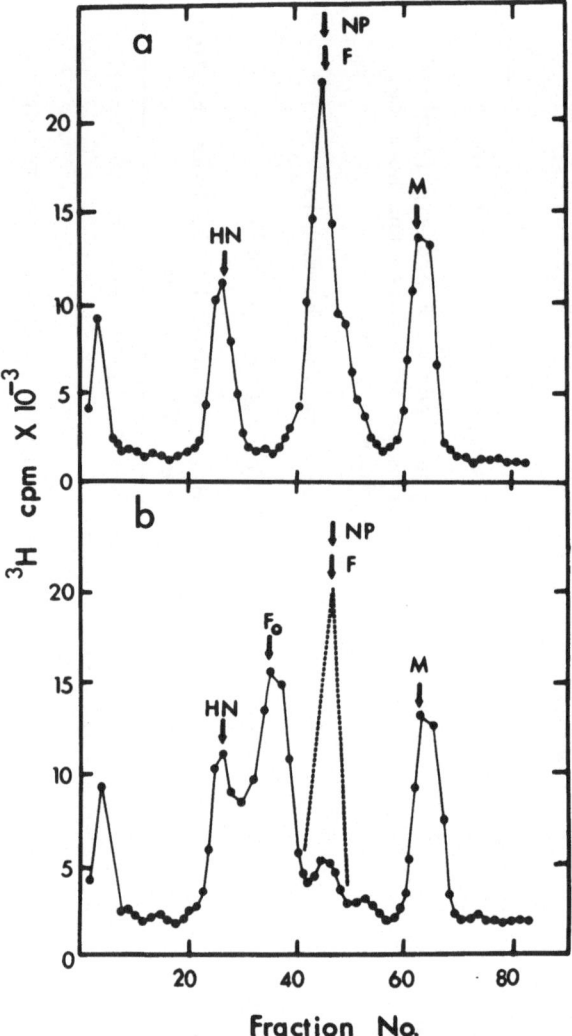

Fig. 5. a; Polyacrylamide gel electrophoresis of the polypeptides of purified virions of NDV strain Texas harvested from infected chick embryo cells. Virions were labelled with ^{3}H-leucine, disrupted with sodium dodecyl sulphate and mercaptoethanol and electrophoresed for 14 hours on 10 cm. gels of 7.5% acrylamide using the techniques described by Caliguiri et al. (1969). Arrows indicate the position of the haemagglutinin neuraminidase (HN) protein, the F envelope protein, which migrates to the same point as the nucleocapsid (NP) protein, and the membrane (M) protein. b; polyacrylamide gel electrophoresis of purified virions of NDV strain Texas harvested from infected MDBK cells. The conditions of electrophoresis and the designation of the virus structural proteins are as in Fig. 5a, except for the Fo prote-

also influences the protein composition of the virus envelope. Propagation of NDV strain Texas in several different cells reveals significant host cell-induced variation in the composition of the virus envelope (Table V). As shown in Table V, NDV strain Texas grown in bovine (MDBK) cells contains large amounts of the Fo precursor protein and is deficient in the smaller F glycoprotein. However, brief trypsinization of MDBK-grown virus results in recovery of large amounts of the F protein, presumably as a result of proteolytic cleavage of the Fo precursor (Table V). In contrast, when grown in other mammalian cells and *in ovo* the virions possess large amounts of the F envelope protein (Table V). Since strain Texas and other virulent NDV strains cause marked cell damage when grown in eggs and BHK or CE cells yet produce much less cytopathogenicity when grown in MDBK cells (Reeve *et al.*, 1974), one possible interpretation of the results in Table V is that the level of free lysosomal enzymes in infected cells, which would directly reflect the extent of virus-induced cell damage, might determine whether the Fo protein was cleaved to produce the F protein. This interpretation might also apply to the results in Table IV, since the avirulent strains Ulster and Queensland produce very little cell damage (Reeve and Poste, 1971) or lysosomal enzyme release (Reeve *et al.*, 1972) and this could account for the predominance of the uncleaved Fo precursor protein in these strains.

More definitive evidence that host cell lysosomal enzymes produce cleavage of the Fo protein in infected cells has been obtained by the finding that virulent NDV strains grown in BHK cells pretreated with lysosomal stabilizing agents possess larger amounts of the Fo envelope protein than virions grown in untreated control cells (Table VI).

Proteolytic cleavage of virus envelope proteins has also been described recently in cells infected with Sendai virus (Homma and Ohuchi, 1973; Scheid and Choppin, 1974) and Influenza virus (Lazarowitz *et al.*, 1973a,b) and a similar role for host cell lysosomal enzymes in this process has been proposed (Lazarowitz *et al.*, 1973b).

---

in which represents a precursor for the F protein (shown by the dotted line). Treatment of MDBK-grown virions with twice-crystallized trypsin (1 μg trypsin per 75 μg viral protein for 10 minutes at 37°C) results in cleavage of the Fo protein and the appearance of the F protein at the point indicated by the dotted line. The NP protein is not shown in this figure and its position is merely marked by an arrow.

In the electrophoregrams in both a and b the origin is on the left and the anode on the right.

TABLE IV

CHARACTERIZATION OF ENVELOPE GLYCOPROTEINS IN NDV VIRIONS HARVESTED FROM BHK CELLS

Virus strain	Virulence	Envelope Proteins[a]			Agglutination of infected cells by Con A (500 µg/ml)[b]
		HN	Fo	F	
Herts	high	+	-	+	++++
Texas	high	+	-	+	++++
Warwick	high	+	-	+	++++
Beaudette C	medium	+	-	+	+++
F	vaccinal	+	+	+	+
Ulster	low	+	+	-	0
Queensland	low	+	+	-	0

[a]Virions were purified and envelope proteins identified by polyacrylamide gel electrophoresis using the methods given in the legend to Fig. 5. + = present; - = absent.

[b]Cell agglutination scored as given in the legend to Table II.

TABLE V

HOST - CELL INDUCED VARIATION IN THE ENVELOPE GLYCOPROTEINS OF NDV STRAIN TEXAS

Host Cell	Envelope Proteins[b]			Agglutination of infected cells by Con A (500 μg/ml)[c]
	HN	Fo	F	
Embryonated egg	+	–	+	N.D.
BHK	+	–	+	++++
CE	+	–	+	++++
MDBK	+	+	–	o/+
MDBK (trypsin)[a]	+	–	+	N/D.

[a] purified virions (75 μg protein) treated with 1 μg crystalline trypsin for 10 minutes at 37°C.

[b] virions were purified and envelope proteins identified by polyacrylamide gel electrophoresis using the methods given in Fig. 5. + = present; – = absent.

[c] cell agglutination scored as given in the legend to Table II.

N.D. = not done.

TABLE VI

MODIFICATION OF NEWCASTLE DISEASE VIRUS ENVELOPE PROTEINS AND CON A AGGLUTINABILITY IN CELLS TREATED WITH LYSOSOMAL STABILIZING AGENTS AND PROTEASE INHIBITORS

Treatment	% envelope proteins[c]			Con A agglutination[d]
	HN	Fo	F	
untreated control[a]	100	0	100	++++
5x10^{-5}M diphenhydramine HCl[b]	95	65	35	+(+)
30 µg/ml chloroquine diphosphate[b]	95	80	18	o/+
100 µg/ml soybean trypsin inhibitor[b]	95	85	10	o/+

[a]BHK cells were infected with NDV strain Texas (20 EID$_{50}$ per cell).

[b]diphenhydramine and chloroquine were added to cells 2 hours before infection and cells were also incubated in media containing these compounds after infection. Soybean trypsin inhibitor was added to infected cell cultures 4 hours after infection.

[c]determined from purified virion preparations and percentage values calculated from areas under peaks of radioactive electrophorogram profiles.

[d]cellular susceptibility to agglutination by Con A (500 µg/ml) measured 8 hours after infection and scored as in the legend to Table II.

The cleavage of the Fo precursor envelope protein in NDV-infected cells may therefore provide a sensitive indicator of enzymic modification of the cell periphery. It is of interest to note that the cleaved F protein is found in virions of virulent NDV strains harvested from cells which show a significantly increased susceptibility to agglutination by Con A (Table V). In contrast, virions from avirulent strains (Table IV) or virulent strains grown in cell types such as MDBK in which they produce only limited cytopathic effects (Table V) both contain large amounts of the uncleaved Fo protein and it is perhaps not insignificant that the cells from which these virions were obtained do not show an increased susceptibility to agglutination by Con A (Tables IV and V).

In summary, it is proposed that modification of the cell periphery by lysosomal proteases and/or glycosidases may be a common factor in determining the increased susceptibility to Con A agglutination found in transformed cells, normal mitotic cells, and normal cells infected with non-oncogenic viruses or treated with other agents causing lysosomal enzyme release such as vitamin A.

D. Modification of Virus Release and Cytopathogenicity in Cells Treated with Con A:

In addition to the use of Con A as a probe to monitor virus-induced alterations in cell surface properties, Con A and other lectins can also be used to experimentally modify the intracellular replication of viruses.

E. Effect of Con A on Virus Release:

Saturation binding of Con A to the surfaces of cells infected with myxoviruses and paramyxoviruses causes complete inhibition of virus release (Rott et al., 1972; Poste et al., 1974; Reeve et al., 1974).

Treatment of cells infected with NDV strain Herts with 50 µg/ml Con A immediately after virus adsorption and penetration causes complete inhibition of virus release (Table VII). Similar results have also been obtained by binding of Con A to cells infected with a range of other NDV strains (Poste et al., 1974). The rate of viral RNA synthesis in Con A-treated infected cells is similar to that in untreated infected control cells (Fig. 6), indicating that the failure to detect new infective virus is due to a block in virus release rather than a direct inhibition of virus replication.

It has been suggested (Poste et al., 1974) that inhibition of the release of enveloped viruses at the cell surface in Con A-treated cells results from the ability of Con A to act as a ligand by cross-

TABLE VII

THE RELEASE OF VIRAL HAEMAGGLUTININ FROM BABY HAMSTER KIDNEY (BHK), CHICK EMBRYO (CE) AND BOVINE KIDNEY (MDBK) CELLS INFECTED WITH NEWCASTLE DISEASE VIRUS STRAIN HERTS (20 EID_{50} PER CELL) AFTER TREATMENT WITH CONCANAVALIN A (50 µg/ml)[a]

Time of removal of Con A (hours after infection)[b]	Released Haemagglutinin (HAU/0.2 ml)[c]		
	BHK	CE	MDBK
untreated control	16	64	64
1	2	2	2
2	2	2	2
3	2	2	2
4	2	2	2
5	2	2	2
6	2	2	2
Con A + 0.1 M α-methyl-D-glucopyranoside[d]	16	64	64
Con A + 0.1 M α-methyl-mannoside[d]	16	64	64

[a] data taken in part from Poste et al., (1974).

[b] Con A was added at the end of the virus adsorption period and removed at the times shown.

[c] measured 16 hours after infection; HAU = viral hemagglutination units.

[d] Con A was preincubated with inhibitor for 1 hour at 37°C, added to cells at the end of virus adsorption and removed after 2 hours.

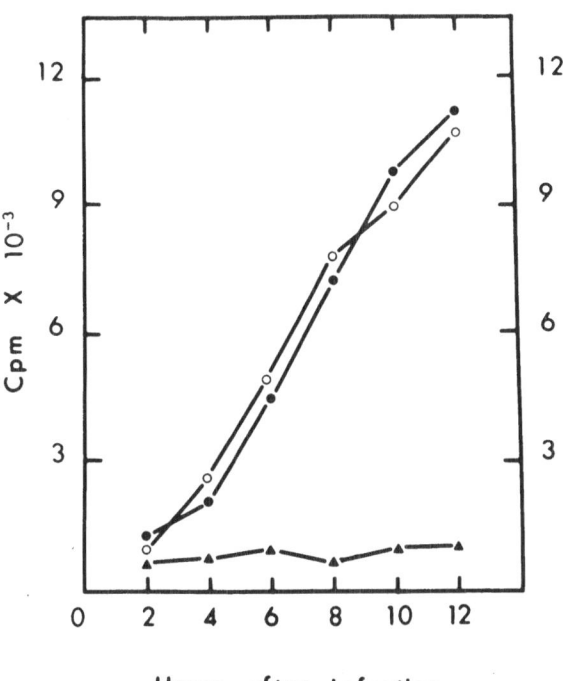

Fig. 6. The accumulative incorporation of ^{3}H-uridine into an acid-insoluble fraction in the presence of actinomycin in BHK cells infected with Newcastle disease virus strain Herts (20 EID_{50} per cell) and similarly infected cells treated with 50 µg/ml Con A added 1 hour after infection. Actinomycin D (5 µg/ml) and ^{3}H-uridine (1 µCi/ml; specific activity 30.8 Ci/mmole) were added 30 minutes after infection. (0―――0) = specific incorporation into Con A treated cells; (●―――●) = virus-infected control cells without Con A; and (▲―――▲) = uninfected control cells without Con A. (Reproduced with permission, Poste et al., 1974).

linking adjacent Con A-binding sites on the cell surface. Binding of multivalent Con A molecules to the surface of infected cells could impair virus release in two ways. Firstly, the cross-linking and lattice formation between surface molecules induced by Con A would reduce the mechanical flexibility and deformability of the plasma membrane which may well hinder the budding process by which new virions are released. Secondly, binding of Con A molecules to the cell surface might interfere directly with the assembly of the virus envelope. The envelope proteins of myxo- and paramyxoviruses are first synthesized in association with intracellular membranes, inserted subsequently into the plasma membrane and finally concentrated in areas of the plasma membrane from which virus budding occurs (Compans, 1973). The mechanism by which virus envelope proteins are transported from intracellular membranes onto the cell surface is presently unclear. However, it has been proposed (Allison, 1971) that once the virus envelope proteins are introduced into the plasma membrane, they are concentrated in specific areas by lateral diffusion within the fluid lipid matrix of the membrane. If this proposal is correct, then it is likely that binding of Con A to saccharide residues on both viral and host cell proteins in the plasma membrane would limit the movement and aggregation of viral proteins within the membrane, thereby frustrating the assembly of the virus envelope even before budding of new virions could take place.

The inhibition of virus release by Con A bound to the cell surface bears certain similarities to the inhibition of the release of myxoviruses from cells treated with antibodies prepared against viral envelope proteins such as neuraminidase and the hemagglutinin (Compans *et al.*, 1969; Dowdle *et al.*, 1974). Although this phenomenon has been interpreted as indicating that viral neuraminidase is involved in virus release (Seto and Chang, 1969), it has been shown recently that multivalent antibody molecules are necessary to block virus release (Becht *et al.*, 1971) and monovalent Fab antineuraminidase fragments are ineffective despite being able to inhibit the viral enzyme. These results are therefore similar to the effects of Con A on virus release, in which native tetravalent molecules inhibit release while divalent succinyl-Con A and trypsin-cleaved Con A have no effect (Poste *et al.*, 1974; Reeve *et al.*, 1974).

F. Effect of Con A on Virus-Induced Cell Fusion:

NDV and other large RNA-containing lipid-enveloped viruses induce cell fusion by two distinct mechanisms (Poste, 1972b). Some confusion still persists in the literature regarding this concept and it is perhaps pertinent to outline the differences between the two forms before discussing the effect of Con A on this aspect of NDV-induced cytopathology.

The first type of NDV-induced cell fusion, called fusion from without (FFWO), occurs independently of virus replication and is induced equally well by infective or non-infective inactivated viruses. Since FFWO can be produced by inactivated viruses, the synthesis of new virus-specific material is not required. This type of cell fusion is completed within 1 to 3 hours of treating cells with very high multiplicities of virus and cell fusion is caused by fusion of the virus envelope with the plasma membrane of the adjacent cells. The second type of NDV-induced cell fusion, called fusion from within (FFWI), occurs after infection at moderate or low virus multiplicities. Intracellular replication of the virus and the synthesis of virus-specific proteins are necessary to induce cell fusion, though the production of new infective virions is not obligatory and only the "early" sequences involving synthesis of virus-specific proteins need be completed (Poste, 1970).

Incubation of CE or BHK cells with Con A (50 µg/ml) for 30 minutes before exposure to high multiplicities (1500 EID_{50} per cell) of virulent NDV strains causes complete inhibition of virus-induced FFWO (Poste *et al.*, 1974). More recent studies in this laboratory have shown that binding of Con A to the cell surface also inhibits FFWO induced by inactivated Sendai virus. However, preincubation of cells of Con A prevents the attachment of both NDV and Sendai virus to their receptors on the cell surface (Okada and Kim, 1972; Poste, *et al.*, 1974), and this alone dictates that FFWO cannot occur since, as indicated above, this type of cell fusion requires fusion of the virus envelope with the plasma membrane of the host cell.

Treatment of BHK or CE cells with 50 µg/ml Con A up to 6 hours after infection with the virulent NDV strains Herts, Texas and Warwick and the mesogenic strain Beaudette C causes marked inhibition of the FFWI which normally accompanies infection by these strains (Table VIII). When added later than 6 hours after infection Con A is increasingly ineffective in inhibiting FFWI (Table VIII).

The mechanism by which Con A inhibits NDV-induced FFWI may be similar to its inhibitory effect on virus release. It has been shown that NDV-induced FFWI requires the insertion of virus-coded proteins into the plasma membrane of infected cells (Reeve *et al.*, 1972). Con A might therefore inhibit FFWI by impairing the correct association of these viral proteins with the plasma membrane by creating cross-linkages and lattice formation between adjacent cellular and viral proteins on the cell surface along the lines proposed for the inhibition of virus released by Con A.

G. <u>Alteration of Viral Cytopathogenicity by Con A</u>:

Under normal conditions of infection the ability of NDV strains to damage cells cultured *in vitro* is in general related directly to

TABLE VIII

THE EFFECT OF CONCANAVALIN A (50 μg/ml) ON CELL FUSION FROM WITHIN INDUCED BY NEWCASTLE DISEASE VIRUS STRAINS HERTS, TEXAS, WARWICK AND BEAUDETTE C[c]

Time of Con A addition[b] (hours after infection)	Uninfected Control	% Polykaryocytosis[c]			
		Herts	Texas	Warwick	Beaudette C
untreated control	2.5	75.6	68.7	65.2	43.9
1	2.5	3.7	2.8	1.9	1.6
2	2.8	2.6	1.5	3.3	2.9
4	2.3	3.1	2.6	3.4	3.2
6	3.4	11.8	9.3	13.8	12.5
8	1.7	39.7	46.3	29.9	21.6
12	2.9	70.4	60.3	45.7	46.2
Con A + 0.1M α-methyl-D-glucopyranoside[d]	1.6	65.7	66.4	70.3	45.9

[a]data taken in part from Poste et al. (1974).

[b]Con A was removed after 2 hours.

[c]measured 16 hours after infection.

[d]Con A was preincubated with inhibitor for 1 hour at 37°C, added to cells 1 hour after infection and removed after 2 hours.

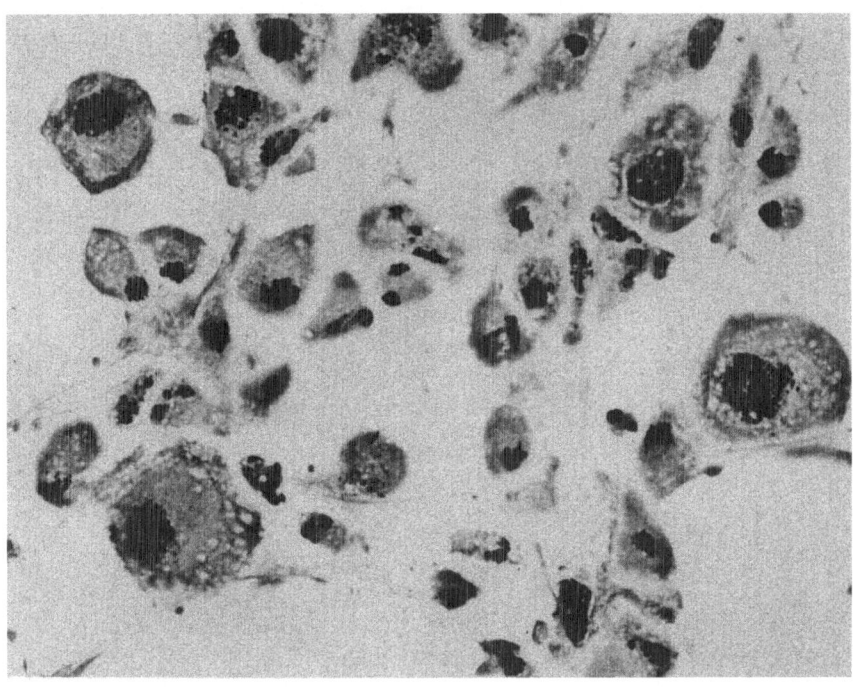

Fig. 7. Con A-treated BHK cells 16 hours after infection with Newcastle disease virus strain Herts (30 EID_{50} per cell) showing extensive cell damage and vacuolation. There is considerable variation in cell size and many cells show significant nuclear damage. Con A (50 µg/ml) was added 1 hour after infection and removed after 2 hours. Stain, May-Grunwald-Giemsa; magnification x450. (Reproduced with permission, Poste et al., 1974).

their virulence for chick embroys and chickens in vivo (Reeve and Poste, 1971; Reeve et al., 1971, 1972). For example, the normal cytopathic effect (CPE) induced by virulent or mesogenic NDV strains in BHK or CE cell cultures involves extensive cell damage, inhibition of host cell macromolecular metabolism and the formation of large polykaryocytes by cell fusion (FFWI). However, as described in the previous section, treatment of cells infected with virulent strains with Con A modifies this type of CPE by inhibiting cell fusion. Instead, Con A-treated cells infected with virulent or mesogenic NDV strains display a new type of CPE which is characterized by extensive cell damage, including marked fragmentation and destruction of cell nuclei (Fig. 7). This type of CPE is not found in uninfected cells treated with identical concentrations of Con A for the same time or in infected control cells without Con A or in cells treated with

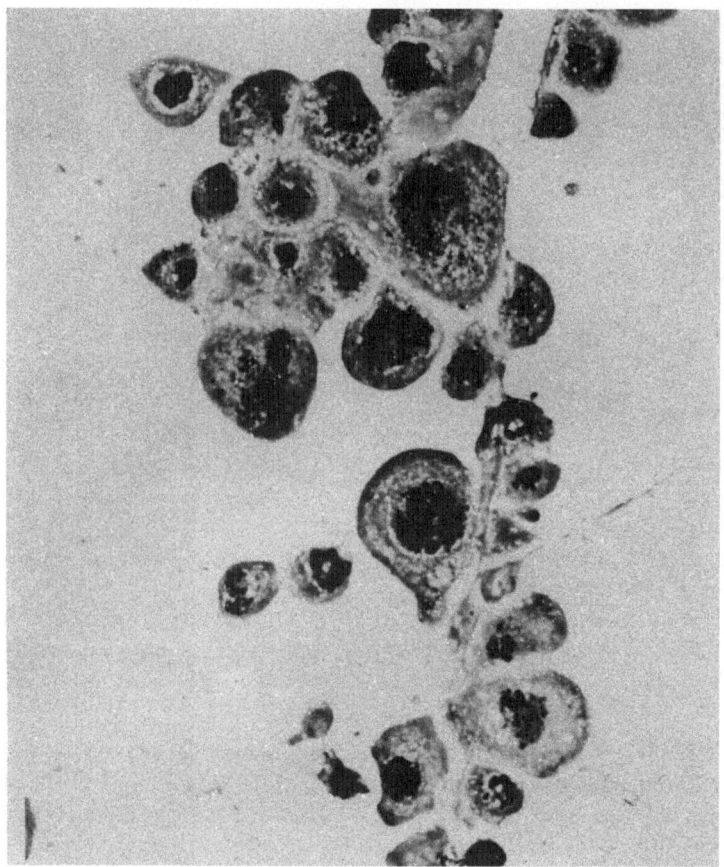

Fig. 8. Con A-treated BHK cells 16 hours after infection with Newcastle disease virus strain Queensland (50 EID_{50} per cell) showing cellular vacuolation and extensive distortion and fragmentation of cell nuclei. The cells are not firmly attached to the glass and show a tendency to round up and form clumps. Con A (50 µg/ml) was added 1 hour after infection and removed 2 hours later. Stain, May-Grunwald-Giemsa; magnification x450. (Reproduced with permission, Poste et al., 1974).

Con A preincubated with α-methylmannoside. Although the CPE shown in Fig. 7 is not normally found in NDV-infected cells, an even more striking modification of NDV-induced CPE is found in Con A-treated cells infected with the avirulent strains F and Queensland. Under normal conditions cells infected with strain F show only a limited degree of FFWI and little other damage, while Queensland-infected cells show no obvious morphological change at the light microscope level. However, treatment of BHK cells infected with these strains with Con A results in extensive cell damage (Fig. 8) similar to that found in Con A-treated cells infected with virulent strains (Fig. 7).

Poste et al. (1974) have suggested that the increased cytopathogenicity of avirulent NDV strains in Con A-treated cells results from the inhibition of virus release imposed by Con A, which leads to intracellular accumulation of viral subunits and inhibition of host cell macromolecular synthesis. Previous studies on the replication of NDV strains of differing virulence have shown that under normal conditions of infection the differing ability of strains to damage cells does not result simply from differences in their rate of replication and the production of new infective virions. A more important factor in determining their cytopathogenicity appears to be the balance between the rate at which viral subunits are synthesized intracellularly and the rate of their subsequent assembly and release as new virions at the cell surface (Reeve et al., 1971; Alexander et al., 1973a,b). For example, normal infection by virulent strains causing extensive CPE is characterized by very high intracellular titers of virus subunits and the rate of subunit synthesis exceeds the rate at which new virions can be assembled and released from the plasma membrane. Conversely, in cells infected under normal conditions with low virulence vaccine strains or naturally avirulent strains, the intracellular titer of viral subunits remains consistently low, but the number of new virus particles released from cell is equal to that in cells infected with virulent strains. Thus, in cells infected with these latter strains there is an efficient balance between subunit synthesis in the cytoplasm and the assembly and release of new virions at the cell surface. However, when virus release is blocked from cells infected with avirulent strains by binding of Con A molecules at the cell surface, this balance is disturbed and viral subunits begin to accumulate intracellularly resulting in inhibition of host cell protein synthesis and death of the cell (Poste et al., 1974), similar to the sequence of events occurring normally in cells infected with virulent strains.

REFERENCES

1. Alexander, D. J., and Reeve P. (1972)."The Proteins of Newcastle disease virus. 2. Virus-induced proteins." Microbios 5, 247.

2. Alexander, D. J., Reeve, P., and Allan, W. H. (1970). "Characterization and biological properties of the neuraminidase of strains of Newcastle disease virus which differ in virulence." Microbios 2, 155.

3. Alexander, D. J., Reeve, P., and Poste, G. (1973a). "Studies on the cytopathic effects of Newcastle disease virus: RNA synthesis in infected cells." J. Gen. Virol. 18, 369.

4. Alexander, D. J., Hewlett, G., Reeve, P., and Poste, G. (1973b). "Studies on the cytoplasmic effects of Newcastle disease virus: the cytopathogenicity of strain Herts 33 in five cell types." *J. Gen. Virol.* **21**, 323.

5. Allison, A. C. (1969). "Lysosomes and Cancer." In: *Lysosomes in Biology and Pathology*, eds. J. T. Dingle and H. B. Fell, **2**, 178. North-Holland, Amsterdam.

6. Allison, A. C. (1971). "The role of membranes in the replication of animal viruses." *Int. Rev. Exp. Pathol.* **10**, 181.

7. Becht, H., Hammerling, U., and Rott, R. (1971). "Undisturbed release of influenza virus in the presence of univalent antineuraminidase antibodies." *Virology* **46**, 337.

8. Becht, H. A., Rott, R., and Klenk, H-D. (1972). "Effect of concanavalin A on cells infected with enveloped RNA viruses." *J. Gen. Virol.* **14**, 1.

9. Birdwell, C. R., and Strauss, J. H. (1973). "Agglutination of sindbis virus and of cells infected with sindbis virus by plant lectins." *J. Virol.* **11**, 502.

10. Borek, C., Grob, M., and Burger, M. M. (1973). "Surface alterations in transformed epithelial and fibroblastic cells in culture: a disturbance of membrane degradation versus biosynthesis?" *Exp. Cell Res.* **77**, 207.

11. Bosmann, H. B., Lockwood, T., and Morgan, H. R. (1974). "Surface biochemical changes accompanying primary infection with Rous sarcoma virus. II. Proteolytic and glycosidase activity and sublethal autolysis." *Exp. Cell Res.* **83**, 25.

12. Burger, M. (1969). "A difference in the architecture of the surface membrane of normal and virally transformed cells." *Proc. Nat. Acad. Sci. U.S.A.* **62**, 994.

13. Burger, M. (1970). "Changes in the chemical architecture of transformed cell surfaces." In: *Permeability and Function in Biological Membranes*, eds. L. Bolis, A. Katchalsky, R. Keynes, and W. Loewenstein, p. 107. North-Holland, Amsterdam.

14. Burger, M. M. (1971). "Forssman antigen exposed on surface membrane after viral transformation." *Nature New Biol.* **231**, 125.

15. Burger, M. M. (1973). "Surface changes in transformed cells detected by lectins." *Fed Proc.* **32**, 91.

16. Caliguiri, L. A., Klenk, H-D., and Choppin, P. W. (1969). "The proteins of the parainfluenza virus SV 5. 1. Separation of virion polypeptides by polyacrylamide gel electrophoresis." *Virology* 39, 460.

17. Compans, R. W. (1973). "Distinct carbohydrate components of influenza virus glycoproteins in smooth and rough cytoplasmic membranes." *Virology* 55, 541.

18. Compans, R. W., Dimmock, N. J., and Meir-Ewert, H. (1969). "Effect of antibody to neuraminidase on the maturation and hemagglutinating activity of an influenza A 2 virus." *J. Virol.* 4, 528.

19. Dowdle, W. R., Downie, J. C., and Laver, W. G. (1974). "Inhibition of virus release by antibodies to surface antigens of influenza viruses." *J. Virol.* 13, 269.

20. Edelman, G. M., Yahara, I., and Wang, J. L. (1973). "Receptor mobility and receptor-cytoplasmic interactions in lymphocytes." *Proc. Nat. Acad. Sci. U.S.A.* 70, 1442.

21. Guerin, C., Zachowski, A., Prigent, B., Paraf, A., Dunia, I., Diawara, M. A., and Benedetti, E. L. (1974). "Correlation between the mobility of inner plasma membrane structure and agglutination by concanavalin A in two cell lines of MOPC 173 plasmacytoma cells." *Proc. Nat. Acad. Sci. U.S.A.* 71, 114.

22. Gunther, G. R., Wang, J. L., Yahara, I., Cunningham, B. A., and Edelman, G. M. (1973). "Concanavalin A derivatives with altered biological activities." *Proc. Nat. Acad. Sci. U.S.A.* 70, 1012.

23. Homma, M., and Ohuchi, M. (1973). "Trypsin action on the growth of Sendai virus in tissue culture cells. III. Structural difference of Sendai viruses grown in eggs and tissue culture cells." *J. Virol.* 12, 1457.

24. Hynes, R. O. (1973). "Alteration of cell-surface proteins by viral transformation and by proteolysis." *Proc. Nat. Acad. Sci. U.S.A.* 70, 3170.

25. Inbar, M., and Sachs, L. (1969). "Interaction of the carbohydrate-binding protein concanavalin A with normal and transformed cells." *Proc. Nat. Acad. Sci. U.S.A.* 63, 1418.

26. Inbar, M., and Sachs, L. (1973). "Mobility of carbohydrate-containing sites on the surface membrane in relation to the control of cell growth." *FEBS Letters* 32, 124.

27. Ji, T. H., and Nicolson, G. (1974). *Proc. Nat. Acad. Sci. U.S.A.*

In press.

28. Latner, A. L., Longstaff, E., and Pradhan, K. (1973). "Inhibition of malignant cell invasion *in vitro* by a proteinase inhibitor." *Brit. J. Cancer* 27, 460.

29. Lazarowitz, S. G., Compans, R. W., and Choppin, P. W. (1973a). "Proteolytic cleavage of the hemagglutinin polypeptide of influenza virus. Function of the uncleaved polypeptide HA." *Virology* 52, 199.

30. Lazarowitz, S. G., Goldberg, A. R., and Choppin, P. W. (1973b). "Proteolytic cleavage by plasmin of the HA polypeptide of influenza virus: Host cell activation of serum plasinogen." *Virology* 56, 172.

31. Lis, H., and Sharon, N. (1973). "The biochemistry of plant lectins (phytohemagglutinins)." *Ann. Rev. Biochem.* 42-541.

32. Mosser, A. G., Janoff, A., and Blondin, J. (1973). "Increased concanavalin A-dependent agglutinability of mouse fibroblasts after treatment with leukocyte lysosomal preparations." *Cancer Res.* 33, 1092.

33. Nicolson, G. L. (1971). "Difference in topology of normal and tumour cell membranes shown by different surface distributions of ferritin-conjugated concanavalin A." *Nature New Biology* 233, 244.

34. Nicolson, G. L. (1972). "Topography of membrane concanavalin A sites modified by proteolysis." *Nature New Biology* 239, 193.

35. Nicolson, G. L. (1973a). *Personal Communication*.

36. Nicolson, G. L. (1973b). "The relationship of a fluid membrane structure to cell agglutination and surface topography." *Ser. Haematol.* 6, 275.

37. Nicolson, G. L. (1973c). "Temperature-dependent mobility of concanavalin A sites on tumour cell surfaces." *Nature New Biology* 243, 218.

38. Nicolson, G. L. (1975). This book p. 153.

39. Nicolson, G. L. (1974). *Int. Rev. Cytol.* In press.

40. Nicolson, G. L., and Painter, R. G. (1973). "Anionic sites of human erythrocyte membranes. II. Antispectrin-induced transmembrane aggregation of the binding sites for positively charged colloidal particles." *J. Cell Biol.* 59, 395.

41. Noonan, K. D., and Burger, M. M. (1973). "Binding of (3H)-concanavalin A to normal and transformed cells." *J. Biol. Chem.* 248, 4286.

42. Okada, Y., and Kim, J. (1972). "Interaction of concanavalin A with enveloped viruses and host cells." *Virology* 50, 507.

43. Ossowski, L., Quigley, J. P., Kellerman, G. M., and Reich, E. (1973). "Fibrinolysis associated with oncogenic transformation: Requirements of plasminogen for correlated changes in cellular morphology, colony formation in agar, and cell migration." *J. Exp. Med.* 138, 1056.

44. Ozanne, B., and Sambrook, J. (1971). "Binding of radioactively labelled concanavalin A and wheat germ agglutinin to normal and virus transformed cells." *Nature New Biology* 232, 156.

45. Poste, G. (1970). "Virus-induced polykaryocytosis and the mechanism of cell fusion." *Adv. Virus Res.* 16, 303.

46. Poste, G. (1971). "Sub-lethal autolysis. Modification of cell periphery by lysosomal enzymes." *Exp. Cell Res.* 67, 11.

47. Poste, G. (1972a). "Changes in susceptibility of normal cells to agglutination by plant lectins following modification of cell coat material." *Exp. Cell Res.* 73, 319.

48. Poste, G. (1972b). "Mechanisms of virus-induced cell fusion." *Int. Rev. Cytol.* 33, 157.

49. Poste, G., and Newhouse, P. (1974). Unpublished observations.

50. Poste, G., and Reeve, P. (1972). "Agglutination of normal cells by plant lectins following infection with nononcogenic viruses." *Nature New Biology* 237, 113.

51. Poste, G., and Reeve, P. (1974). "Increased mobility and redistribution of concanavalin A receptors on cells infected with Newcastle disease virus." *Nature* 247, 469.

52. Poste, G., Reeve, P., Alexander, D. J., and Terry, G. (1972a). "Studies on the cytopathogenicity of Newcastle disease virus: Effect of lectins on virus infected cells." *J. Gen. Virol.* 17, 81.

53. Poste, G., Waterson, A. P., Terry, G., Alexander, D. J., and Reeve, P. (1972b). "Cell fusion by Newcastle disease virus." *J. Gen. Virol.* 16, 95.

54. Poste, G., Alexander, D. J., Reeve, P., and Hewlett, G. (1974).

"Modification of Newcastle disease virus release and cytopathogenicity in cells treated with plant lectins." *J. Gen. Virol.* 23, 255.

55. Reeve, P., and Poste, G. (1971a). "Cell fusion by Newcastle disease virus in the absence of RNA synthesis." *Nature New Biology* 229, 157.

56. Reeve, P., and Poste, G. (1971b). "Studies on the cytopathogenicity of Newcastle disease virus: relation between virulence, polykaryocytosis and plaque size." *J. Gen. Virol.* 11, 17.

57. Reeve, P., Alexander, D. J., Pope, G., and Poste, G. (1971). "Studies on the cytopathic effects of Newcastle disease virus: metabolic requirements." *J. Gen. Virol.* 11, 25.

58. Reeve, P., Poste, G., Alexander, D. J., and Pope, G. (1972). "Studies on the cytopathic effects of Newcastle disease virus: cell surface changes." *J. Gen. Virol.* 15, 219.

59. Reeve, P., Poste, G., and Alexander, D. J. (1974). In: *Negative Strand Viruses*, eds. R. Barry and B. W. J. Mahy. Academic Press, New York and London. In press.

60. Rosenblith, J. Z., Ukena, T. E., Yin, H. H., Berlin, R. D., and Karnovsky, M. J. (1973). "A comparative evaluation of the distribution of concanavalin A-binding sites on the surfaces of normal, virally-transformed, and protease-treated fibroblasts." *Proc. Nat. Acad. Sci. U.S.A.* 70, 1625.

61. Rott, R., Becht, H., Klenk, H-D., and Scholtissek, C. (1972). "Interactions of concanavalin A with the membrane of influenza virus infected cells and with envelope components of the virus particle." *Z. Naturforsch.* 27B, 227.

62. Rubin, H. (1970). "Overgrowth stimulating factor released from Rous sarcoma cells." *Science* 167, 1271.

63. Sachs, L. (1974). *Personal communication*.

64. Samson, A. C. R., and Fox, C. F. (1973). "Precursor protein for Newcastle disease virus." *J. Virol.* 12, 579.

65. Scheid, A., and Choppin, P. W. (1973). "Isolation and purification of the envelope proteins of Newcastle disease virus." *J. Virol.* 11, 263.

66. Scheid, A., and Choppin, P. W. (1974). "Identification of biological activities of paramyxovirus glycoproteins. Activation of cell fusion, hemolysis, and infectivity by proteolytic

cleavage of an inactive precursor protein of Sendai virus." *Virology* 57, 475.

67. Schnebli, H. P., and Burger, M. M. (1972). "Selective inhibition of growth of transformed cells by protease inhibitors." *Proc. Nat. Acad. Sci. U.S.A.* 69, 3825.

68. Sefton, B. M., and Rubin, H. (1970). "Release from density dependent growth inhibition by proteolytic enzymes." *Nature, London* 227, 843.

69. Seto, J. T., and Chang, F. S. (1969). "Functional significance of sialidase during influenza virus multiplication: an electron microscope study." *J. Virol.* 4, 58.

70. Shapiro, S. C., and Bratt, M. S. (1971). "Proteins of four biologically distinct strains of Newcastle disease virus." *Proc. Soc. Exp. Biol. Med.* 136, 834.

71. Sheppard, J. R. (1972). "Difference in the cyclic adenosine 3', 5'-monophosphate levels in normal and transformed cells." *Nature New Biology* 236, 14.

72. Taylor, J. M., and Lambach, K. J. (1973). *Personal communication.*

73. Taylor, J. C., Hill, D. W., and Rogolsky, M. (1972). "Detection of caseinolytic and fibrinolytic activities of BHK-21 cell strains." *Exp. Cell Res.* 73, 422.

74. Tarro, G. (1973). "Appearance in trypsinized normal cells of reactivity with antibody presumably specific for malignant cells." *Proc. Nat. Acad. Sci. U.S.A.* 70, 325.

75. Tevethia, S. S., Lowry, S., Rawls, W. E., Melnick, J. L., and McMillan, V. (1972). "Detection of early cell surface changes in herpes simplex virus infected cells by agglutination with concanavalin A." *J. Gen. Virol.* 15, 93.

76. Unkeless, J. C., Tobia, A., Ossowski, L., Quigley, J. P., Rifkin, D. B., and Reich, E. (1973). "An enzymatic function associated with transformation of fibroblasts by oncogenic viruses. I. Chick embryo fibroblast cultures transformed by avian RNA tumor viruses." *J. Exp. Med.* 137, 85.

77. Weiss, L. (1967). *The Cell Periphery, Metastasis and Other Contact Phenomena.* North-Holland, Amsterdam.

78. Whur, P., Robson, R. T., and Payne, N. E. (1973). "Effect of a protease inhibitor on the adhesion of Ehrlich ascites cells to

host cells in vivo." *Brit. J. Cancer* 28, 417.

79. Wickus, G. G., and Robbins, P. W. (1973). "Plasma membrane proteins of normal and Rous sarcoma virus-transformed chick-embryo fibroblasts." *Nature New Biology* 245, 65.

80. Yunis, E. J., and Donnelly, W. H. (1969). "The ultrastructure of replicating Newcastle disease virus in the chick embryo chorioallantoic membrane." *Virology* 39, 352.

81. Zarling, J. M., and Tevethia, S. S. (1971). "Expression of concanavalin A binding sites in rabbit kidney cells infected with vaccinia virus." *Virology* 45, 313.

8

CONCANAVALIN A AS A QUANTITATIVE AND ULTRASTRUCTURAL

PROBE FOR NORMAL AND NEOPLASTIC CELL SURFACES*

Garth L. Nicolson
The Salk Institute for Biological Studies
San Diego, California, U. S. A.

ABSTRACT

Concanavalin A (Con A) has been popularly used as a cell surface probe for normal and neoplastic cells. Differences between normal fibroblasts and their transformed derivatives were examined using Con A agglutination, quantitative labeling with ^{125}I-Con A and ultrastructural labeling with fluorescent- or ferritin-Con A. Con A agglutinates confluent-SV3T3 and 3T3 cells at midpoints of 20-60 and >1,500-2,000 μg/ml, respectively, and sparse cells at 5-15 and 1,200-1,500 μg/ml, respectively. Quantitative binding of ^{125}I-Con A at 4°C (10 or 15 min) with saturating lectin concentrations does not indicate a difference in the number of Con A receptors on sparse or confluent 3T3 and SV3T3 cells similar to many publications, but contrary to Noonan and Burger (1973). Under these conditions of labeling, ferritin-Con A is not internalized, indicating absence of endocytosis. The lateral mobility of Con A receptors and their relative ability to be aggregated on the cell surface by Con A was investigated with fluorescent- and ferritin-Con A. The initial distribution of Con A receptors on 3T3, SV3T3 and MSV3T3 cells under conditions of labeling at low temperature (0-5°C) or to fixed cells was essentially randomly dispersed, but changes quickly to aggregated on SV3T3 and MSV3T3 (but not 3T3) after shifting the temperature to 20-37°C, indicating, in general, a greater relative mobility of Con A receptors on SV3T3 and MSV3T3 cells. The aggregated Con A receptors seem to be directly involved in cell agglutination because they are usually found at the

---

*Supported by N.C.I. contract CB-33879 from the Tumor Immunology Program and grant CA-75122, NSF grant P4B1719 from the Human Cell Biology Program and a grant from the Cancer Research Institute, Inc.

sites of cell-to-cell contact during 10 min agglutination experiments with ferritin-Con A. When confluent-3T3 cells are labeled on monolayer with ferritin-Con A at 0-4°C, washed and then shifted to 20-37°C for 10-15 min prior to *in situ* embedding, two classes of Con A receptors can be identified. One class appears to have low relative mobility and is associated with the 3T3 cell's extensive sub-plasma membrane microfilament network, while the other is capable of being aggregated and eventually endocytosed. On confluent-SV3T3 cells, only the latter class of receptors appears to be present, indicating a possible loss of cytoplasmic control over the distribution and mobility of lectin-binding sites on transformed cell surfaces.

I. INTRODUCTION

Plant lectins such as concanavalin A (Con A) have been used to distinguish differences between normal and transformed cells grown *in vitro* (reviews: Burger, 1973; Nicolson, 1974a,b,c,; Lis and Sharon, 1973). Normal cells are characterized by their ability to show density-dependent inhibition of growth, and this property is lost after transformation concomitantly with the appearance of enhanced cell tumorigenicity. Inbar and Sachs (1969a,b) first used Con A to show that a variety of transformed cells agglutinate at much lower concentrations than their normal counterparts. The enhanced agglutinability of transformed cells was thought to be due to the unmasking of "cryptic" lectin-binding sites because normal cells briefly treated with proteolytic enzymes were found to be as agglutinable as transformed cells (Burger, 1969).

The first report on the number of Con A-binding sites on normal and transformed cells utilized ^{63}Ni-Con A, and the results indicated that transformed cells have more Con A sites than normal cells (Inbar and Sachs, 1969a), although the non-specific labeling was very high in these experiments, and several investigators have found ^{63}Ni-Con A to be unsuitable as a quantitative labeling reagent. Using ^{3}H- and ^{125}I-labeled Con A, other investigators subsequently found that there is little or no difference between the number of Con A sites on normal or transformed cells (Arndt-Jovin and Berg, 1971; Cline and Livingston, 1971; Ozanne and Sambrook, 1971; Inbar *et al.*, 1971; Nicolson, 1973a; Barbarese *et al.*, 1973; Rosenblith *et al.*, 1973; Phillips *et al.*, 1974). Also, quantitative studies using lectins from *Glycine max* (soy bean) (Sela *et al.*, 1971), *Ricinus communis* (Nicolson, 1973a) and *Pisum sativum* (pea) (Trowbridge, 1974) indicate that other lectins with differing saccharide specificities bind equivalently to normal and transformed cells, although in each case these lectins show the usual differential agglutination characteristics with transformed cells (Sela *et al.*, 1970; Tomita *et al.*, 1970; Nicolson and Blaustein, 1972; Vesely *et al.*, 1972). Not all types of transformed cells show these differential agglutination properties (Liske and Franks, 1968;

Gantt et al., 1969), and many normal cells are readily agglutinable by plant lectins (Liske and Franks, 1968, Moscona, 1971; Nicolson and Yanagimachi, 1972; Uhlenbruck and Herrmann, 1972), particularly during non-oncogenic virus infection (Poste and Reeve, 1972; Becht et al., 1972; Tevethia et al., 1972; Poste, 1975).

Since it appeared that the number of Con A-binding sites was not the determining factor in tumor cell aggregation, the distribution of Con A sites was investigated using ferritin-Con A (Nicolson, 1971, 1972) or peroxidase-Con A techniques (Bretton et al., 1972; Martinez-Palomo et al., 1972). In these experiments unfixed normal and transformed cells were labeled at room temperature, and the resulting distributions of Con A-binding sites on these cells were random dispersed for normal fibroblasts, but clustered for transformed or trypsinized fibroblasts. One exception has been found. Bretton et al. (1972) and Garrido et al. (1974) have reported that the distribution of Con A-peroxidase on rat embryo fibroblasts could not be distinguished from that on spontaneously transformed cells.

It is now well known that cell surface components are capable of fairly rapid lateral diffusion in the membrane plane under physiological conditions (Frye and Edidin, 1970; Taylor et al., 1971; Davis, 1972; Pinto da Silva, 1972; de Petris and Raff, 1972; Edidin and Weiss, 1972; Karnovsky et al., 1972; Edidin and Fambrough, 1973), so the experiments on the distribution of Con A receptors on normal and transformed cells had to be repeated with this in mind. Using fluorescent labeling techniques, Inbar and Sachs (1973) and Nicolson (1973b) demonstrated that the initial distribution of Con A-binding sites on transformed fibroblasts is uniform (as indicated by ring fluorescence when unfixed cells are labeled at low temperature or formaldehyde-fixed cells are labeled at room temperature), but this distribution changes rapidly to a patchy distribution within a few minutes at 37°C. Rosenblith et al. (1973) used hemocyanin-ferritin labeling techniques to demonstrate that Con A receptors are initially randomly dispersed on normal, protease-treated normal and transformed cells labeled at low temperature, but elevation of the temperature to 37°C for a few minutes permitted rapid Con A-induced rearrangement of its receptors to a clustered distribution on the protease-treated and transformed cells. These experiments indicate that the *relative* mobility of lectin receptors on the transformed cells is higher than on normal cells. De Petris et al. (1973) have questioned whether there is a difference in the mobility of Con A receptors on normal and transformed cells. They did not find dramatic differences in Con A receptor distributions using ferritin-lectin techniques. However, their results may have been due to the different cell lines used, conditions of labeling, etc.

In the present paper Con A agglutinability and Con A quantitative labeling of 3T3 and SV40-transformed 3T3 cells will be investigated as a function of cell culture density, since it has been demonstrated that cell-contact in culture can lead to surface modifications (Hakomori, 1970; Kijimoto and Hakomori, 1972; Roth and White, 1972). Also, the relationship of cell cytoplasmic structures such as microfilaments to Con A surface receptors will be examined.

II. METHODS

A. Cells

Mouse Balb/c 3T3 (clone A31) fibroblasts and their transformed derivatives, SV3T3 and MSV3T3, were obtained from Dr. S. Aaronson and were grown and harvested as described (Nicolson and Lacorbiere, 1973).

B. Lectins

Concanavalin A was affinity purified by the procedures of Agrawal and Goldstein (1967) and *Ricinus communis* agglutinin (RCA_I) by the procedures of Nicolson and Blaustein (1972). Ferritin conjugation of Con A and affinity purification of ferritin-Con A was performed as described (Nicolson and Singer, 1974). Lectins were radioisotope-labeled with ^{125}I as in previous studies (Nicolson, 1973a; Nicolson and Lacorbiere, 1973) and were repurified by affinity before use.

C. Quantitative Labeling

Cells were grown in Dulbecco's modified Eagle's medium containing 10% calf serum and were removed from Falcon Petri dishes with 0.2% buffered EDTA. The cells were washed and suspended in tissue culture phosphate-buffered saline (TC-PBS) or Hank's buffered balanced salt solution at 0-4°C. To cell samples in plastic culture tubes (precoated with 1% bovine serum albumin and washed) at a concentration of $1-2 \times 10^6$ cells/ml (0.3 ml) affinity purified ^{125}I-Con A was added (0.1 ml). After 10 min at 0-4°C the cells were quickly centrifuged and resuspended in fresh buffer. This was repeated once again, and the resulting pellet was transferred to another plastic tube and counted in a Packard gamma scintillation counter. Triplicate samples were used in each experiment, and controls were run identically except that α-methyl-D-mannoside (0.1 M, final) was present in the incubation and wash solutions.

D. Fluorescent-Labeling

Cells were labeled with Con A and fluorescent-anti-Con A as previously described (Nicolson, 1973b).

E. Ferritin-Labeling

Cells were grown to confluency on 60 mm Petri dishes coated with a thin layer of 'Epon 812' (the 'Epon 812' coated dishes were incubated in medium and then TC-PBS for 1-2 days before the cells were seeded) and washed with TC-PBS at 0-4°C. Cold ferritin-Con A (∼500 µg/ml protein) was added and the labeling continued for 15 min on ice. At that time the dishes were washed once with cold TC-PBS or Hank's solution and completely flooded with cold 1% glutaraldehyde, or they were incubated 15-20 min further at 37°C in a CO_2 incubator. After the 37°C incubation, the dishes were flooded with 1% glutaraldehyde as before. *In situ* fixation continued for >60 min at 0-4°C or 30 min at 37°C. The fixed cell samples were washed and post-fixed in 1% osmium tetroxide, washed again and dehydrated in ethanol. The cells were embedded on the 'Epon' layer forming a sandwich which was later stripped from the plastic dish. Sections were cut perpendicular to the plane of growth with a diamond knife and were stained with uranyl acetate prior to examination in an Hitachi HU-12 electron microscope.

III. RESULTS AND DISCUSSION

When fibroblasts are cultured *in vitro*, their lectin-mediated agglutinability can vary with the cell density. This property was first noted by Goto *et al.* (1972). They found that spontaneously-transformed 3T6 cells (3T6 is a tumorigenic, non-density-dependent inhibited cell line) were less agglutinable with Con A as they grew from sparse to confluent culture densities. This characteristic is shared with 3T3 and SV3T3 cells; that is, these cell lines also become less Con A agglutinable after cell-contact (Table I). The decrease in agglutinability at cell contact is also noticeable if *R. communis* agglutinin (RCA_I) is used instead of Con A (Nicolson, 1974a); thus, it may be a general property of fibroblasts grown in culture. When cells are replated immediately after their removal from dishes with trypsin, their agglutinability is low and develops with time after subculturing (Inbar *et al.*, 1971).

The quantity of Con A-binding sites does not appear to be sensitive to cell culture density. Sparse and contacting 3T3 and SV3T3 cells were labeled at low temperature for 10 min with various amounts of ^{125}I-Con A (Fig. 1). The results of this experiment indicate that (a) the saturation curves for 3T3 and SV3T3 cells are nearly identical and (b) there is no significant difference in the number of Con A-binding sites on sparse and contacting cells. These results are consistent with other laboratories (Arndt-Jovin and Berg, 1971; Cline and Livingston, 1971; Ozanne and Sambrook, 1971; Inbar *et al.*, 1971; Barbarese *et al.*, 1973; Rosenblith *et al.*, 1973; Phillips *et al.*, 1974) but differ from those of Noonan and Burger (1973). These latter authors claim that differences cannot be found between normal and

TABLE I

CELL DENSITY-DEPENDENT CON A AGGLUTINATION OF
3T3 and SV3T3 FIBROBLASTS*

Cell	Cell Density**	Saccharide Inhibitor	Concentration of Con A (µg/ml)	
			Mid-Point Agglutination†	End-Point Agglutination†
3T3	sparse-growing	-	1,200-1,500	400-1,000
		αMM	-	-
	contacting-growing	-	>2,000	1,200-2,000
		αMM	-	-
	confluent-quiescent	-	>2,000	1,200-2,000
		αMM	-	-
SV3T3	sparse-growing	-	8-16	<4
		αMM	-	-
	contacting-growing	-	30-60	8-16
		αMM	-	-
	confluent-growing	-	30-60	8-16
		αMM	-	-

*Data adopted, in part, from Nicolson (1974a).

**Cell densities were as follows: sparse, $3-6 \times 10^3/cm^2$; contacting, $4-9 \times 10^4/cm^2$; confluent, $>0.8-2 \times 10^5/cm^2$. Cells were removed with 0.02% buffered EDTA after subculturing for at least two days.

0.05 M α-methyl-D-mannopyranoside present.

†Agglutination was scored after a 15-20 min incubation at room temperature in 16 mm wells of Linbro FP-54 trays rotating at 1-2 Hz (Nicolson and Lacorbiere, 1973). Mid-point agglutination equals ++ by the method of Sela et al. (1970).

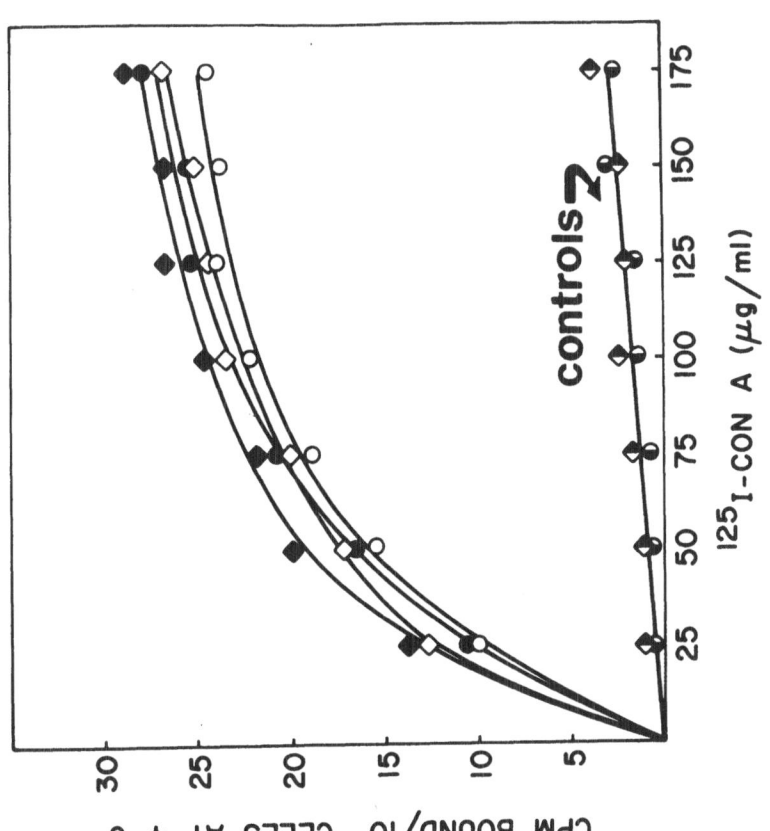

Fig. 1. Binding of ^{125}I-Con A to sparse and confluent 3T3 and SV3T3 fibroblasts during a 10 min incubation at 0-4°C (see Methods for details). ◆, confluent SV3T3; ◇, sparse SV3T3; ●, confluent 3T3; ○, sparse 3T3; ◆, SV3T3 control (α-methyl-D-mannoside); ◉, 3T3 control (α-methyl-D-mannoside).

transformed cells if endocytosis is permitted. That endocytosis does not occur under the conditions we chose for the ^{125}I-Con A binding assay was shown by incubating the cells at 0-4°C for 10 min with ferritin-Con A. After the incubation, the cells were fixed and embedded. An analysis of thin sections of the ferritin-Con A labeled cells did not reveal endocytosis of the ferritin conjugate. Although some patching occurred on a few of the cells, there was no evidence of ferritin-Con A inside the cells. Thus, it appears from the data here and elsewhere that there is no difference in the total number of Con A receptors on 3T3 and SV3T3 cells; however, the Con A receptor structure or environment could be quite different on these cells. One explanation for the results that Noonan and Burger (1973) obtained is that they selectively saturated a minor class of Con A-receptors characterized by a high binding constant and that transformed cells have more of these high avidity sites (Nicolson, 1974c). High avidity Con A receptors have been found by Cuatrecasas (1973) on fat cells where it appears that heterogeneity of Con A receptors exists. Quantitative labeling studies on normal and transformed cells with other lectins have yielded results consistent with a majority of the publications on Con A labeling. Soy bean agglutinin (Sela *et al.*, 1971), pea lectin (Trowbridge, 1974) and *Ricinus communis* (RCA$_I$) agglutinin (Nicolson, 1973a) bind equivalently to confluently grown normal and transformed fibroblasts, although the latter lectin can distinguish between sparse and contacting 3T3 cells in the binding assay (Nicolson and Lacorbiere, 1973).

Several techniques have been used to study the distributions of Con A receptors on normal, protease-treated and transformed cells. These studies were necessary when it became obvious that the number of Con A receptors had a tenuous relationship to Con A-mediated agglutinability. Using techniques such as ferritin-Con A labeling to briefly formalinized, mounted plasma membranes of 3T3 and SV3T3 cells, a difference in the topography of Con A sites was first noted (Nicolson, 1971). SV3T3 cells had on the average a more clustered distribution of Con A receptors. Similar results using the Con A-peroxide technique were reported by Martinez-Palomo *et al.* (1972) and Bretton *et al.* (1972). The clustered Con A-binding sites were reasoned to be directly involved in cell agglutination because SV3T3 or trypsinized 3T3 cells, specifically agglutinated with ferritin-Con A into small cell aggregates, showed high concentrations of ferritin between adjacent agglutinated cells (Nicolson, 1972, 1974a). From several studies it is clear that the clustered lectin sites arise from ligand-induced clustering of inherently dispersed receptor sites. This was demonstrated by labeling aldehyde-fixed cells or cells at low temperature (at or near 0°C) with fluorescent-labeled Con A (Nicolson, 1973b; Inbar and Sachs, 1973; Inbar *et al.*, 1973), hemocyanin-labeled Con A (Rosenblith *et al.*, 1973) or peroxidase-labeled Con A (Garrido *et al.*, 1974). All cells had uniform, dispersed distributions of Con A receptors if labeled at low temperature or if fixation was performed before labeling, but the Con A receptors on unfixed protease-

TABLE II

DISTRIBUTION OF CONCANAVALIN A RECEPTORS ON NORMAL AND SV40- AND MSV-TRANSFORMED MOUSE FIBROBLASTS BY IMMUNOFLUORESCENCE*

Cell	First Incubation (5–20 μg/ml Con A)		Second Incubation			Third Incubation (FITC-anti-Con A)		% Distribution of Fluorescent Anti-Con A	
	Prior Fixation[†]	Temp (°C)	Prior Fixation	Time (Min)	Temp (°C)	Prior Fixation	Temp (°C)	Uniform	Patchy
3T3	2% FA; 15 min	0,20	–	–	–	–	0,20	>95	<5
	–	0	–	15	0	–	0	>90	<10
	–	0	2% FA; 20 min	15	0,20	–	0,20	>90	<10
	–	0	–	10,15	20,37	2% FA; 15 min	0,20	>85	<15
	–	0	–	15	20	–	37	>50	<50
SV3T3	2% FA; 15 min	20	–	–	–	–	20	>95	<5
	–	0	–	15	0	–	0	>85	<15
	–	0	2% FA; 20 min	15	0,20	–	0,20	>85	<15
	–	0	–	10,15	20,37	2% FA; 15 min	0,20	<10	>90
	–	0	–	–	–	–	37	<5	>95
MSV3T3	2% FA; 15 min	0,20	–	–	–	–	0,20	>95	<5
	–	0	–	15	0	–	0	>85	<15
	–	0	2% FA; 20 min	15	0,20	–	0,20	>85	<15
	–	0	–	10,15	20,37	2% FA; 15 min	0	<10	>90
	–	0	–	–	–	–	37	<5	>95

*Cells were treated with 5–20 μg/ml Con A for 30–60 min (First Incubation). Then they were washed once and incubated for various times at various temperatures (Second Incubation). Cells were treated with fluorescein isothiocyanate (FITC)-conjugated anti-Con A, washed and subsequently fixed in 2% buffered formaldehyde for 15 min at the incubation temperature and observed in a Leitz Ortholux microscope with UV illumination. Data, in part, from Nicolson (1973b).

[†]Fixation in 2% buffered formaldehyde.

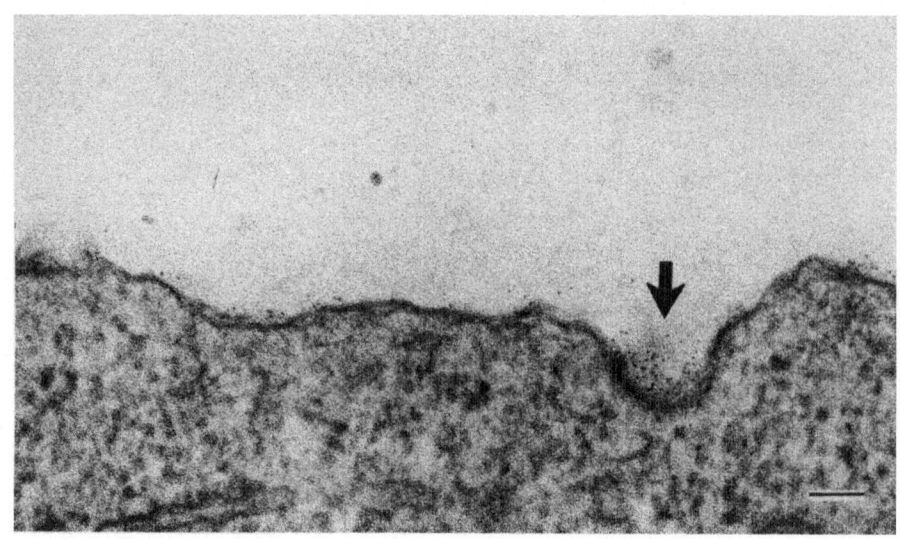

Fig. 2. Distribution of Con A receptors on a region of a 3T3 cell lacking a subplasmalemma membrane-associated microfilament structure. Cells were labeled with ferritin-Con A at 0-4°C, washed, and then incubated at 20°C for 30 min prior to fixation. The ferritin-Con A molecules are clustered in certain regions and endocytosis is starting (arrow). Bar equals 0.1 μm.

treated and transformed cells changed quickly to clustered distributions after elevating the temperature to 20 or 37°C for a few minutes (Table II). These experiments suggest that a temperature-dependent difference in the relative mobility of Con A receptors exists on transformed cells (Inbar and Sachs, 1973; Rosenblith et al., 1973; Nicolson, 1973b, 1974a). Similar results have been obtained using low concentrations of RCA_I (Nicolson, 1974a).

If the mobility of Con A receptors determines, in part, the agglutinability of cells, what is responsible for the decreased Con A agglutinability after cell contact? It is known that structures within the cell (membrane-associated contractile components such as microfilaments and microtubules [Nicolson, 1974c]) reorient and appear to be closely interacting with the plasma membrane after cell contact (McNutt et al., 1971; Perdue, 1973). To examine the possible role of these structures in determining the topography and mobility of Con A receptors we labeled confluent 3T3 cells with ferritin-Con A in situ, so the intracellular microfilament/microtubule system could be seen as well as the distribution of Con A-binding sites. When confluent 3T3 cells are labeled for short periods with ferritin-Con A at 0-4°C, washed, fixed and embedded, the Con A re-

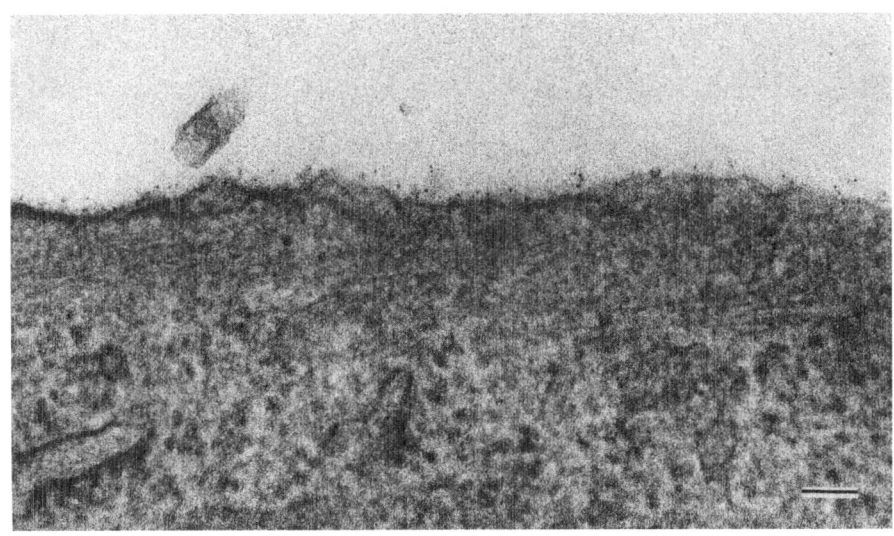

Fig. 3. Distribution of Con A receptors on a region of a 3T3 cell with membrane-associated microfilaments closely opposed to the plasma membrane. Cells were labeled with ferritin-Con A at 0-4°C, washed, and then incubated at 20°C for 30 min prior to fixation. The ferritin-Con A molecules remain relatively dispersed. Bar equals 0.1 µm.

ceptors are found in a dispersed distribution over the entire cell surface. However, when confluent cells are labeled with ferritin-Con A at 0-4°C, washed once, and then incubated at 20°C for several (15-30) minutes, receptor site rearrangement occurs. On the transformed cell surfaces ferritin-Con A-induced clustering occurs within minutes, and the label is distributed in a patchy manner. Under these conditions, normal cells show two types of Con A receptor distributions. In regions where membrane-associated structures such as microfilaments could not be seen (usually the sides or undersides of the cells on substrate), the Con A receptors were redistributed, and in a few cases endocytotic vesicles could be seen forming (Fig. 2). But on the upper cell surfaces where the membrane-associated microfilament systems are extensive and closely opposed to the plasma membrane, Con A receptor distributions generally remained dispersed (Fig. 3). The exception to this was on the surfaces of the microvilli where the quality of ferritin-Con A seemed to decrease after the temperature elevation similar to the findings of Rosenblith et al. (1973). The details of these experiments will be published elsewhere (G. L. Nicolson, in preparation).

The Con A-binding sites that are relatively less mobile than their counterparts on transformed cells may be linked differently with other membrane components and/or membrane-associated components. In this case there would be no need to invoke the synthesis of completely different new plasma membrane proteins after viral transformation. Evidence in favor of such a proposal are the findings of Hogg (1974) and Hynes (1973) where transformation of murine fibroblasts by SV40 virus resulted in the deletion of one major component from the purified plasma membranes. Here it appears that the transformed cells have lost some degree of control over the topography and mobility of their Con A receptors, and this control may be determined, in part, by the extensive microfilament systems under the normal cell membrane. It is tempting to speculate that the deleted membrane component is involved in this control, but further analysis of the membranes of normal and transformed cells is necessary.

There is recent evidence that supports the notion that cell surface receptors can be controlled by structures within the cell cytoplasm. Berlin and Ukena (1972) and Yin et al. (1972) found that microtubule-disrupting drugs such as colchicine and vinblastine affect Con A-mediated agglutination. These drugs also prevent Con A inhibition of anti-Ig-induced capping on lymphocytes (Yahara and Edelman, 1972, 1973) and allow membrane transport sites and lectin receptors to be internalized during phagocytosis. These sites are normally not internalized but remain on the surface during phagocytosis (Berlin and Ukena, 1972; Oliver et al., 1974). Dibutyryl-cyclic AMP converts hamster CHO cell morphology to a fibroblastic form and lowers the agglutinability by wheat germ agglutinin. These effects are reversed by microtubule disrupting drugs (Hsie and Puck, 1971; Hsie et al., 1971). Cytochalasin B, a microfilament disrupting drug (Wessels et al., 1971) has been reported to modify the lectin agglutinability of certain cells (Kaneko et al., 1973), although this may not be a general phenomenon (de Petris et al., 1973; Nicolson, 1973c).

These findings suggest that there are several interrelated properties that can affect cell agglutination (Fig. 4) (Nicolson, 1974c). These include the characteristics of the agglutination molecule such as its valency, charge, size, binding constant, etc., the number, location, distribution and mobility of the cell agglutination sites, cell rigidity, and surface charge density (zeta potential). Surface structures such as microvilli may, under certain circumstances, aid in agglutination by presenting specialized surfaces to other cells or by trapping them more effectively at long range. Finally, structures within the cell, such as peripheral and membrane-associated components, may affect agglutination by restraining the mobility and controlling the topography of receptors or by increasing cell rigidity. These factors may favor or oppose agglutination, but agglutination in any given system should depend on the sum of the aggregation forces minus the sum of the repulsive forces. Thus, certain of these factors may not favor agglutination, but can-

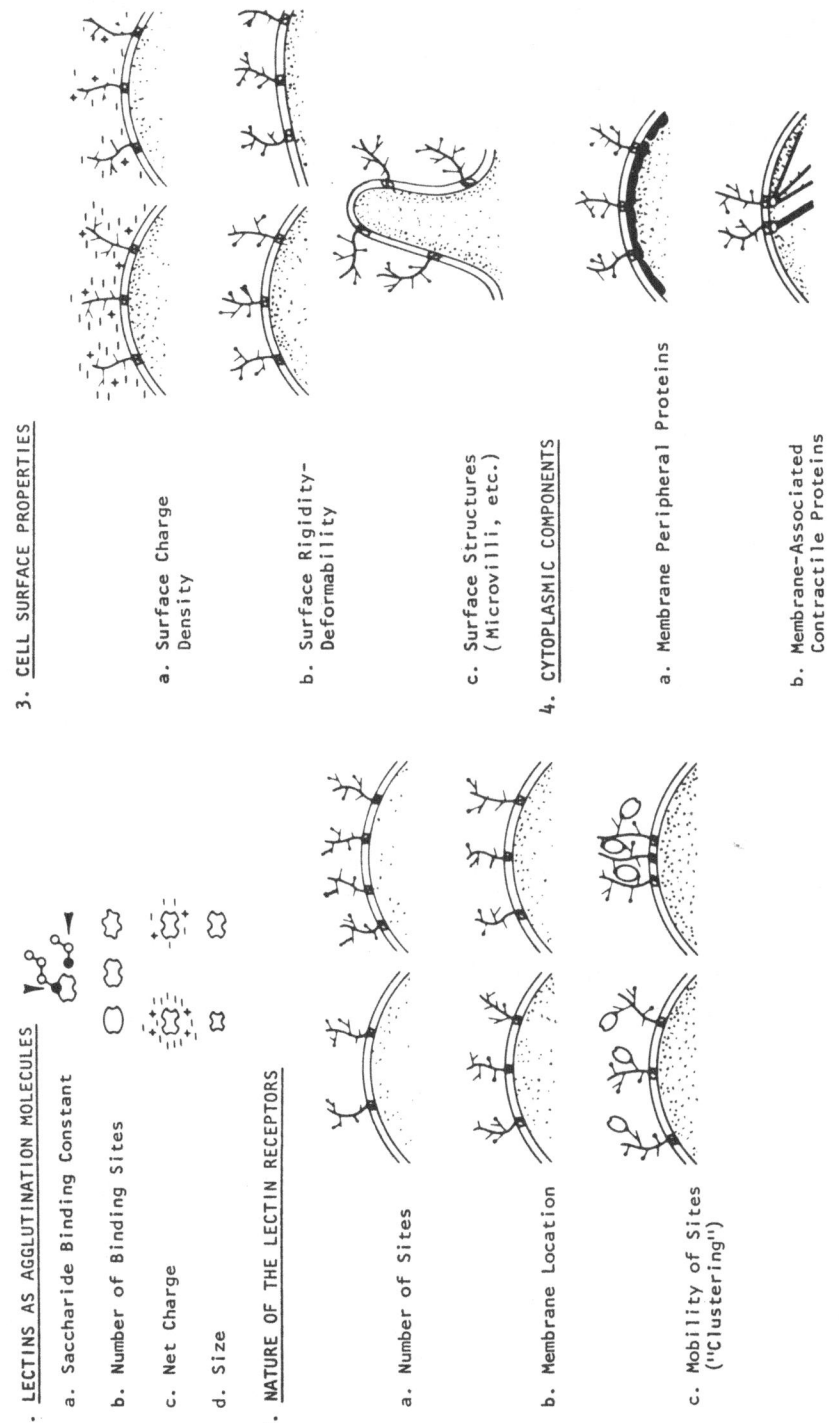

Fig. 4. Factors affecting cell agglutination by lectins.

not prevent it, if the aggregation forces prevail over the repulsive forces.

REFERENCES

1. Agrawal, B. B. L., and Goldstein, I. J. (1967). "Protein-carbohydrate interaction. VI. Isolation of concanavalin A by specific adsorption on cross-linked dextran gels." *Biochim. Biophys. Acta.* 147, 262.

2. Arndt-Jovin, D. J., and Berg, P. (1971). "Quantitative binding of ^{125}I-concanavalin A to normal and transformed cells." *J. Virol.* 8, 716.

3. Barbarese, E., Sauerwein, H., and Simkins, H. (1973). "Alteration in the surface glycoproteins of chick erythrocytes following transformation with erythroblastosis strain R virus." *J. Mem. Biol.* 13, 129.

4. Becht, H., Rott, R., and Klenk, H. D. (1972). "Effect of concanavalin A on cells infected with enveloped RNA viruses." *J. Gen. Virol.* 14, 1.

5. Berlin, R. D., and Ukena, T. E. (1972). "Effect of colchicine and vinblastine on the agglutination of polymorphonuclear leucocytes by concanavalin A." *Nature New Biol.* 238, 120.

6. Bretton, R., Wicker, R., and Bernhard, W. (1972). "Ultrastructural localization of concanavalin A receptors in normal and SV40-transformed hamster and rat cells." *Int. J. Cancer* 10, 397.

7. Burger, M. M. (1969). "A difference in the architecture of the surface membrane of normal and virally transformed cells." *Proc. Nat. Acad. Sci. U. S. A.* 62, 994.

8. Burger, M. M. (1973). "Surface changes in transformed cells detected by lectins." *Fed. Proc.* 32, 91.

9. Cline, M. J., and Livingston, D. C. (1971). "Binding of ^{3}H-concanavalin A by normal and transformed cells." *Nature New Biol.* 232, 155.

10. Cuatrecasas, P. (1972). "Interaction of wheat germ agglutinin and concanavalin A with isolated fat cells." *Biochemistry* 12, 1312.

11. Davis, W. C. (1972). "H-2 antigen on cell membranes: An explanation for the alteration of distribution by indirect label-

ing techniques." *Science* 175, 1006.

12. De Petris, S., and Raff, M. C. (1972). "Distribution of immunoglobulin on the surface of mouse lymphoid cells as determined by immunoferritin electron microscopy. Antibody-induced, temperature-dependent redistribution and its implications for membrane structure." *Europ. J. Immunol.* 2, 524.

13. De Petris, S., Raff, M. C., and Mallucci, L. (1973). "Ligand-induced redistribution of concanavalin A receptors on normal, trypsinized and transformed fibroblasts." *Nature New Biol.* 244, 275.

14. Edidin, M., and Fambrough, D. (1973). "Fluidity of the surface of cultured cell muscle fibers. Rapid lateral diffusion of marked surface antigens." *J. Cell Biol.* 57, 27.

15. Edidin, M., and Weiss, A. (1972). "Antigen cap formation in cultured fibroblasts: A reflection of membrane fluidity and of cell motility." *Proc. Nat. Acad. Sci. U. S. A.* 69, 2456.

16. Frye, L. D., and Edidin, M. (1970). "The rapid inter-mixing of cell surface antigens after formation of mouse-human heterokaryons." *J. Cell Sci.* 7, 319.

17. Gantt, R. R., Martin, J. R., and Evans, V. J. (1969). "Agglutination of *in vitro* cultured neoplastic and non-neoplastic cell lines by a wheat germ agglutinin." *J. Nat. Cancer Inst.* 42, 369.

18. Garrido, J., Burglen, M.J., Samolyk, D., Wicker, R., and Bernhard, W. (1974). "Ultrastructural comparison between the distribution of concanavalin A and wheat germ agglutinin cell surface receptors of normal and transformed hamster and rat cell lines." *Cancer Res.* 34, 230.

19. Goto, M., Kataoka, Y., Goto, K., Yodoyama, T., and Sato, H. (1972). "Decrease in agglutinability of cultured tumor cells to concanavalin A at the plateau of cell growth." *Gann* 63, 505.

20. Hakomori, A. (1970). "Cell density-dependent changes of glycolipids in fibroblasts and loss of this response in the transformed cells." *Proc. Nat. Acad. Sci. U. S. A.* 67, 1741.

21. Hogg, N. M. (1974). "A comparison of membrane proteins of normal and transformed cells by the lactoperoxidase method." *Proc. Nat. Acad. Sci. U. S. A.* 71, 489.

22. Hsie, A. W., and Puck, T. T. (1971). "Morphological transformation of Chinese hamster cells by dibutyryl adenosine cyclic

3':5'-monophosphate and testosterone." *Proc. Nat. Acad. Sci. U. S. A.* **68**, 358.

23. Hsie, A. W., Jones, C., and Puck, T. T. (1971). "Further changes in differentiation state accompanying the conversion of Chinese hamster cells of fibroblastic form by dibutyryl adenosine cyclic 3':5'-monophosphate and hormones." *Proc. Nat. Acad. Sci. U. S. A.* **68**, 1648.

24. Hynes, R. O. (1974). "Alteration of cell-surface proteins by viral transformation and by proteolysis." *Proc. Nat. Acad. Sci. U. S. A.* **70**, 3170.

25. Inbar, M., and Sachs, L. (1969a). "Structural differences in sites on the surface membrane of normal and transformed cells." *Nature* **233**, 710.

26. Inbar, M., and Sachs, L. (1969b). "Interaction of the carbohydrate-binding protein concanavalin A with normal and transformed cells." *Proc. Nat. Acad. Sci. U. S. A.* **63**, 1418.

27. Inbar, M., and Sachs, L. (1973). "Mobility of carbohydrate containing sites on the surface membrane in relation to the control of cell growth." *FEBS Lett.* **32**, 124.

28. Inbar, M., Ben-Bassat, H., and Sachs, L. (1971). "A specific metabolic activity on the surface membrane in malignant cell-transformation." *Proc. Nat. Acad. Sci. U. S. A.* **68**, 2748.

29. Inbar, M., Ben-Bassat, H., Huet, C., Oseroff, A. R., and Sachs, L. (1973). "Inhibition of lectin agglutinability by fixation of the cell surface membrane." *Biochim. Biophys. Acta* **311**, 594.

30. Kaneko, I., Sato, H., and Ukita, T. (1973). "Effect of metabolic inhibitors on the agglutination of tumor cells by concanavalin A and *Ricinus communis* agglutinin." *Biochim. Biophys. Res. Comm.* **50**, 1087.

31. Karnovsky, M. J., Unanue, E. R., and Leventhal, M. (1972). "Ligand-induced movement of lymphocyte membrane macromolecules. II. Mapping of surface moieties." *J. Exp. Med.* **136**, 907.

32. Kijimoto, S., and Hakomori, S. I. (1972). "Contact-dependent enhancement of net synthesis of Forssman glycolipid antigen and hematoside in NIL cells at the early stage of cell-to-cell contact." *FEBS Lett.* **25**, 38.

33. Lis, H., and Sharon, N. (1973). "The biochemistry of plant lectins (phytohemagglutinins)." *Ann. Rev. Biochem.* **43**, 541.

34. Liske, R., and Franks, D. (1968). "Specificity of the agglutinin in extracts of wheat germ." *Nature* 217, 860.

35. McNutt, N. S., Culp, L. A., and Black, P. H. (1971). "Contact-inhibited revertant cell lines isolated from SV40-transformed cells. II. Ultrastructural study." *J. Cell Biol.* 50, 691.

36. Martinez-Palomo, A., Wicker, R., and Bernhard, W. (1972). "Ultrastructural detection of concanavalin A surface receptors in normal and in polyoma-transformed cells." *Int. J. Cancer* 9, 676.

37. Moscona, A. A. (1971). "Embryonic and neoplastic cell surfaces: Availability of receptors for concanavalin A and wheat germ agglutinin." *Science* 171, 905.

38. Nicolson, G. L. (1971). "Difference in the topology of normal and tumor cell membranes as shown by different distributions of ferritin-conjugated concanavalin A on their surfaces." *Nature New Biol.* 233, 244.

39. Nicolson, G. L. (1972). "Topography of cell membrane concanavalin A-sites modified by proteolysis." *Nature New Biol.* 239, 193.

40. Nicolson, G. L. (1973a). "Neuraminidase 'unmasking' and the failure of trypsin to 'unmask' β-D-galactose-like sites on erythrocyte, lymphoma and normal and virus-transformed fibroblast cell membranes." *J. Nat. Cancer Inst.* 50, 1443.

41. Nicolson, G. L. (1973b). "Temperature-dependent mobility of concanavalin A sites on tumour cell surfaces." *Nature New Biol.* 243, 218.

42. Nicolson, G. L. (1973c). "The relationship of a fluid membrane structure to cell agglutination and surface topography." *Ser. Haematol.* 6, 275.

43. Nicolson, G. L. (1974a). "Factors influencing the dynamic display of lectin-binding sites on normal and transformed cell surfaces." *In Control of Proliferation in Animal Cells* (B. Clarkson and R. Baserga, eds.), Cold Spring Harbor Laboratory, New York, in press.

44. Nicolson, G. L. (1974b). "Cell-contact and transformation-induced changes in the dynamic organization of normal and neoplastic cell plasma membranes and their role in lectin-mediated toxicity toward tumor cells." *In Biology and Chemistry of Eucaryotic Cell Surfaces,* Proc. of the 6th Miami Winter Symposia (E. Y. C. Lee and E. E. Smith, eds.), Academic Press, New York, in press.

45. Nicolson, G. L. (1974c). "The interactions of lectins with animal cell surfaces." *Int. Rev. Cytol.*, in press.

46. Nicolson, G. L., and Blaustein, J. (1972). "Interaction of *Ricinus communis* agglutinin with normal and tumor cell surfaces." *Biochim. Biophys. Acta* 266, 543.

47. Nicolson, G. L., and Lacorbiere, M. (1973). "Cell contact-dependent increase in membrane D-galactopyranosyl-like residues on normal, but not virus- or spontaneously-transformed murine fibroblasts." *Proc. Nat. Acad. Sci. U. S. A.* 70, 1672.

48. Nicolson, G. L., and Singer, S. J. (1974). "The distribution and asymmetry of saccharides on mammalian cell membrane surfaces utilizing ferritin-conjugated plant agglutinins as specific saccharide stains." *J. Cell Biol.* 60, 236.

49. Nicolson, G. L., and Yanagimachi, R. (1972). "Terminal saccharides on sperm plasma membranes: Identification by specific agglutinins." *Science* 177, 276.

50. Noonan, K. D., and Burger, M. M. (1973). "The relationship of concanavalin A binding to lectin-initiated cell agglutination." *J. Cell Biol.* 59, 134.

51. Oliver, J. M., Ukena, T. E., and Berlin, R. D. (1974). "Effects of phagocytosis and colchicine on the distribution of lectin-binding sites on cell surfaces." *Proc. Nat. Acad. Sci. U. S. A.* 70, 394.

52. Ozanne, B. and Sambrook, J. (1971). "Binding of radioactively labeled concanavalin A and wheat germ agglutinin to normal and virus-transformed cells." *Nature New Biol.* 232, 156.

53. Perdue, J. F. (1973). "The distribution, ultrastructure, and chemistry of microfilaments in cultured chick embryo fibroblasts." *J. Cell Biol.* 58, 265.

54. Phillips, P. G., Furmanski, P., and Lubin, M. (1974). "Cell surface interactions with concanavalin A: Location of bound radiolabeled lectin." *Exptl. Cell Res.*, in press.

55. Pinto da Silva, P. (1972). "Translational mobility of the membrane intercalated particles of human erythrocyte ghosts. pH-dependent, reversible aggregation." *J. Cell Biol.* 53, 777.

56. Poste, G., and Reeve, P. (1972). "Agglutination of normal cells by plant lectins following infection with non-oncogenic viruses." *Nature New Biol.* 237, 113.

57. Poste, G. (1975). This book, p. 117.

58. Rosenblith, J. Z., Ukena, T. E., Yin, H. H., Berlin, R. D., and Karnovsky, M. J. (1973). "A comparative evaluation of the distribution of concanavalin A-binding sites on the surfaces of normal, virally-transformed, and protease-treated fibroblasts." *Proc. Nat. Acad. Sci. U. S. A.* 70, 1625.

59. Roth, S., and White, D. (1972). "Intercellular contact and cell-surface galactosyl transferase activity." *Proc. Nat. Acad. Sci. U. S. A.* 69, 485.

60. Sela, B., Lis, H., Sharon, N., and Sachs, L. (1970). "Different locations of carbohydrate-containing sites in the surface membrane of normal and transformed mammalian cells." *J. Mem. Biol.* 3, 267.

61. Sela, B., Lis, H., Sharon, N., and Sachs, L. (1971). "Quantitation of N-acetyl-D-galactosamine-like sites on the surface membrane of normal and transformed mammalian cells." *Biochim. Biophys. Acta* 249, 564.

62. Taylor, R., Duffus, P., Raff, M., and de Petris, S. (1971). "Redistribution and pinocytosis of lymphocyte surface immunoglobulin molecules induced by anti-immunoglobulin antibody." *Nature New Biol.* 233, 225.

63. Tevethia, S., Lowry, S., Rawls, W., Melnick, J., and McMillan, V. (1972). "Detection of early cell surface changes in Herpes simplex virus by agglutination with concanavalin A." *J. Gen. Virol.* 15, 93.

64. Tomita, M., Osawa, T., Sakurai, Y. and Ukita, T. (1970). "On the surface of murine-ascites tumors. I. Interactions with various phytoagglutinins." *Int. J. Cancer* 6, 283.

65. Trowbridge, I. (1974). Personal communication.

66. Uhlenbruck, G., and Herrmann, W. P. (1972). "Agglutination of normal, coated and enzyme-treated human spermatozoa with heterophile agglutinins." *Vox Sang.* 23, 444.

67. Vesely, P., Entlicher, G., and Kocourek, J. (1972). "Pea phytohemagglutinin selective agglutination of tumour cells." *Experientia* 15, 1085.

68. Wessells, N. K., Spooner, B. S., Ash, J. F., Bradly, M. O., Luduena, M. A., Taylor, E. L., Wrenn, J. T., and Yamada, K. M. (1971). "Microfilaments in cellular and developmental processes." *Science* 171, 135.

69. Yahara, I., and Edelman, G. M. (1972). "Restriction of the mobility of lymphocyte immunoglobulin receptors on concanavalin A." *Proc. Nat. Acad. Sci. U. S. A.* **69**, 608.

70. Yahara, I., and Edelman, G. M. (1973). "Modulation of lymphocyte receptor redistribution by concanavalin A, anti-mitotic agents and alterations of pH." *Nature* **236**, 152.

71. Yin, H. H., Ukena, T. E., and Berlin, R. D. (1972). "Effect of colchicine and vinblastine on the agglutination of virus-transformed cells by concanavalin A." *Science* **178**, 867.

9

MICROTUBULAR PROTEINS AND CONCANAVALIN A RECEPTORS

Richard D. Berlin
University of Connecticut
　Health Center
Farmington, Connecticut, U. S. A.

ABSTRACT

　　The inherent topographical distribution of Con A binding sites (CABS) is disperse or random in all cell types studied using hemocyanin to mark CABS in surface replicas. In virally transformed cells, the addition of Con A leads to the formation of clusters (CABS). A role for microtubules is suggested in this process since colchicine treatment of transformed cells and Con A addition lead to the aggregation of Con A into a "cap." During phagocytosis CABS are selectively removed from the surface. This selective movement is abolished by drugs that disrupt microtubules. Binding of Con A or RCA to intact cells at 37°C leads to the removal of their receptors from the surface, presumably by "micropinocytosis."

I. INTRODUCTION

　　Important differences in the biological behavior of tumor cells and normal cells are reflected in altered surface properties. For tumor cells *in vitro*, these include the loss of contact inhibition of cell movement and division, altered membrane transport, and decreased surface adhesiveness (Hakamori, 1973). Since the work of Aub (1963), it is recognized that in general transformed cells are considerably more agglutinable by Con A and other lectins than their normal counterparts. Studies of temperature sensitive mutants have shown an intimate correlation between the transformed state (established at permissive temperatures) and agglutinability by lectins. It is assumed that agglutination results from the cross-linking of receptors on adjacent cells by the multivalent lectins. Initially it was supposed that transformed cells had increased numbers of receptors, but it was found that the numbers of receptors are essentially identical (Ozanne, 1971; Cline, 1971). This does not necessarily signify, of course, that the lectin receptors are chemically identical. However, it suggests that factors other than available binding sites determine agglutinability. Nicolson (1971), using ferritin-conjugated antibody to Con A, found that the topographical distribution of Con A binding sites (CABS) was clustered in the transformed cell surface and yet disperse (or random) in the normal cell surface. Presumably the clustering of sites would facilitate the formation of multiple linkages between cells at points of contact.

This paper presents a review of the evidence developed in our laboratory concerning the topographical distribution of Con A binding sites and the mechanisms that control topography, with particular emphasis on the possible role of microtubular proteins.

II. METHODS

The morphological distribution of CABS was determined by the technique of Smith and Revel (1972) in which Keyhold limpet hemocyanin (a high molecular weight blood pigment) that contains carbohydrate receptors for Con A is used to identify Con A adsorbed to the cell surface. CABS were enumerated by saturation with ^{125}I-Con A.

Phagocytosis and membrane isolation in rabbit polymorphonuclear leukocytes have been described recently in detail (Oliver *et al.*, 1974).

III. RESULTS

In accord with Singer and Nicolson (and using an entirely different technique which allows an analysis of larger areas of the surface) we found that CABS are dispersed in normal fibroblasts (Fig. 1), whereas in transformed fibroblasts CABS are drawn into clusters (Fig. 3). Moreover, the clusters in turn assume a characteristic distribution; they are withdrawn from the edges and pseudopods of the cell (Fig. 4) (Rosenblith *et al.*, 1973). These experiments emphasize the possibility that it is the topographical arrangement of membrane constituents and not their chemical identities - or perhaps factors in addition to them - which differentiate normal from transformed cells.

The clustered topography is closely correlated with increased agglutinability and associated biological properties. Thus, brief trypsinization of normal fibroblasts increases their agglutinability and at the same time alters the topography of CABS from dispersed to

MICROTUBULAR PROTEINS AND RECEPTORS 175

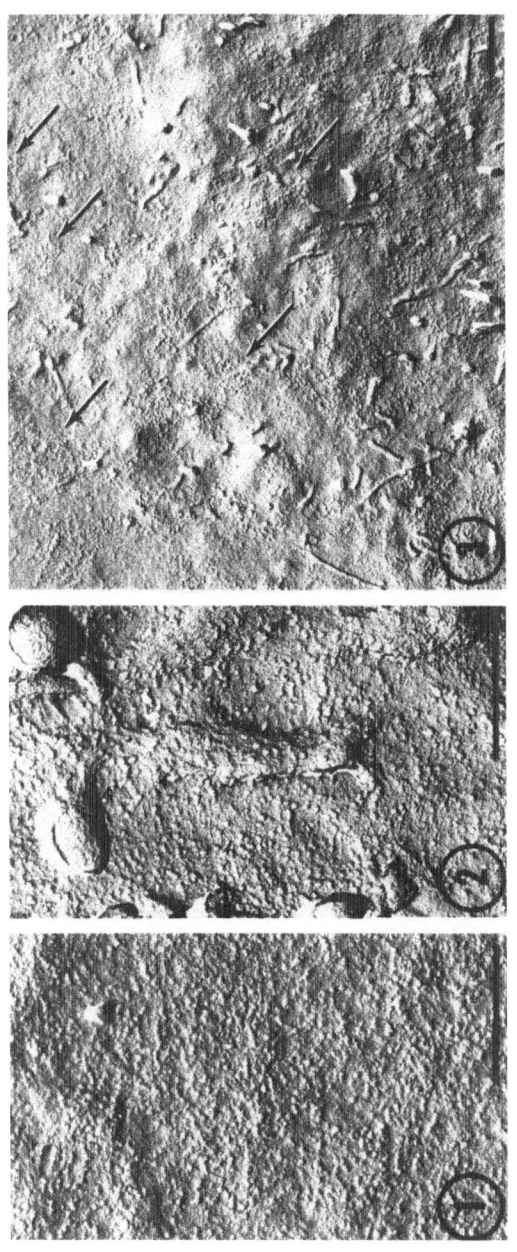

Figs. 1-3. All figures are shadow-cast replicas; direction of shadow is bottom to top. Bar = 1 μm. (From Rosenblith et al., 1973)

1. 3T3 cell showing dispersed hemocyanin labeling after Con A/Hemocyanin treatment and incubation at 37°C; x 15,300.

2. SV3T3 cell with dispersed hemocyanin labeling after Con A/H treatment and incubation at 4°C; x 15,300.

3. SV3T3 cell showing clusters (arrows) of hemocyanin after Con A/H treatment at 4°C followed by incubation at 37°C; x 6,100.

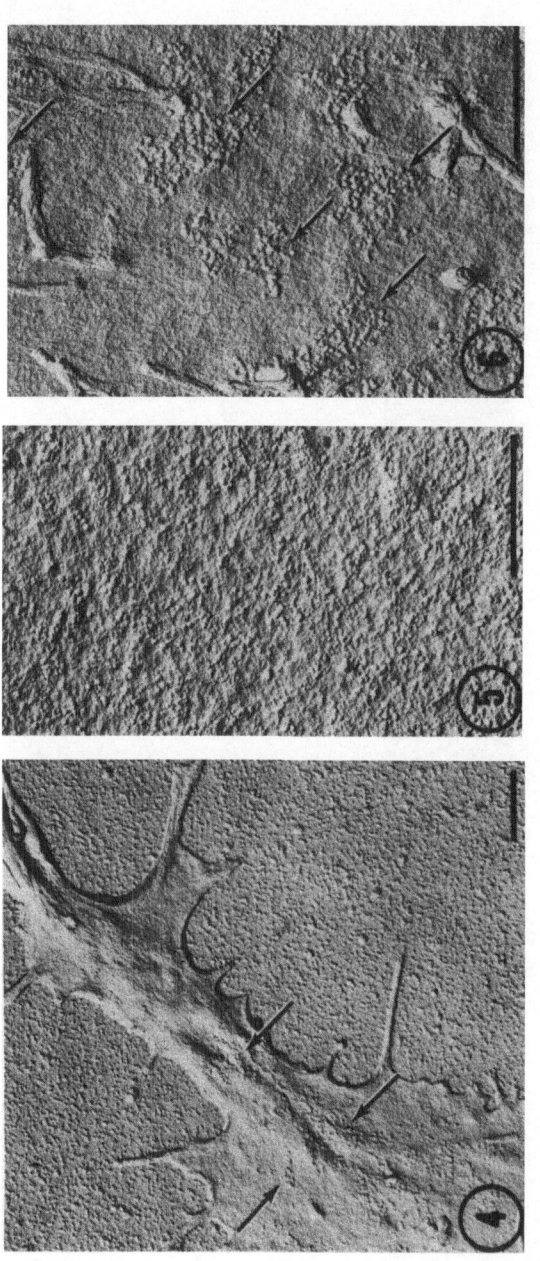

Figs. 4-6. Shadow-cast replicas; direction of shadow is bottom to top. Bar = 1 μm. (From Rosenblith et al., 1973)

4. Pseudopod of SV3T3 cell treated as in Fig. 3. Clusters (arrows) of hemocyanin occur in proximal part of pseudopod close to main body of cell, while distal aspect of pseudopod is devoid of clusters; x 7,650.

5. Trypsinized 3T3 cell maintained and Con A/H-labeled at 4°C showing dispersed hemocyanin labeling; x 15,300.

6. Trypsinized 3T3 cell incubated and Con A/H-labeled at 37°C, showing hemocyanin clusters; x 15,300.

TABLE I

CON A/H TREATMENT OF 3T3 AND SV3T3 CELLS

Cells	Con A/H* labeling °C	Incubation† post Con A/H °C	CABS§
3T3	4	4	Dispersed
	4	37	Dispersed
	37	37	Dispersed
3T3 Prefix‡	37	37	Dispersed
SV3T3	4	4	Dispersed
	4	37	Clustered
	37	37	Clustered
	37	4	Clustered
SV3T3 Prefix‡	37	37	Dispersed

*Cells were treated with Con A (10 min), rinsed three times in PBS, treated with hemocyanin (10 min), and rinsed and
†Monolayers were then incubated for an additional 10 min.
‡Fixation before Con A/H labeling was in 1% paraformaldehyde for 10 min.
§CABS = Con A-binding-site topography after Con A/H labeling.

clustered (Fig. 6). Conversely 30-60 min treatment of transformed fibroblasts with cyclic AMP restores normal growth characteristics to a large degree; in these cells surface topography of CABS is dispersed (Ukena et al., unpublished). In neither case (0.01% trypsin or cyclic AMP for 30') can it be supposed that significant alterations in the composition of the membrane can have occurred.

Additional experiments showed that the clustered topography of transformed cells was actually induced by the addition of exogenous lectin (Table I). When cells are held in the cold, or prefixed with paraformaldehyde, a random or homogeneous distribution is seen on all cells (normal and transformed) (Fig. 2). Thus, an essential difference between normal and transformed cells is the fact that, although the inherent distribution of CABS is random, only in the transformed cells are the binding sites induced to cluster by exogeneous Con A. One interpretation of this result is that the membrane of transformed cells is more fluid, allowing CABS to be cross-linked by the multivalent Con A. Very limited information is available on the viscosity of membranes. There are suggestions of a decreased viscosity of membranes of transformed cells, but quantitatively the measured differences are small. In addition, from the numbers and dimensions of CABS and approximations of membrane area and viscosity, one can readily calculate (Poo and Cone, 1974) the approximate collision frequency of CABS. The value is sufficiently high that it is reasonable to suppose that, within the incubation period of Con A labelling, the bulk of the CABS would have approached each other sufficiently to permit cross-linking to occur - assuming that CABS mobility was random.

An alternative hypothesis is that CABS are restrained by forces which arise extrinsic to the membrane proper, and that these forces are diminished in the transformed state. Although we are far from understanding the full spectrum of such hypothetical forces or their mechanisms of action, there is some evidence which was developed in our laboratory which indicates a role for microtubules in the organization and movement of CABS. These studies have relied on several drugs, particularly colchicine, which at low concentrations specifically disrupts microtubular structures, and its photochemically derived analog lumicolchicine which has many similar physical properties, but does not interact with microtubules.

If 3T3 fibroblasts are pretreated with colchicine the distribution of CABS is unaltered, i.e., remains random. However, if SV40 transformed fibroblasts are pretreated with colchicine, CABS topography is profoundly altered: instead of dispersed clusters, CABS are drawn into a large aggregate or "cap" (Fig. 7).

The cells which have proved most useful in our analyses thus far are phagocytes-alveolar macrophages and polymorphonuclear leukocytes (PMN). Although these are normal cells, they are also highly agglutinable by Con A. Like other cells, the distribution of CABS on prefixed cells is random (Borysenko, 1974). However, without fixation, at 37°C, clusters are formed and these progress to the formation of caps (Ryan *et al.*, 1974). It was also shown that colchicine altered the agglutinability of PMN by Con A (Berlin and Ukena, 1972) and further studies were directed at an analysis of CABS on PMN surfaces during phagocytosis.

Some years ago we began studies of the effect of phagocytosis on the membrane transport of non-electrolytes (Tsan and Berlin, 1971). During phagocytosis the particles are enveloped in plasma membrane which is then lost from the surface by internalization. By suitable choice of particle a large part of the membrane is internalized. Despite this extensive internalization, membrane transport (representing membrane proteins) does not change (Table II). The insertion of new membrane, containing transport sites, was ruled out. However, when phagocytes were treated with colchicine and then allowed to phagocytize, transport was reduced - and reduced in proportion to the amount of phagocytosis (internalization) (Ukena and Berlin, 1972) (Table III). Since this effect was obtainable with low concentrations of colchicine and vinblastine, but not by lumicolchicine, we inferred that it was mediated by microtubules. These studies have been further amplified by using CABS and *Ricinus communis* agglutinin binding sites

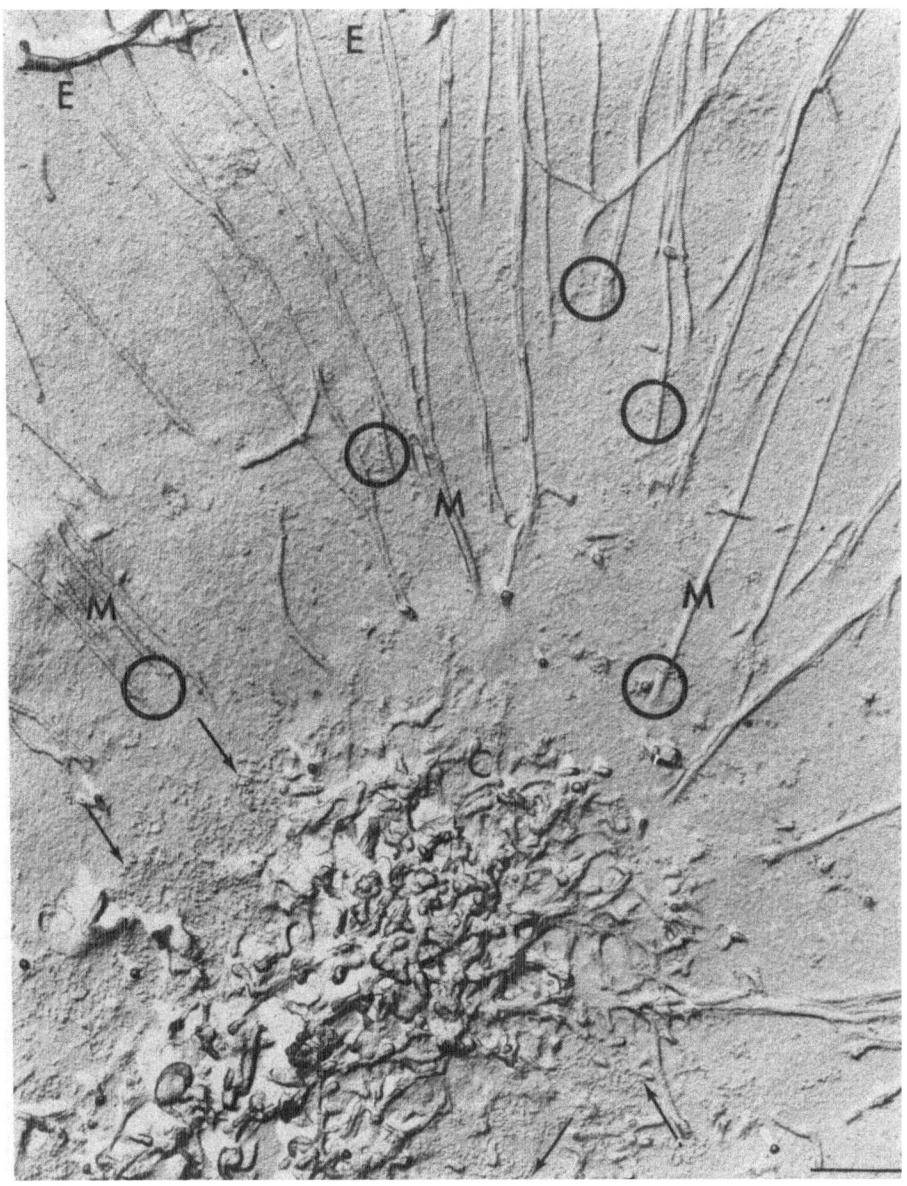

Fig. 7. SV3T3 cell treated with 10^{-6} M colchicine before Con A/H labeling. Microspikes (M) extend from the central area of the cell (C) past the cell edges (E). Large aggregates of hemocyanin (arrows) are seen in the center of the cell, and small clumps (circles) are often associated with the points of origin of the microspikes. Bar = 1 µm. x 11,700 (From Ukena et al., 1974).

TABLE II

EFFECT OF PHAGOCYTOSIS ON MEMBRANE TRANSPORT IN THE RABBIT ALVEOLAR MACROPHAGE AND POLYMORPHONUCLEAR LEUKOCYTE

Transport Systems	Rate of Transport pmoles per 45 sec					$p^{\ddagger}$
Alveolar macrophages*						
Lysine	133	± 27	(6)	126 ± 21	(6)	> 0.2
Adenosine	35	± 14	(4)	34 ± 14		> 0.25
Polymorphonuclear[†] Leukocytes						
Adenosine	11.4 ±	2.1	(3)	10.5 ± 0.14	(3)	> 0.2
Adenine	6.0 ±	1.5	(7)	6.3 ± 1.3		> 0.2

*0.3 million cells. Number of observations in parentheses.
[†]2.0 million cells.
[‡]P values determined from paired differences.
(From Tsan and Berlin, 1971).

TABLE III

EFFECT OF PHAGOCYTOSIS BY POLYMORPHONUCLEAR LEUKOCYTES ON MEMBRANE TRANSPORT OF NONELECTROLYTES WITH OR WITHOUT COLCHICINE PRETREATMENT

Transport System	Percent depression relative to nonphagocytizing monolayers*	
	No colchicine	Colchicine
Adenine	2.5 ± 2.3 (14) P[‡] > 0.9	42.8 ± 5.0 (14) P <0.001
Lysine	6.3 ± 3.9 (5) P > 0.8	43.5 ± 6.3 (5) P <0.001

*Values are percent depression ± standard error of the mean. Number of observations in parentheses.
[‡]P values refer to significance of depression from control after phagocytosis.
(From Ukena and Berlin, 1972).

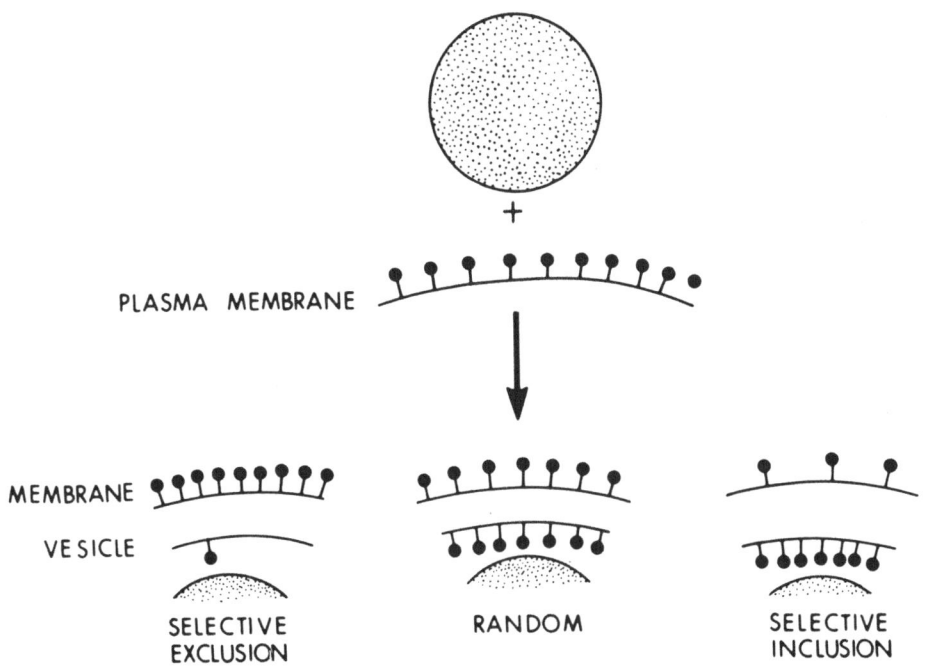

Fig. 8. Diagrammatic representation of probing of surface by particle (top) for surface topography of lectin binding sites. Binding sites are disperse on plasma membrane. Hypothetical effects of membrane internalization on the specific activities (sites per unit area) of membrane are shown below. At center, random inclusion of sites leaves specific activity unaffected, whereas selective exclusion (left) leaves membrane enriched and selective inclusion makes the membrane impoverished (low specific activity) for receptors.

as surface markers (Oliver et al., 1974).

The experimental advantage of using CABS as surface markers in these studies is that (unlike the transport sites) the distribution of CABS prior to phagocytosis is known - and is disperse. Again, the strategy here was to allow the phagocytized particle to probe the surface. We developed a method for isolating cell membranes - before and after phagocytosis - and for titrating the CABS on the isolated membrane. No Con A was added to the membranes that could induce clustering or other deviations from the random CABS distribution. Labelled Con A was used simply to titrate CABS after membrane isolation. The CABS titrations are expressed in µg Con A bound per 100 µg of membrane protein. The rationale of the approach which involves

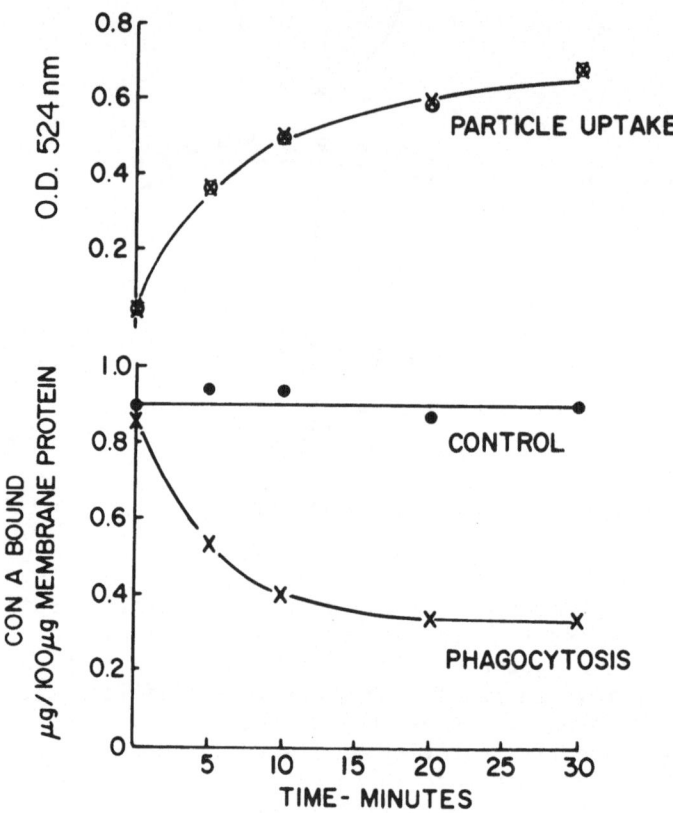

Fig. 9. Top: A plot of the extent of phagocytosis as a function of time with (0——0) or without (X——X) colchicine. Cells were treated with the alkaloid at 5 μM for 30 min before initiation of phagocytosis. Bottom: Effect of phagocytosis on specific activity of CABS of isolated membranes. 4×10^6 cells/ml of medium were incubated without (control) and with (phagocytosis) emulsion for the times indicated. Cells were then centrifuged, washed, and resuspended in modified Hanks' medium containing 7 μg/ml of labeled Con A and incubated at 4°C for 30 min. Membranes were then isolated, and the specific activity of Con A bound was determined (From Oliver et al., 1974).

essentially the determination of the specific activities of lectin binding sites is best understood by reference to Fig. 8. If a given membrane constituent is internalized in parallel with other membrane proteins, its specific activity will not be affected by phagocytosis. However, if a constituent is selectively excluded from the internalized membrane, its specific activity will be increased in the residual plasma membrane, and decreased if it is selectively included in the internalized membrane.

In Fig. 9 it is seen that CABS are selectively removed from the surface during and in parallel with phagocytosis. Similar findings were made when using *Ricinus communis* agglutinin (RCA) to analyze for a second membrane receptor.

One of the more interesting findings to emerge from these studies was the observation that the addition of Con A to intact cells resulted in the nearly quantitative internalization of CABS. When Con A at high concentration was added at 37°C and the membrane isolated, the amount of radioactive Con A that could be bound was reduced by more than 90%. Moreover, the amount bound was not significantly increased when the membranes were first treated with high concentrations of α-methyl-mannose (αMM) to elute Con A previously bound. By contrast when Con A was bound at 4°C the capacity to label isolated membranes pretreated with αMM was not different from contols. This result indicates that Con A induces the internalization of its own receptors. Similar results were obtained using RCA. In addition it was found that Con A induced the internalization of 50% of RCA binding sites and vice versa. In our view the most likely interpretation of this result is that there exists a class of molecules or molecular complexes that has receptors for both lectins. It was shown that phagocytosis results in the selective internalization of this class. Alternatively, the internalization of membrane associated with phagocytosis could lead to the exhaustion of a membrane element essential for linkage between separate Con A and RCA receptors. When cells were treated with colchicine prior to phagocytosis the selective internalization of CABS or RCA receptors did not occur (Fig. 10a,b). Fig. 10c shows that the derivative lumicolchicine (inactive against microtubules) has no effect on the selective loss of CABS associated with phagocytosis.

We conclude that under appropriate stimulation (phagocytosis) certain CABS are directed from their random distribution into specific internalized regions. This directed movement is in turn dependent on colchicine sensitive structures - probably microtubules.

IV. DISCUSSION

The foregoing observations indicate that CABS, while distributed relatively homogeneously over the cell surface, are in a highly dynamic

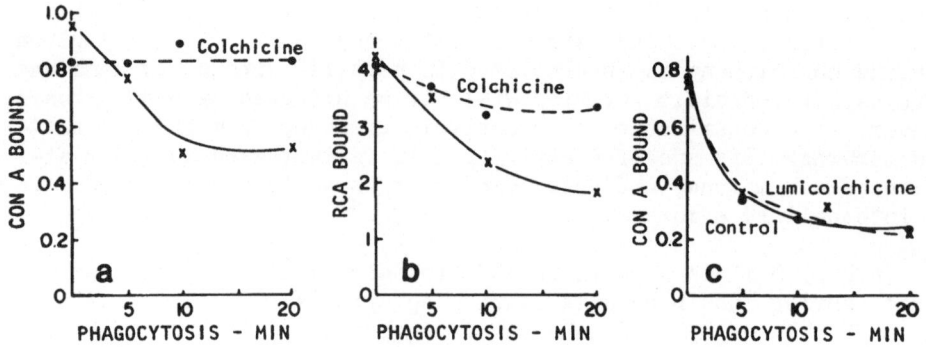

Fig. 10. Effect of colchicine alkaloids on changes in membrane lectin binding sites produced by phagocytosis. Cells were preincubated with 5 µM alkaloid (dashed line) or modified Hanks' medium alone (solid line) for 30 min before addition of emulsion and during phagocytosis. (a) Effect of colchicine on CABS. Labeled with 7.5 µg/ml of RCA. (c) Effect of lumicolchicine on CABS. Labeled with 6.3 µg/ml of Con A (from Oliver et al., 1974).

state. Under certain conditions, they may be induced to aggregate into clusters by the addition of Con A. Although this is a property of the transformed mouse fibroblast, as opposed to the normal, it is probably not a reflection of differences in the biochemical composition of cells, since short treatment with cyclic AMP abolishes the transformed characteristic.

In certain cell types, such as the PMN and lymphocyte, binding of Con A leads to cap formation - a further cross-linking of CABS. Cap formation also occurs in transformed fibroblasts first treated with colchicine. This process of aggregation effected by crosslinking of multivalent ligands, is not unexpected, provided there is some mobility of CABS within the plane of the membrane (the absence of cross-linking in membranes of certain cells is in this context the more remarkable). On the other hand, the selective movement of CABS into the phagocytic vesicle in the absence of apparent exogenous cross-linking agents suggests a different mechanism whereby CABS undergo translational movement. This mechanism must include microtubules as an essential element, since selective inclusion of CABS into phagocytic vesicles does not occur after colchicine treatment. We believe that CABS are normally disperse in all cells, for lymphocyte binding of CABS by Con A may induce the association of CABS with microtubules that limits their movement (Edelman et al., 1973). In the case of phagocytes contact with a particle may induce such an association. Reaven and Axline (1973) have demonstrated a network of microfilaments and microtubules in the areas of contact between cultured macrophages and their substrate or particulates. In phagocytes, however, the evidence goes beyond the notion of "fixation." It is suggested that a highly directional movement of CABS occurs with endocytosis that is dependent on the microtubule.

REFERENCES

1. Aub, J. C., Tieslau, C., and Lankester, H. (1963). "Reactions of normal and tumour cell surfaces to enzymes. I. Wheat-germ lipase and associated mucopolysaccharides." *Proc. Nat. Acad. Sci. U. S. A.* 50, 613.

2. Berlin, R. D., and Ukena, T. E. (1972). "Effect of colchicine and vinblastine on the agglutination of polymorphonuclear leukocytes by Con A." *Nature New Biology* 238, 120.

3. Borysenko, J. Z. (1974). Unpublished data.

4. Cline, M. J., and Livingston, D. C. (1971). "Binding of ^{3}H-concanavalin A by normal and transformed cells." *Nature New Biology* 232, 155.

5. Edelman, G. M., Yahara, I., and Wang, J. L. (1973). "Receptor mobility and receptor-cytoplasmic interactions in lymphocytes." *Proc. Nat. Acad. Sci. U. S. A.* 70, 1442.

6. Hakomori, S. (1973). "Glycolipids of tumor cell membrane." *Adv. in Cancer Res.* 18, 265.

7. Nicolson, G. L. (1971). "Difference in topology of normal and tumour cell membranes shown by different surface distributions of ferritin-conjugated concanavalin A." *Nature New Biology* 233, 244.

8. Oliver, J. M., Ukena, T. E., and Berlin, R. D. (1974). "Effects of phagocytosis and colchicine on the distribution of lectin-binding sites on cell surfaces." *Proc. Nat. Acad. Sci. U. S. A.* 71, 394.

9. Ozanne, B., and Sambrook, J. (1971). "Binding of radioactively labeled concanavalin A and wheat germ agglutinin to normal and virus-transformed cells." *Nature New Biology* 232, 156.

10. Poo, M., and Cone, R. A. (1974). "Lateral diffusion of rhodopsin in the photoreceptor membrane." *Nature* 247, 438.

11. Reaven, E. P., and Axline, S. G. (1973). "Subplasmalemmal microfilaments and microtubules in resting and phagocytizing cultivated macrophages." *J. Cell Biol.* 59, 12.

12. Rosenblith, J. Z., Ukena, T. E., Yin, H. H., Berlin, R. D., and Karnovsky, M. J. (1973). "A comparative evaluation of the distribution of concanavalin A-binding sites on the surfaces of normal, virally transformed, and protease-treated fibroblasts." *Proc. Nat. Acad. Sci. U. S. A.* 70, 1625.

13. Ryan, G. B., Borysenko, J. Z., and Karnovsky, M. J. "Capping of concanavalin A on human polymorphonuclear leukocytes: factors affecting redistribution of surface receptors." *J. Cell Biol.* In press.

14. Smith, S. B., and Revel, J. P. (1972). "Mapping of concanavalin A binding sites on the surface of several cell types." *Develop. Biol.* 27, 434.

15. Ukena, T. E., and Berlin, R. D. (1972). "Effect of colchicine and vinblastine on the topographical separation of membrane functions." *J. Exp. Med.* 136, 1.

16. Ukena, T. E., Borysenko, J. Z., Black, P. H., Karnovsky, M. J., and Berlin, R. D. "Effects of dibutyryl cyclic AMP and theophylline on concanavalin A binding site distribution on normal and transformed cells, and the distribution of concanavalin A binding sites on cells resistant to concanavalin A." (Submitted for publication).

17. Ukena, T. E., Borysenko, J. Z., Karnovsky, M. J., and Berlin, R. D. (1974). "Effects of colchicine, cytochalasin B, and 2-deoxyglucose on the topographical organization of surface-bound concanavalin A in normal and transformed fibroblasts." *J. Cell Biol.* 61, 70.

18. Yahara, I., and Edelman, G. M. (1972). "Restriction of the mobility of lymphocyte immunoglobulin receptors by concanavalin A." *Proc. Nat. Acad. Sci. U. S. A.* 69, 608.

10

EFFECTS OF CONCANAVALIN A ON CELLULAR DYNAMICS AND MEMBRANE TRANSPORT*

Tushar K. Chowdhury
University of Oklahoma Health Sciences Center
Oklahoma City, Oklahoma, U.S.A.

ABSTRACT

How does Con A affect the cellular dynamics and cellular transport of various substances? It has been observed that following a 30 minute incubation with Con A (20-66 µg/ml) the tumor cells (Anaplastic carcinoma Type 15091A), unlike the normal cells, gradually retract their pseudopodia, take on a rounded shape and lose their motility. The maximum rounding of tumor cells takes place during the period of 200-250 min following the withdrawal of Con A from the incubating medium. All of these rounded cells later develop their regular pseudopodia. A partial synchrony of cell division is also observed. Associated with the rounding effect, there is a decrease in the rate of uptake of some specific amino acids by the tumor cells. The maximum reduction in the uptake coincides with the stage of maximum rounding. Con A also affects some of the electrical properties of tumor cells. In addition, the incorporation of radioactive thymidine into the nuclei is reduced to a much larger extent than can be accounted for by the reduction in the rate of entry of thymidine into the cytoplasm. The saturation densities of both normal and tumor cells are not significantly altered by Con A. Tumor cells treated with Con A (chronically or acutely) are equally as potent as the untreated cells in yielding solid tumors when the host animals are injected with these cells subcutaneously. In conclusion, it is suggested that while Con A does not revert or destroy this particular line of tumor cells, this lectin has some interesting selective effects on the morphology and transport properties of tumor cells.

---

*This work has been supported by NSF grant No. GB18726 and by an American Cancer Society grant No. 5-69.

I. INTRODUCTION

In recent years particular interest has centered around concanavalin A (Con A) because of its effects on cancer cells. Many of the effects are found in cancer cells in distinction from normal cells. Con A, for example, has been reported to agglutinate leukemic cells as well as cells transformed by polyoma virus, Simian virus, chemical carcinogens, and X-irradiation. It does not agglutinate normal cells, however, unless prior to Con A treatment the cell surface is significantly modified by the proteolytic enzyme trypsin (Inbar and Sachs, 1969). Most of the transformed or tumor cells lack the so-called "contact inhibited movement" and "density dependent growth" (Abercrombie, 1967; Foster and Pardee, 1969; Martz and Steinberg, 1972). Under some special culture conditions, certain normal cells also behave like tumor cells, with regard to the above characteristics (Aaronson and Todaro, 1968). These tumor cells grow *in vitro* beyond confluency and pile over each other to form multiple layers. Working with one such line of transformed cells, e.g., polyoma virus-transformed 3T3 fibroblasts (Py-3T3), Burger and Noonan (1970) reported that by treating the transformed cells with monovalent Con A, the growth pattern of these cells is restored to that of normal cells. The main growth characteristic studied above was the surface density of cells grown *in vitro*. It would have been desirable, of course, to investigate whether the Con A-treated transformed cells compare with the normal cells with regard to some other physiological properties, e.g., electrophysiology, transport activity, etc.

The purpose of this investigation is to study the effects of Con A on some of the physiological characteristics of a tumor cell line which exhibits on the one hand the "contact inhibited movement" and yet, on the other hand, lacks the "density dependent growth" (Chowdhury and Chaudhuri, 1974). The specific questions posed in this study are:

1. Does Con A exhibit an effect on the morphology of the tumor cells as well as on the corresponding normal cells? Is the effect on tumor cells different from that on normal cells?

2. Are the permeability and transport characteristics of tumor cells different from those of normal cells? How does Con A affect these properties?

3. Does Con A act only at the level of the plasma membrane or does its effect extend also to the nucleus?

4. Are the electrical properties of tumor cells different from those of normal cells? If so, does Con A treatment of tumor cells revert these properties to the values observed in normal cells?

II. METHODS OF PROCEDURE

A. Cells

Normal fibroblasts were grown out of the primary explants derived from the subcutaneous tissue of normal A/J mice (Jackson Laboratory, Bar Harbor, Maine). The growth medium was a modified Eagle's minimum essential medium supplemented with dextrose, asparagine, $CaCl_2$, glutamine, NCTC-135, penicillin, streptomycin, fungizone, $NaHCO_3$, and 10% fetal calf serum. This medium will be referred to as B_1D medium (Chou and Chowdhury, 1974). The tumor cells originated from explant cultures of the solid subcutaneous tumor (type 15091A, Jackson Laboratory) grown in A/J host mice. Normal, as well as the tumor cells, were grown in plastic flasks with a 25 sq cm surface area. The cells were fed with fresh B_1D medium at two-day intervals and on the day preceding their use in the transport study. Serial subcultures of both lines of cells were started before the cells reached confluency. In an attempt to ensure that the cells grown *in vitro* were not too far removed from the original *in vivo* cell lines, primary explant cultures of normal as well as tumor cells were started every six to eight weeks (tumor cell line from the progressively maintained solid tumor and normal cell line from the normal tissue). The normal cells grow only up to a confluent monolayer. This particular line of tumor cells also exhibits contact-inhibited movement in the formation of a confluent monolayer. However, the cells continue multiplying without any sign of density-dependent growth controls (Chowdhury and Chaudhuri, 1974). Only a single layer of cells remains attached to the substratum and all other cells remain floating. The floating tumor cells are capable of attaching back to the substratum only if there is some empty space on it. Secondary cultures can be started quite effectively by transferring these floating cells into new flasks. The present study was conducted primarily on attached cells.

B. Concanavalin A

Grade III, highly purified Con A was obtained from Sigma Chemical Company (Saint Louis, Missouri). It contains approximately 15% protein and 85% NaCl and is substantially free from carbohydrates. Two different types of stock solution of Con A were used. The first type was a solution of intact Con A in a phosphate buffer solution. The second solution consisted of the so-called monovalent forms of Con A which were prepared by the method described by Burger and Noonan (1970). Each of the stock solutions of Con A was diluted with the specific incubation medium so that the final concentration of Con A in the incubation medium ranged from 20 µg/ml to 66 µg/ml.

C. Transport Study

In some previous experiments (Chou and Chowdhury, 1974) it has been observed that a number of amino acids are taken up into the present cell lines, presumably by active transport. From them only three amino acids have been selected for this particular investigation. These are (1) the simplest and smallest amino acid, glycine; (2) a basic amino acid, histidine; and (3) an aromatic amino acid, phenylalanine. The uptake of these amino acids was studied by using tritium- or ^{14}carbon-labelled forms of the amino acids in tracer quantities in the incubating medium and following their entry into the cells. The general procedure adopted for the amino acid transport study was as follows: Flasks with sparse cells population were used two to seven days after seeding and they were always arranged in pairs; one flask of cells was used for Con A experiments, whereas the other flask of cells served as control. All flasks of cells were preincubated in a phosphate buffer solution (PBS) at 37°C for one hour. Following this preincubation, the incubation medium in the experimental flask was replaced by a Con A-containing PBS (concentration of Con A = 66 µg/ml), whereas the incubation medium in the control flask was replaced by fresh PBS. Thirty minutes later all flasks were washed five times with fresh PBS at 37°C and finally the cells were incubated with fresh PBS. The end of the Con A incubation period will be referred to as the zero time. On the first hour, second hour, fourth hour and sixth hour, groups of paired flasks were chosen in which the simple PBS was replaced with PBS containing trace quantities of a specific radioactive amino acid. Following an incubation period of one hour with the isotope-containing medium, each flask was rinsed for a total of 30 seconds with six lots of 2 ml volumes of non-radioactive PBS. Each flask was then scraped with clean soft-bristled pipe cleaners in 1.5 ml PBS and the cell suspension transferred to a graduated conical centrifuge tube. A milliliter of this cell suspension was immediately removed for counting cells in a hemocytometer. The cell suspension was treated with 2.5 ml of 0.1 N NaOH, thereby lysing the cells. Aliquots of this lysate were used in determining the radioactivity and the protein content. The cellular uptake of amino acid was expressed as counts per minute of the lysate radioactivity per unit specific activity of the incubation medium per microgram of cellular protein (CPM/µg).

In the study of the uptake of thymidine the procedure was very similar with the exception that the incubation medium used in this case was a conditioned culture medium instead of PBS. The conditioned medium was collected daily from a number of flasks in which the cells were in their logarithmic phase of growth.

D. Electrophysiological Measurements

Intracellular electrical potentials of both normal and tumor cells were determined with the aid of fine tipped glass microelectrodes having diameters ranging between 100 Å and 700 Å (Chowdhury, 1969). The reference electrode was placed in the culture medium bathing the cells. The membrane resistance of the cells was estimated by passing through the intracellular microelectrode several short pulses of electrical current and monitoring the corresponding pulses of deflection in the microelectrode potential (Chowdhury and Chou, 1973).

III. RESULTS

A. Effects of Con A on Cellular Morphology and Motility

Figure 1A illustrates the gross morphological features of A/J tumor cells. These cells have multiple pseudopodia. They exhibit a considerable degree of contact inhibition of movement, i.e., they do not pile over each other (Chowdhury and Chaudhuri, 1974). Incubation of these tumor cells with Con A (20 to 66 µg/ml) in PBS or B_1D medium at 37°C for a period of as little as 30 min produced a noticeable degree of retraction of pseudopodia and cell rounding. The cell rounding was maximum at about 4 hours following the withdrawal of Con A (Fig. 1B).

With time lapse cinematography, it was observed that prior to treatment with Con A tumor cells exhibited a considerable degree of surface motility; numerous pseudopodia erupted at various points along the edge of the cells. None of these pseudopodia is of a permanent nature; all of them continuously keep changing their locations, sometimes even appearing to vanish within the cell. Immediately following Con A treatment, some of the pseudopodia began to retract but quite sluggishly and the cells gradually became rounded. This rounding effect continued for a few hours, even following the replacement of the Con A solution with fresh PBS or B_1D medium. By the fourth hour of post-Con A incubation, 80 to 90 per cent of the cells became rounded with sharp, smooth boundaries and their normal motility was fully arrested. A question may arise regarding the viability of these cells. To test this point the fully rounded cells were watched for eight hours while they were in the PBS or fed with B_1D culture medium equilibrated with 5% CO_2 at 37°C. The rounded cells began to develop pseudopodia within six hours following the withdrawal of Con A-containing incubation media. By the eighth hour every single rounded cell regained its regular multiple pseudopodia and the associated motility (Fig. 1C). For the cells to survive beyond this period and to grow properly, it was necessary to feed them with B_1D medium. At the above concentrations of the lectin, the cellular dynamics was hindered by both types of Con A solutions. The effect of the so-called monovalent form of Con A was identical to that of intact Con A.

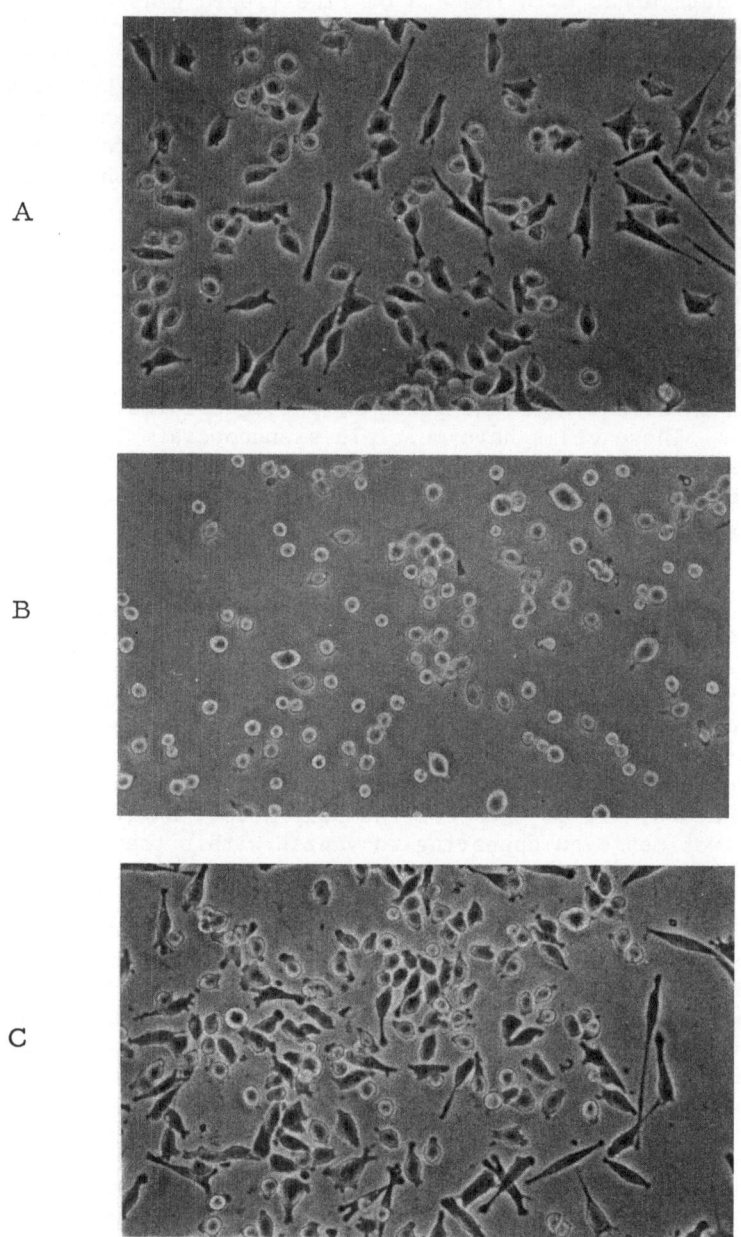

Fig. 1. Phase contrast micrographs of A/J tumor cells (x290).
A : Prior to Con A treatment. B : Four hours following a 30 min treatment with Con A. C : Eight hours after Con A treatment

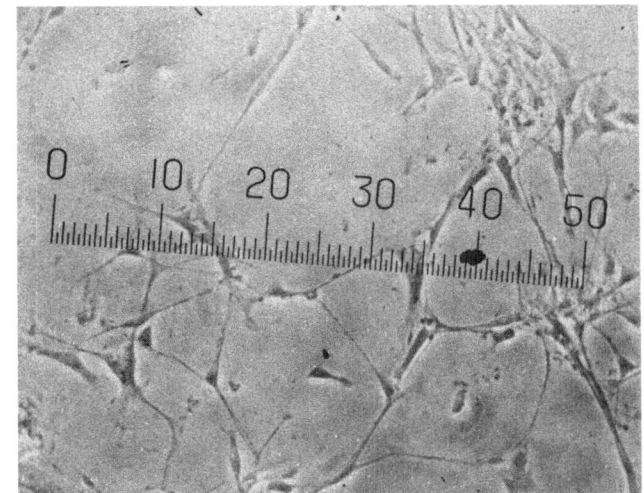

Fig. 2. Phase contrast micrographs of normal fibroblasts (x290). A : Prior to Con A treatment. B : Four hours after a 30 min treatment with Con A.

Figure 2A represents the gross morphology of the normal A/J fibroblasts. Under time lapse cinematography it was observed that each of the cells in the fibrous network was associated with pseudopodia and while they had a noticeable degree of motility it was less than the tumor cell motility. Treatment of these cells with Con A (20 to 66 μg/ml) did not produce either any retraction of pseudopodia

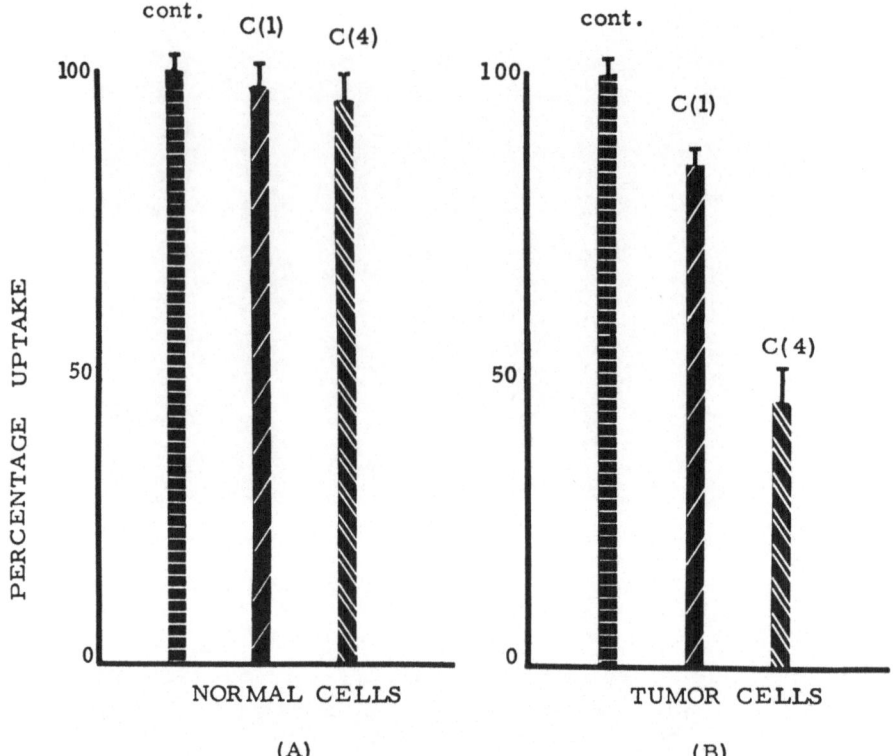

Fig. 3. Effect of Con A on the cellular uptake of glycine.
A : Normal cells. The first bar graph represents the uptake of glycine by untreated (control) cells, whereas the second and third bar graphs represent the respective relative uptake of glycine by normal cells in the first and fourth hours following Con A treatment. B : Tumor cells. The bar graph representations are similar to those in A. The standard error of the mean is represented by the vertical line above each bar graph.

or any sluggishness in their motility (Fig. 2B). Even at Con A concentrations of as high as 100 µg/ml no effect on the above dynamic properties of the normal cells could be detected.

B. <u>Cellular Uptake of Amino Acids</u>

In our study of the transport of amino acids into the cells, the parameter under investigation was not the rate of transport itself, but the maximum amount of amino acid taken up within the cells at a

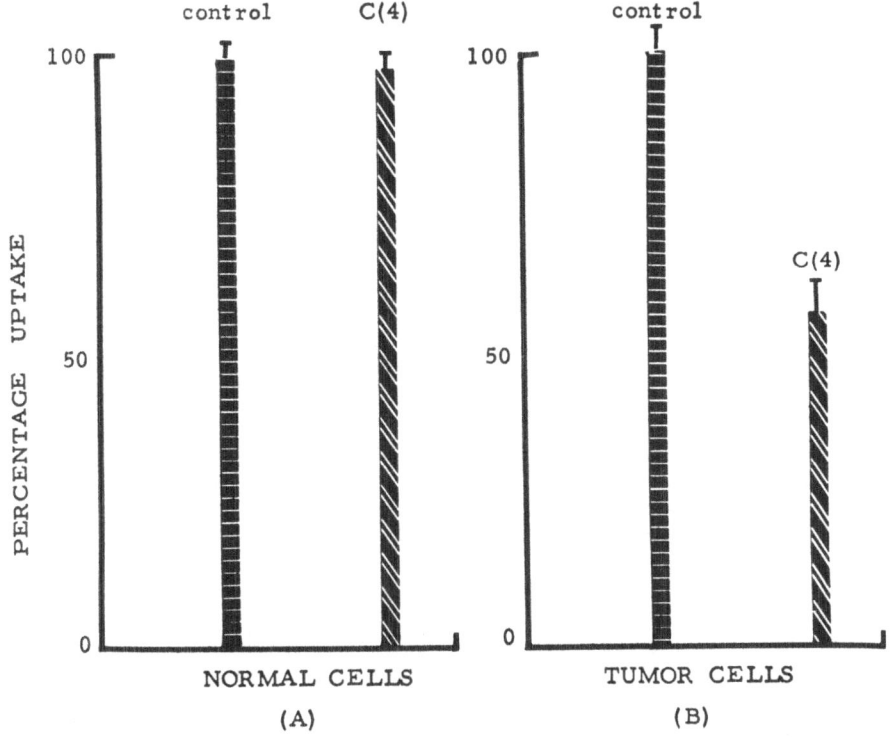

Fig. 4. Effect of Con A on the cellular uptake of histidine. A : Normal cells. The first bar graph represents the uptake of histidine by untreated cells, whereas the second bar graph represents the relative uptake of histidine by normal cells in the fourth hour following Con A treatment. B : Tumor cells. The bar graph representations are similar to those in A. The standard error of the mean is represented by the vertical line above each bar graph.

steady state. Sixty minutes of incubating the cells with a radioactive amino acid were sufficient for the isotopic amino acids to reach an equilibrium distribution with the non-radioactive amino acids (Chou and Chowdhury, 1974). The maximum amount of amino acid taken up by the cells is expressed as counts per minute of radioactivity per microgram of cellular proteins (CPM/µg).

Figure 3 shows our findings on the uptake of glycine by normal A/J fibroblasts prior to and following treatment with Con A (66 µg/ml). The first bar graph in Fig. 3A depicts the average glycine uptake in eight flasks of normal cells without any Con A treatment.

The second and third bar graphs in Fig. 3A represent the data on Con A treated normal cells. More specifically, the second bar graph corresponds to the uptake of glycine by normal cells during the first hour following a 30-minute period of incubation with Con A. The effect of Con A on the glycine uptake seems to be of an inhibitory nature even though it is not significant from a statistical point of view. The uptake of glycine during the fourth hour following Con A incubation remains also practically unaltered (the third bar graph in Fig. 3A).

Figure 3B represents the uptake of glycine by A/J tumor cells prior to and following treatment with Con A. The first bar graph corresponds to the average glycine uptake in twelve flasks of tumor cells in the absence of Con A. One hour following the Con A incubation, the uptake of glycine by the tumor cells is significantly reduced (second bar graph). The maximum inhibition of Con A on glycine transport occurs during the fourth hour following Con A incubation (third bar graph). During this period the tumor cells also exhibited the maximum degree of rounding and the least degree of motility. In the second as well as in the sixth hour the inhibitory effect of Con A on glycine transport was less than during the fourth hour. The data on the uptake of histidine are shown in Fig. 4. The effect of Con A on the histidine uptake by normal cells is almost insignificant (Fig. 4A). On the other hand, the uptake of histidine by Con A-treated tumor cells during the fourth hour is considerably reduced from the control value of histidine uptake by untreated A/J tumor cells (Fig. 4B).

The effects of Con A on the uptake of the aromatic amino acid, phenylalanine, is presented in Fig. 5. As was to be expected, Con A has no significant effect on the uptake of phenylalanine by normal cells (Fig. 5A). However, unlike the inhibitory effect of Con A on tumor cells in the uptake of glycine and histidine, this lectin fails to inhibit the uptake of phenylalanine by tumor cells (Fig. 5B). It is apparent, therefore, that Con A does not necessarily block the transport of all amino acids.

C. Nuclear Incorporation of Thymidine

This group of experiments was designed to ascertain the effect of Con A (1) on the entry of thymidine, particularly into tumor cells, and (2) on the incorporation of thymidine into the nuclei of these tumor cells. For these experiments paired flasks of tumor cells with identical cell densities were chosen. Following a one hour incubation period of the cells in filtered conditioned medium, Con A was added into the experimental flask at a final concentration of 66 µg/ml, while the control flask contained only the conditioned medium. The Con A incubation lasted for 30 min following which the flask was rinsed several times with fresh conditioned medium at

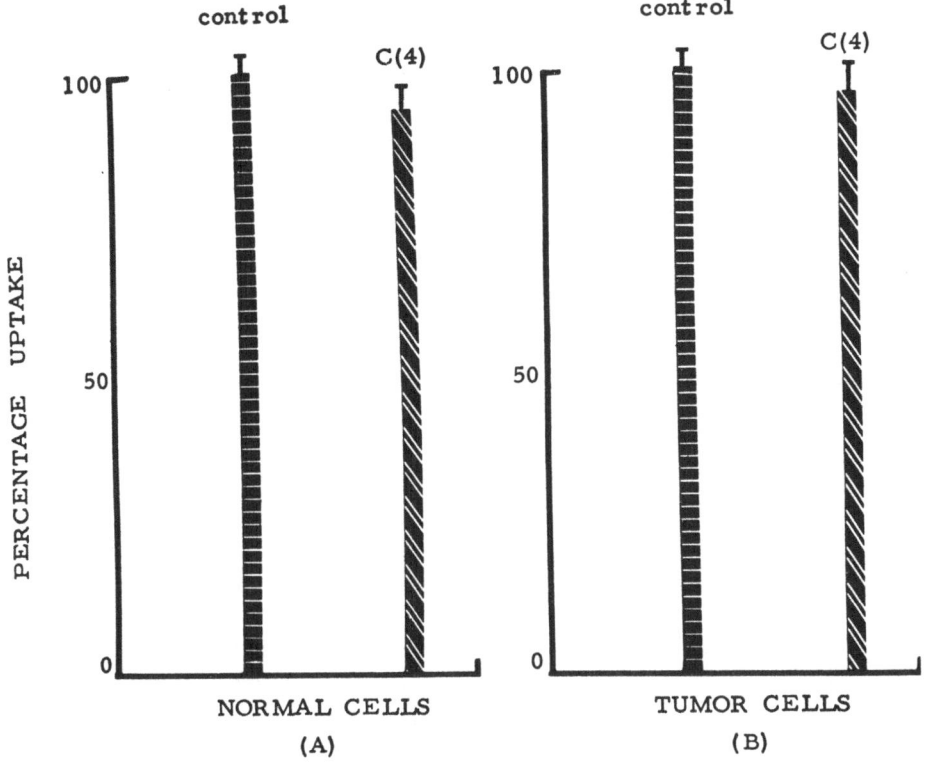

Fig. 5. Effect of Con A on the cellular uptake of phenylalanine. The bar graph representations are similar to those in Fig. 4.

controlled pH and temperature. Radioactive thymidine-containing conditioned medium was placed in the paired control and experimental flasks either during the first or fourth hour after the termination of incubation of the tumor cells with Con A. The isotopic thymidine was used in tracer amounts and care was taken to avoid any thymidine blockage of nuclear DNA synthesis. Following a 45 min incubation of the cells in the radioactive medium, the flasks were washed with six lots of non-radioactive PBS over a period of 30 seconds and a final volume of 1.5 ml of PBS at 37°C was placed in each flask. The cells were then detached from the substratum by scraping the surface with clean fine-bristled pipe cleaners. The suspended cells were then centrifuged at 500 x g for five minutes and resuspended in 1.0 ml of a hypotonic tris buffer solution which caused the cells to become swollen. These swollen cells were subjected to a mild homogenization in a Dounce homogenizer and the homogenate was centrifuged at 850 x g

TABLE I

EFFECT OF CON A ON THYMIDINE INCORPORATION IN TUMOR CELLS

Cells	Relative Thymidine Radioactivity in	
	Cytoplasm (CPM/μg cell protein)	Nucleus (CPM/μg cell protein)
Untreated (control) tumor cells	100 ± 6	400 ± 10
Con A-treated tumor cells	70 ± 5	180 ± 12

for ten minutes. The supernatant was saved for the determination of thymidine radioactivity. The pellet of nuclei was resuspended in 1 ml of isotonic buffer solution. The yields and appearance of the nuclear preparations from the two flasks were identical. The nuclear suspensions were then treated with 1.5 ml of 0.1 N NaOH whereby the nuclei were lysed. The lysate was then analyzed for thymidine radioactivity. The protein contents of this lysate, as well as of the cytoplasmic supernatant fraction, were determined by the standard method of Lowry et al. (1951). As shown in Table I, the content of radioactive thymidine in the cytoplasm is much less for the Con A-treated tumor cells than for the control tumor cells. Likewise, the nuclear content of radioactive thymidine is less in the Con A-treated tumor cells. However, the reduction in nuclear thymidine content is proportionately more pronounced in the Con A-treated cells than in the untreated tumor cells. It appears, therefore, that the Con A effect is also extended to the nuclei of the tumor cells.

D. Electrical Properties of Cells

Two of the electrical properties of cells studied are the intracellular electrical potential and the transmembrane resistance. The magnitudes of these two parameters somewhat reflect the permeability and transport characteristics of the cells with respect to electrolytes in particular. The intracellular electrical potentials of all cells were negative with respect to the reference electrode placed in the B_1D culture medium (Table II). The average intracellular potential of normal control cells was -0.6 ± 0.2 mV, whereas that in the Con A-treated normal cells was -0.7 ± 0.2 mV. Con A did not alter the magnitude of the intracellular potential in normal cells. In the case of tumor cells, however, Con A had a significant effect on the intracellular potential. The potential of the Con A-treated tumor cells was -8.0 ± 0.4 mV, whereas that in the untreated tumor cells was -4.5 ± 0.2 mV. Almost the same pattern of effect is reflected with the transmembrane resistance. In the case of normal cells the average membrane resistance of the Con A-treated cells was

TABLE II

EFFECT OF CON A ON INTRACELLULAR ELECTRICAL POTENTIAL
AND MEMBRANE RESISTANCE

Cells	Treatment	Intracellular Potential (mV)	Membrane Resistance (MΩ)
Normal cells	Untreated	-0.6 ± 0.2	2.3 ± 0.4
Normal cells	Con A	-0.7 ± 0.2	2.4 ± 0.4
Tumor cells	Untreated	-4.5 ± 0.2	3.4 ± 0.4
Tumor cells	Con A	-8.0 ± 0.4	2.6 ± 0.2

The reference electrode was placed in the culture medium.

2.4 ± 0.4 MΩ, as compared to 2.3 ± 0.4 MΩ for the control cells. Again in tumor cells the control value of 3.4 ± 0.4 MΩ was significantly reduced by Con A to 2.6 ± 0.2 MΩ. It appears, therefore, that particularly with regard to the transmembrane resistance, the magnitude for Con A-treated tumor cells tends to be closer to that of untreated normal cells.

E. Effects of Con A on the Growth of Tumor Cells

For this group of experiments tumor cells in the early logarithmic phase of growth were chosen. In the absence of Con A the doubling time, T_d, of tumor cells was estimated to be 32 hours (Fig. 6A). However, when the cells were treated chronically with Con A, i.e., daily for a period of one hour, the growth rate was significantly diminished. The doubling time of the Con A-treated cells was estimated to be 50 hours (Fig. 6B). Interest developed at this time to find an answer to the question of what happens to the growth rate of Con A-treated tumor cells when the Con A treatment is terminated entirely. Interestingly enough, it was observed that such tumor cells grow at a rate almost identical with that of control tumor cells. These results suggest, therefore, that the effects of Con A on the growth of tumor cells are not permanent. A reduction in the growth rate of tumor cells is achieved particularly when the cells are treated chronically or intermittently with Con A.

F. Effects of Con A on the Potency of Tumor Transplantation

This group of experiments was designed to test whether Con A-treated tumor cells were capable of producing a solid tumor in the

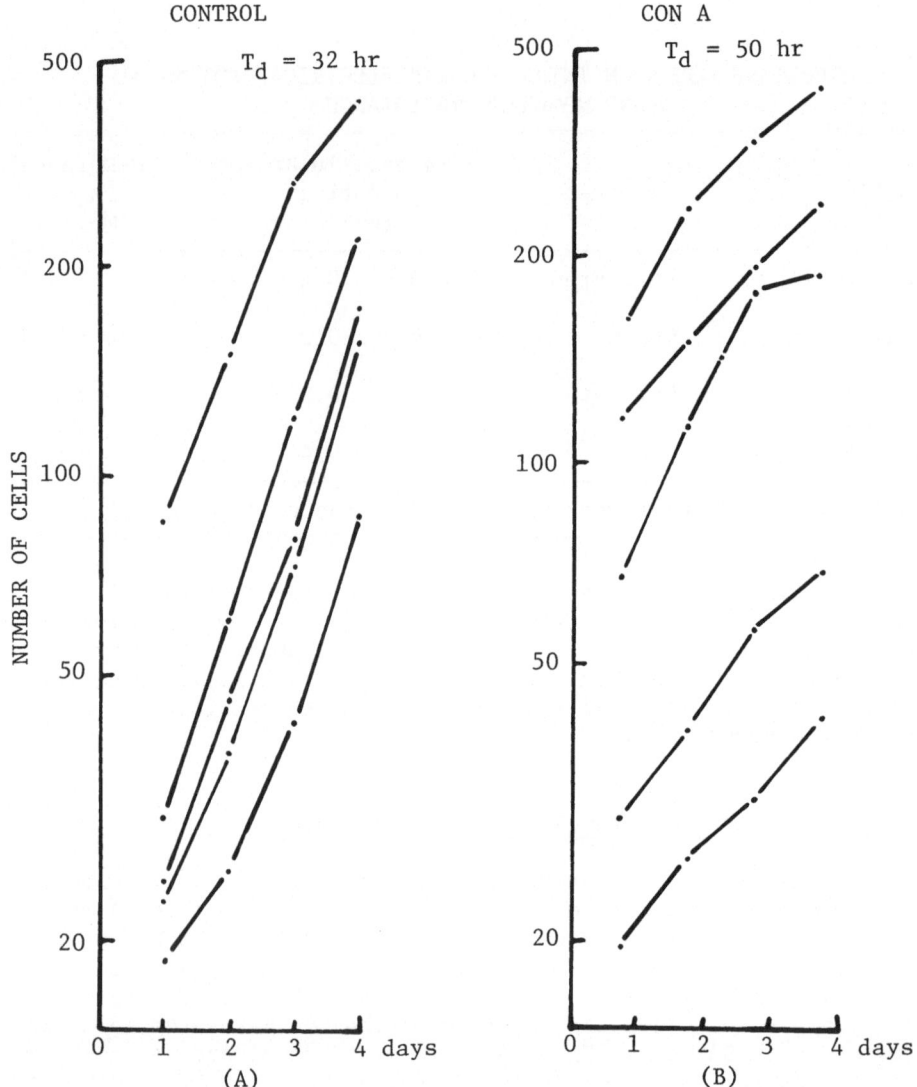

Fig. 6. Growth rate of tumor cells. A : Untreated tumor cells. Ordinate: Number of cells per 0.1 sq cm. Abscissa: Number of days following seeding. B : Tumor cells treated with Con A for one hour daily.

host A/J mice. Two procedures of Con A treatment were used. The first procedure (acute) involved a single treatment of the cells with Con A (66 μg/ml) for a period of one hour, following which the cells were washed several times with fresh B_1D medium and were ready for inoculation. The second procedure (chronic) involved daily treatment

TABLE III

TUMORIGENICITY OF CON A-TREATED TUMOR CELLS*

	Untreated (Control) Tumor Cells	Acute Con A-treated Cells	Chronic Con A-treated Cells
Palpable	6 days	8 days	6 days
Survival period	21-25 days	25-31 days	20-27 days

*Each animal was inoculated with a million tumor cells. There were ten animals in each of the above three groups.

of the tumor cells with Con A (66 μg/ml) for one hour over a period of four months. During this four month period, a number of serial subcultures was made in thinning down the cell population. The last treatment with Con A was given 24 hours prior to inoculation of the cells into animals. Three groups of ten mice were inoculated subcutaneously, each mouse getting one million tumor cells in a carrier medium of 0.1 ml. The first group of mice was inoculated with untreated tumor cells, the second group with acute Con A-treated tumor cells and the third group with chronic Con A-treated tumor cells. The results are presented in Table III. The tumor in the first group of mice (control) was palpable within 6 days, whereas in the second and third groups it became palpable on the 8th and 6th day respectively. The survival time of the control group of mice ranged from 21 to 25 days, whereas that of the acute Con A-treated mice ranged from 25 to 31 days. On the other hand, the survival time of animals inoculated with the chronically treated tumor cells was almost identical with that found in animals inoculated with the highly potent control tumor cells.

IV. DISCUSSION

In the present investigation, one of the primary effects of Con A appears to be upon the surface membrane of the cells, resulting in a gradual loss of pseudopodia. However, this effect is quite specific to tumor cells and not to normal cells, at least at the relatively low concentrations of Con A used in this study. The effect of this lectin on tumor cells is not instantaneous, rather there is a fairly large latent period. For example, the maximum rounding of the tumor cells occurs some four hours after the withdrawal of Con A solution. Of course, the withdrawal of Con A from the incubating medium does not necessarily imply a removal of Con A molecules from the cell surface. Nevertheless, it is interesting to note that the adherent Con A molecules produce a rather gradual retraction of the cellular pseudopodia with an ultimate arrest of the tumor cells in a non-motile condition during the fourth hour. There is a general tendency of contraction over the entire surface of the tumor cells and this may cause the formation of clusters of Con A binding sites (Nicolson, 1971) on tumor cell surfaces. The rounded state of tumor cells with practically no visible pseudopodia is associated with the minimum uptake of a number of amino acids (Chou and Chowdhury, 1974). Can the reduction in the uptake of amino acids by Con A-treated cells be attributed to the reduced surface area of the rounded cells? This indeed can be a possibility, especially since the surface area of untreated tumor cells is considerably larger due to the presence of numerous pseudopodia. However, several lines of evidence emerge from this study which may de-emphasize the roll of the above factor. First, the parameter studied here is the maximal steady state level of radioactive amino acid present at the end of a one hour period of incubation. The initial rate of transport of the amino acids may depend significantly upon the surface area of the cell, but the near-maximal concentration of any particular amino acid transported actively into the cells increases asymptotically with time. The three amino acids studied here indeed enter the cells by active transport processes (Chou and Chowdhury, 1974) and the sixty minutes allowed for incubation in the present investigation is a long enough period for the attainment of the maximal concentration. It is assumed, however, that the volume of the tumor cell did not alter significantly with treatment with Con A. The second line of evidence against the concept of sphericity being the cause of reduced amino acid uptake by Con A-treated tumor cells comes from the finding that, while the uptake of glycine and histidine are considerably reduced in the above cells, the uptake of phenylalanine remained practically unaltered. Reduced surface area cannot be regarded as a cause of reduced uptake for glycine and histidine specifically and not for phenylalanine as well. The specific inhibitory effect of Con A on the uptake of amino acids may indeed be related to the lectin's interaction with specific carrier molecules at specific transport sites (Inbar *et al.*, 1971). The results of a related study in this laboratory suggest that, while Con A may block the transport sites for glycine and histidine and not for phenylalanine, the polycation polylysine blocks the transport sites for phenylalanine and histidine (Chou and Chowdhury, 1974). This type of specific inhibitory effect of Con A on cellular transport may also be attributed to a possible blockage of some carrier proteins otherwise extended from the microtubular structures (Berlin, 1975) to the surface of the cells.

The effects of Con A on the morphology and transport of amino acids in tumor cells were quite reversible. All of the fully rounded cells regain their full complements of pseudopodia as well as the regular transport activity within a few hours after the change-over to Con A-free incubation medium. From there on, the morphological characteristics as well as many of the functional characteristics become indistinguishable from those of untreated tumor cells. This particular line of tumor cells exhibits a considerable degree of contact-inhibited movement and so do tumor cells treated with Con A either acutely or chronically. From time lapse cinematographic studies it appears that in the sixth hour following Con A incubation there is a short burst of enhanced mitotic activity. It has been reported with a different cell line that Con A may arrest some cells in the G_2 phase (Collard et al., 1975). Further experiments are needed to establish any such synchronizing activity of Con A on A/J tumor cells.

The effect of Con A on thymidine uptake into tumor cells appears to be a two-fold one. Firstly, Con A exerts an inhibitory effect on the permeation of thymidine through the plasma membrane. Secondly, the effect of Con A is extended even to the nucleus. At the level of the nucleus the effect may be either a reduced permeability of the nuclear membrane to thymidine or a reduction in the de novo synthesis of DNA. The latter possibility is now under investigation.

Two electrophysiological properties studied in the present investigation are particularly useful with reference to the cellular transport of electrolytes. Moreover, it appears that the magnitudes of intracellular potential and transmembrane resistance of tumor cells are significantly different from those of normal cells (Aull, 1967; Borle and Loveday, 1968; Chowdhury and Chou, 1973). Consequently, it was of interest to investigate what effect Con A has on the intracellular electrical potential and transmembrane electrical resistance. The microelectrode technique used here was particularly suitable, since it produces only a very minimal perturbation to a cellular system (Chowdhury and Snell, 1965; Chowdhury and Chou, 1973). The results of the present study indicate that in normal cells Con A has no significant effect on these two electrophysiological properties. However, in tumor cells Con A does alter the magnitudes of these two parameters. It is tempting to postulate that this effect of Con A is perhaps the result of a morphological effect produced by Con A. The fluidity and motility of the plasma membrane may indeed play an important role in the transport characteristics of cells in general. Any tampering with this membrane fluidity is therefore likely to affect the transport of electrolytes as well as of non-electrolytes.

With reference to a recent suggestion that certain Con A-treated transformed cells regain the growth characteristics of normal cells (Burger and Noonan, 1971) an obvious question can be posed: Are the electrophysiological properties of Con A-treated tumor cells reverted back to magnitudes corresponding to normal cells. Obviously in the present line of tumor cells that was not the case. Even the growth characteristics of A/J tumor cells was not altered by Con A. Only in case of routine intermittent use of Con A did the cells exhibit a longer doubling time (a significant change from 32 hours to 50 hours). However, this change is a temporary one. When these Con A-treated A/J tumor cells were allowed to grow in Con A-free culture medium, they exhibited again a faster growth rate.

Lastly, the question of the tumorigenicity of Con A-treated tumor cells was investigated. This was investigated by inoculating host mice under the skin of the hind leg with a fixed number of Con A-treated or untreated tumor cells and then following the survival time of these mice. It was observed that the survival period of the animal was increased, but by a few days only when the inoculant tumor cells received Con A treatment immediately prior to their inoculation into the host mice. In this group of mice the palpability of the subcutaneous tumor was also delayed by two to three days. One possible explanation for this finding may be that many of the freshly Con A-treated tumor cells experienced the delayed morphological effect of Con A while they were within the host animals. Unlike untreated tumor cells or even tumor cells treated with Con A 24 hours earlier, many of these freshly treated cells had lost their motility while they were within the animals. Such non-motile cells may be susceptible to attack by the animal's reticuloendothelial system. A fraction of the inoculated cells escaping this defense attack can keep on growing at their usual growth rate; this may be a possible reason for the two to three day delay in the palpability of the tumor in this particular group of animals. This delay in the initial growth of the tumor can also explain the slightly longer survival time of these host mice.

In conclusion, it can be stated that neither an acute treatment nor a chronic treatment with Con A helps A/J tumor cells to gain the morphological and physiological characteristics of their normal counterpart. Nevertheless, the effects of Con A on some physiological properties of tumor cells are quite specific and these are presumably taking place through some specific biochemical interactions (Goldstein *et al.*, 1973). Thus an understanding of the mechanisms of these effects is likely to help understand some basic features of normal as well as tumor cells.

REFERENCES

1. Aaronson, S. A., and Todaro, G. J. (1968). "Development of 3T3-like lines from Balb/c mouse embryo culture: Transformation susceptibility to SV40." *J. Cell Physiol.* 72, 141.

2. Abercrombie, M. (1967). "Contact inhibition: The phenomenon and its biological implications." *Nat'l Cancer Inst. Monograph* 26, 249.

3. Aull, F. (1967). "Measurement of electrical potential difference across the membrane of Ehrlich mouse ascites tumor cell." *J. Cell Physiol.* 69, 20.

4. Berlin, R. D. (1975). "Microtubular proteins and concanavalin A receptors." This book, p. 173.

5. Borle, A. B., and Loveday, J. (1968). "Effects of temperature, potassium and calcium on the electrical potential difference in Hela cells." *Cancer Res.* $\underline{28}$, 2401.

6. Burger, M. M., and Noonan, K. D. (1970). "Restoration of normal growth by covering of agglutinin sites on tumor cell surface." *Nature* $\underline{228}$, 512.

7. Chou, A. C., and Chowdhury, T. K. (1974). "Effects of Con A and polylysine on the tumor cell membrane." Submitted for publication.

8. Chowdhury, T. K. (1969). "Fabrication of extremely fine glass micropipette electrodes." *J. Sci. Instr.* $\underline{2}$, 1087.

9. Chowdhury, T. K., and Chaudhuri, T. K. (1974). "A tumor cell line which exhibits contact inhibition of movement." Submitted for publication.

10. Chowdhury, T. K., and Chou, A. C. (1973). "Some electrical properties of chemically induced tumor cells." *J. Nat'l Cancer Inst.* $\underline{51}$, 1981.

11. Chowdhury, T. K., and Snell, F. M. (1965). "A microelectrode study of electrical potential in frog skin and toad bladder." *Biochim. Biophys. Acta.* $\underline{94}$, 461.

12. Chowdhury, T. K., Stith, R. D., and Murthy, S. V. K. N. (1974). "Effects of concanavalin A on protein synthesis and DNA synthesis in tumor cells." In preparation.

13. Collard, J. G., Temmink, J. H. M., and Smets, L. A. (1974). "Cell cycle dependent agglutinability, distribution of concanavalin A binding sites and surface morphology of normal and transformed fibroblasts." This book, p. 221.

14. Foster, D. O., and Pardee, A. B. (1969). "Transport of amino acids by confluent and nonconfluent 3T3 and polyoma virus-transformed 3T3 cells growing on glass cover slips." *J. Biol. Chem.* $\underline{244}$, 2675.

15. Goldstein, I. J., Reichert, C. M., Misaki, A., and Gorin, P. A. J. (1973). "An extension of the carbohydrate binding specificity of concanavalin A." *Biochim. Biophys. Acta* $\underline{317}$, 500.

16. Inbar, M., and Sachs, L. (1969). "Interaction of the carbohydrate-binding protein concanavalin A with normal and transformed

cells." *Proc. Nat. Acad. Sci. U. S. A.* <u>63</u>, 1418.

17. Inbar, M., Ben-Bassat, H., and Sachs, L. (1971). "Location of amino acid and carbohydrate transport sites in the surface membrane of normal and transformed mammalian cells." *J. Membrane Biol.* <u>6</u>, 195.

18. Lowry, O. H., Rosebrough, N. J., Farr, A. L., and Randall, R. J. (1951). "Protein measurements with the folin phenol reagent." *J. Biol. Chem.* <u>193</u>, 265.

19. Martz, E., and Steinberg, M. S. (1972). "Contact inhibition of what? An analytical review." *J. Cell Physiol.* <u>81</u>, 25.

20. Nicolson, G. (1971). "Difference in the topology of normal and tumor cell membranes as shown by different distributions of ferritin-conjugated concanavalin A on their surfaces." *Nature New Biol.* <u>233</u>, 244.

11

THE CHARACTERISTICS OF SUCCINYLATED CON A INDUCED GROWTH INHIBITION OF 3T3 CELLS IN TISSUE CULTURE*

Raphael J. Mannino, Jr., and Max M. Burger
Biocenter of the University of Basel
Basel, Switzerland

ABSTRACT

The growth of untransformed 3T3 fibroblasts can be inhibited by dimeric, non-agglutinating concanavalin A prepared by succinylation (Suc-Con A). This growth inhibition is non-toxic, reversible, and specific for Suc-Con A binding; the cell density at which growth terminates is dependent upon the final cell number and independent of the initial (i.e., plating) density. The part of the cell cycle during which Suc-Con A can exert its growth inhibitory effect appears to be restricted to mitosis and/or early G-1 phase.

I. INTRODUCTION

Lectins are molecules that specifically agglutinate animal cells and can be used as tools for probing the cell surface. They are available from many sources, but are most commonly isolated from plant seeds. A number of lectins has been shown to preferentially agglutinate transformed cells over untransformed ones. Among these, two which seem to discriminate between normal and transformed cells quite well although not absolutely, are concanavalin A (Con A), an agglutinin isolated from jack bean, and wheat germ agglutinin (WGA).

In a number of tissue culture cell lines which have been studied a good correlation has been demonstrated between the saturation density and the agglutinability by lectins. For example, Pollack and

---

*This work was supported by the Swiss National Foundation (Grant No. 3.1330.73 SR).

Burger (1969) have shown, using 3T3 mouse fibroblasts and a number of transformed lines derived from the 3T3 cell line, that the higher the saturation density in tissue culture the lower the concentration of WGA necessary for half maximal agglutination, i.e., the higher the saturation density the higher the agglutinability, by certain lectins.

A correlation between *in vitro* saturation density and *in vivo* tumorigenicity has been demonstrated by Aaronson and Todaro (1968). Using tissue culture cells derived from inbred mice they demonstrated that the higher the saturation density in tissue culture the greater the neoplastic potential of a cell when introduced back into a mouse. Thus it has been suggested that, at least for the cell types that have been studied, a lectin-detectable cell surface configuration is associated both with uncontrolled cell growth and with a neoplastic potential.

The relationship between a lectin-detectable cell surface configuration and the transformed state has also been suggested in a variety of studies using cell lines that are temperature-sensitive for transformation (Burger and Martin, 1972; Eckhart *et al.*, 1971; Noonan *et al.*, 1973). At the permissive temperature the cells grow as transformed cells and are highly agglutinable, whereas at the non-permissive temperature the cells return to normal, controlled growth and concomitantly become non-agglutinable with lectins. In other studies, clones of transformed cells selected for their resistance to Con A toxicity were shown to have lost their increased agglutinability and regained the morphological and growth characteristics of untransformed cells.

Studies of the binding of fluorescent-labeled WGA (and more recently of fluorescent-labeled Con A) to normal and transformed cells have demonstrated that untransformed cells in interphase bind very little fluorescent lectin, while transformed cells are highly fluorescent. Unexpectedly, however, untransformed cells in mitosis were shown to bind fluorescent-labeled lectin at levels approximately equal to those of the transformed cells (Fox *et al.*, 1971). Subsequently it has been shown that untransformed cells in mitosis bind approximately as much tritium-labeled Con A as transformed cells, 3-5 fold higher than untransformed interphase cells (Noonan and Burger, 1973a) and that the agglutinability of untransformed cells in mitosis is similar to that of both transformed cells and untransformed interphase cells treated with trypsin (Noonan and Burger, 1973b). Shoham and Sachs (1974) have reported that lectin binding and agglutinability during the cell cycle of transformed cells is the reciprocal of normal cells, i.e., transformed cells in mitosis bind low levels of fluorescent-labeled lectin and are non-agglutinable whereas transformed interphase cells are agglutinable and highly fluorescent.

Since it appeared that a lectin-detectable cell surface configuration was not a peculiar property of transformed cells but was also demonstrable in untransformed mitotic cells, it became of interest to investigate whether this cell surface configuration is only an indication of the "proliferating" state or whether it plays an active role in the mechanisms regulating cellular growth.

Mild proteolytic treatment of the surfaces of untransformed cells in a density-inhibited monolayer results in both the appearance of the agglutinable state and of one round of cell division (Burger, 1969, 1970; Sefton and Rubin, 1970). Other treatments which induce density-inhibited cells to grow, such as insulin, serum stimulation, and urea (Weston and Hendricks, 1972), also induce the agglutinable state. Thus, a treatment which induces the lectin-detectable cell surface change seems to result in at least one round of cell division.

It has also been possible to "cover" this agglutinable cell surface through the binding of the lectin Con A (Burger and Noonan, 1970). Con A is toxic in its native, agglutinable form, a fact which necessitated the preparation of a "monovalent" Con A, a molecule that binds to the cell surface, but one that does not cause agglutination. When "monovalent" Con A prepared by chymotrypsinization (Burger and Noonan, 1970) was added to polyoma virus-transformed 3T3 cells the cells grew to densities similar to those of untransformed cells. Upon removal of the non-agglutinating Con A the cells once again grew to high saturation densities. Therefore the "covering-up" of the lectin sites results in untransformed growth, whereas their subsequent exposure leads to a return to uncontrolled growth.

All of these data have led to the proposal that the lectin-detectable cell surface change accompanying mitosis of untransformed cells is involved in producing the signal which commits a cell to enter the next round of cell division (Fox *et al.*, 1971). Since a transformed cell is constantly in the "mitotic configuration", it would constantly be induced to grow.

II. METHODS

3T3 Swiss mouse embryo fibroblasts were grown in Dulbecco's modified Eagle's medium supplemented with calf serum. The cultures were routinely checked for PPLO infection by autoradiography and fresh cultures were begun every three months from frozen stocks. Succinyl Con A was prepared by the method of Gunther *et al.* (1973). Briefly, Con A (100 mg) was dissolved in saturated sodium acetate (25 ml) and the precipitate was removed by centrifugation (1000 x g, 10 min.). The supernatant was allowed to react with succinic anhydride (30 mg) for one hour at 0°C, then dialyzed against distilled water (3 changes), lyophilized and dissolved in saturated sodium acetate (20 ml). This solution was allowed to react with succinic anhydride (30 mg) for 90 min. at 23°C, and again dialyzed exhaustively against distilled water and lyophilized.

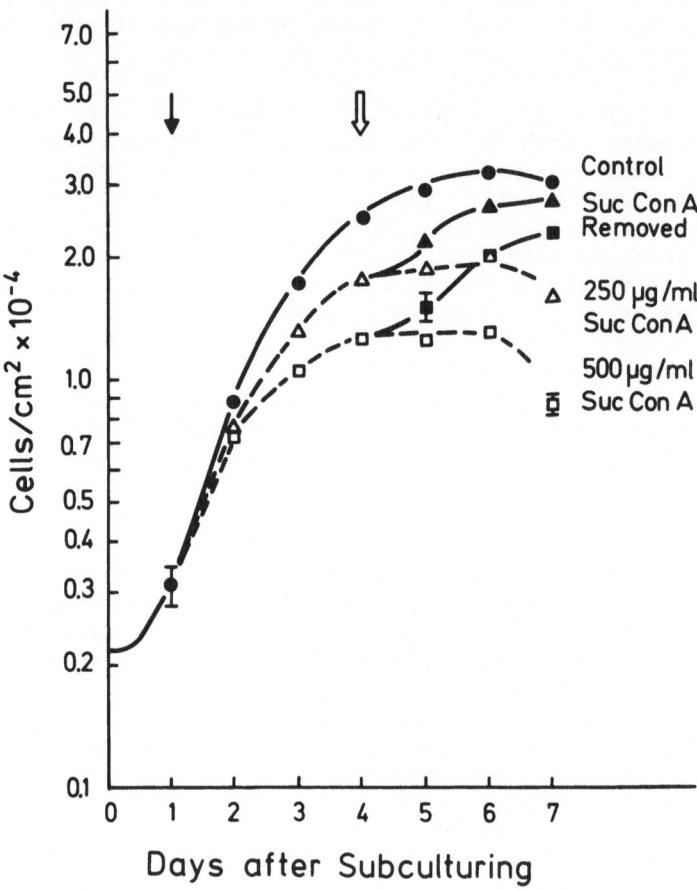

Fig. 1. Suc-Con A induced growth inhibition of 3T3 cells. 3T3 cells were subcultured into 3.5 cm Falcon tissue culture dishes containing Dulbecco's Modified Eagle's medium (DME) containing 10% calf serum. The arrows indicate the times at which the medium was changed to DME + 5% calf serum ± Suc-Con A. Cells were removed from the dishes with EDTA + trypsin solution and counted in a Coulter counter. Duplicate dishes were counted for each point. ●——●, control, no Suc-Con A; △——△ 250 µg/ml Suc-Con A on days 1 and 4; □——□, 500 µg/ml Suc-Con A days 1 and 4; ▲——▲ and ■——■, fresh medium without Suc-Con A on day 4.

III. RESULTS

Proceeding from the hypothesis that the surface configuration permanently exhibited by transformed cells is related to the cell surface change accompanying mitosis of untransformed cells and that the interaction of non-agglutinating Con A with transformed cells results in decreased cell growth, it should follow that high enough concentrations of non-agglutinating Con A would also inhibit the growth of untransformed cells. The results of such experiments are illustrated in Fig. 1.

When 3T3 cells are grown in the presence of a preparation of non-agglutinating Con A, they grow to a density that is below the level of the saturation density of control cells. The extent of this inhibition of growth is dependent upon the concentration of non-agglutinating Con A present, i.e., the higher the concentration of non-agglutinating Con A the lower the density at which cell growth stops. This growth inhibition can be reversed by removing the medium containing the non-agglutinating Con A and replacing it with medium not containing Con A.

The non-agglutinating Con A used in these studies was prepared by succinylation according to Gunther *et al.* (1973) rather than by enzymatic methods. Succinylated-Con A (Suc Con A) sediments as a dimer in the analytical ultracentrifuge (S_{20}^W = 4.0) in contrast to native Con A which is a tetramer (S_{20}^W = 6.1). Of the 13 free amino groups present per Con A monomer approximately 10 are succinylated. Suc Con A retains the Con A sugar binding specificity as demonstrated by its ability to bind to Sephadex and by its specific elution with glucose (0.1 M). In contrast to Con A which precipitates glycogen at levels as low as 0.05 mg/ml, Suc-Con A shows no detectable glycogen precipitation. Different preparations of Suc Con A contain tetrameric Con A contamination ranging from a maximum of 5 to generally less than 0.01 as determined by glycogen precipitation and analytical ultracentrifugation. There are no detectable differences in the growth inhibitory properties of these various preparations. The banding patterns of native Con A and Suc-Con A on SDS-polyacrylamide disc-gel electrophoresis are essentially identical.

An interesting characteristic of Suc-Con A-induced growth inhibition is that, for a given concentration of Suc-Con A, the cell density at which growth terminates is independent of the initial cell density. That is to say, for a given concentration of Suc-Con A the density at which growth ceases is highly dependent upon the final cell density, but is independent of the initial cell density (Fig. 2). Furthermore, growth inhibition by Suc-Con A appears to involve specific binding, since succinylated bovine serum albumin has no inhibitory effect on cell growth and 5×10^{-3} M α-methyl-D-mannopyranoside, a specific inhibitor of Con A-mediated cell agglutination, blocks inhibition of growth by Suc-Con A. It appears, therefore,

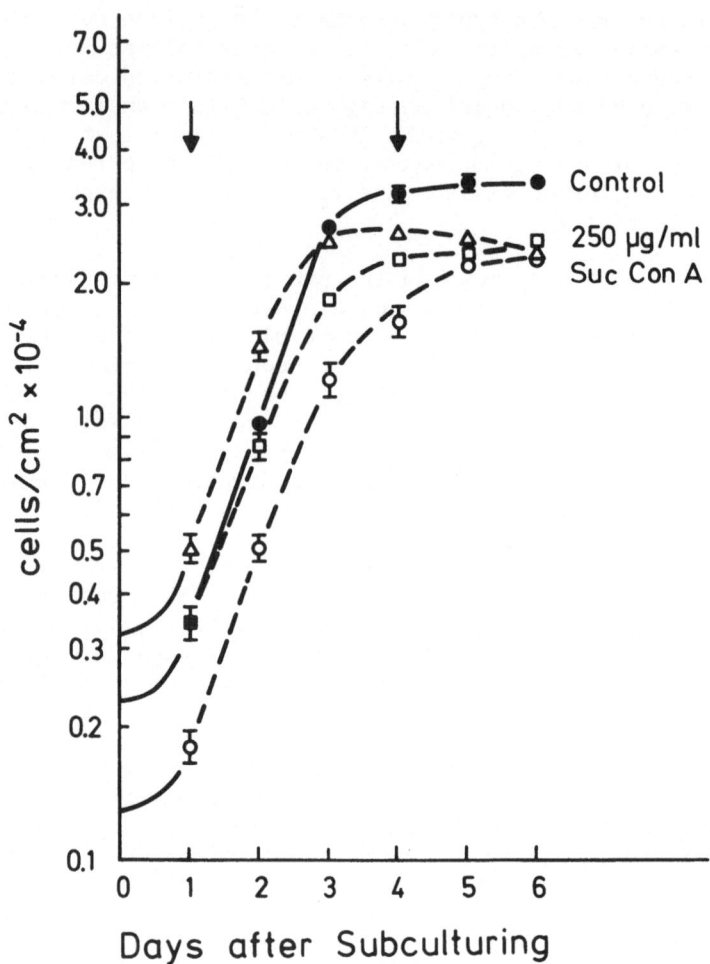

Fig. 2. Suc-Con A-induced growth inhibition is dependent upon final cell density, independent of initial cell density. Growth conditions and cell counting were as described in Fig. 1. ●——●, control, no Suc-Con A; o——o, cell subcultured at ∼ 0.12 x 10^4 cells/cm^2, + 250 µg/ml Suc-Con A; □——□, cells subcultured at 0.23 x 10^4 cells/cm^2, + 250 µg/ml Suc-Con A; △——△, cells subcultured at ∼ 0.32 x 10^4 cells/cm^2, + 250 µg/ml Suc Con A. Note that at this high density the cells grow over the final density, then slowly become less dense.

that Suc-Con A-induced growth inhibition involves some function both of Suc-Con A-cell interactions and of cell-cell interactions. This suggests that the effect of Suc-Con A is neither toxic nor directly inhibitory to some metabolic process.

Growth termination in the presence of Suc-Con A is preceded by a decrease in the rate of ^{3}H-thymidine incorporation per cell. Cultures that have stopped growing under the influence of Suc-Con A show a rate of ^{3}H-thymidine incorporation per cell that is identical to that of a density-inhibited monolayer. Upon removal of the Suc-Con A the observed increase in cell growth is preceded by an increase in the rate of ^{3}H-thymidine incorporation. Measurements of the DNA content per cell through impulse cytophotometry show that Suc-Con A-inhibited cells accumulate in the G-1 phase of the cell-cycle as in the case of density-inhibited cells.

The saturation density of 3T3 cells is dependent on the serum conditions in which the cells are grown; the saturation density can be adjusted upward or downward either by increasing the percentage of serum in the medium or by changing the medium more or less frequently. Cells growing in the presence of Suc-Con A show a serum dependence very similar to that of untreated cells, i.e., the higher the serum concentration, the higher the final density. However it is important to note that, for a given concentration of Suc-Con A, the ratio of the final density of cells whose growth has been inhibited by Suc Con A to the saturation density of control cells grown under identical conditions is constant. Suc-Con A-inhibited cells show the same sensitivity to serum pulses in the presence of Suc-Con A that untreated cells show in the absence of Suc-Con A. We believe this indicates that Suc-Con A does not exert its inhibitory effect by removing necessary growth factors from the serum, but rather that it interacts directly with the cells.

It should be mentioned here that, in our hands, native Con A has never induced a reversible inhibition of growth of 3T3 cells. At low doses (25-50 µg/ml) it appears to slow down the growth of 3T3 cells but never stops it and at higher doses it kills the cells. We observed another qualitative difference between native and Suc-Con A: native Con A, even at low doses, causes the cells to be more firmly attached to the tissue culture dish substratum. This property of strong adhesion is most easily demonstrated after treatment with an EDTA-trypsin (0.05%) solution. Whereas control cells detach within 5-10 min., Con A treated cells will remain attached up to 30 min. and pieces of cells will remain attached even after cell lysis has occurred. In our experience Suc-Con A has never caused cells to become more firmly attached to the growing surface; the adherence of Suc Con A treated cells to the substratum is essentially indistinguishable from that of control cells, as measured by the ease of EDTA-trypsin removal from the substratum.

By all the parameters measured thus far, cells whose growth has been inhibited by Suc-Con A are similar to untreated density-inhibited cells and our working hypothesis is that Suc-Con A interacts with 3T3 cells by a mechanism that mimics the phenomenon of density-dependent inhibition of growth.

At this point, it became of interest to determine whether or not Suc-Con A exerts its growth inhibitory effect at a particular point during the cell cycle. To test this, 3T3 cells were synchronized by plating into medium containing only 2% serum. Under these conditions the cells go through approximately one round of cell division, then shut down in the G-1 phase of growth as demonstrated by impulse cytophotometry. Upon replacement of the old medium with fresh medium containing 10% serum, the resting cells begin to grow and go through mitosis approximately 26-30 hours after the addition of the medium. With such a synchronized population of cells at a low density it was then possible to add Suc-Con A at various times during the cell cycle and to observe the effect of Suc-Con A on growth. The experiment was designed so that the density of the cells at the time of the medium change would be the same as the expected final density of cells grown from day 1 in the chosen concentration of Suc-Con A. Therefore, if Suc-Con A could exert its growth inhibitory effect at the time of the medium change, no growth of the Suc-Con A-treated cells would be expected.

In contrast to what was expected, in the presence of Suc-Con A the cells went through one round of cell division, i.e., through the remainder of G-1, all of S, G-2 and M, before growth was inhibited (Fig. 3). If Suc-Con A was added at the time of the first medium change, but removed just prior to mitosis, growth continued past the first round of cell division. If Suc-Con A was added just before the cells entered mitosis the cells stopped growing after the completion of cell division. These data suggest that Suc-Con A exerts its growth inhibitory effect during the mitotic phase of the cell cycle. Experiments are currently in progress to examine the effect of Suc-Con A on the mitotic phase of the cell cycle in more detail, in order to obtain a better understanding of Suc-Con A-induced growth inhibition.

IV. DISCUSSION

Trowbridge and Hilborn (1974) have reported that Suc-Con A binding to nearly confluent monolayers of 3T3 and SV3T3 cells is saturated at a concentration of 100 µg/ml and they concluded that this concentration has no appreciable effect upon cell growth. Nevertheless, a close examination of their data seems to indicate a growth inhibition of about 30-50% at this concentration of Suc-Con A which, they state, can be abolished by 50 mM α-methyl-D-mannoside. Our findings that concentrations of Suc-Con A higher than 100 µg/ml have a marked, specific and non-toxic inhibitory effect on the growth of

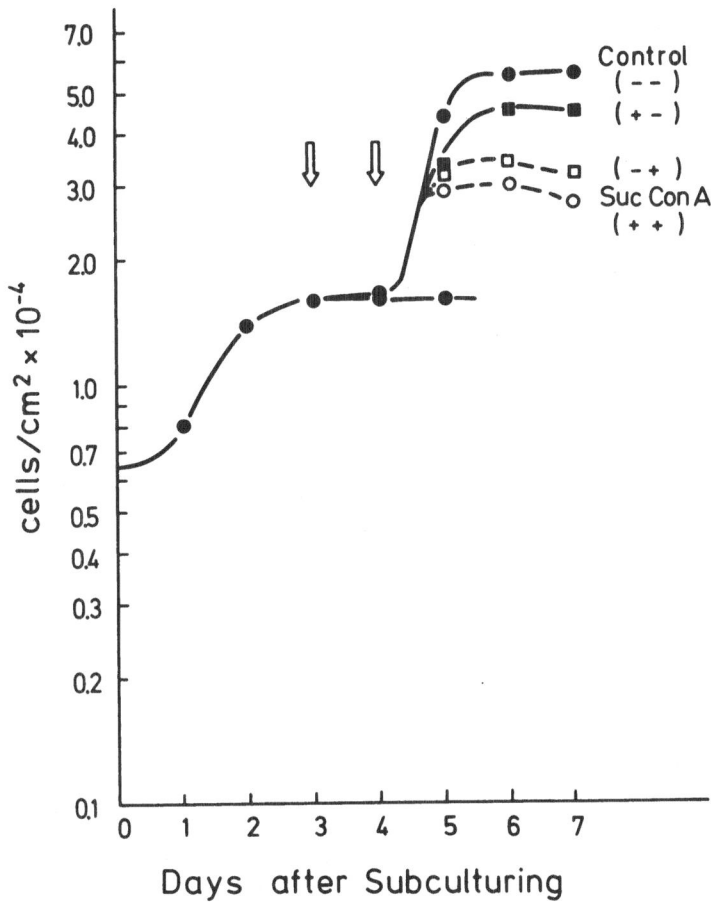

Fig. 3. Suc-Con A can inhibit cell growth only by interacting with cells in mitosis. 3T3 cells were subcultured in DME + 2% calf serum. When growth had become stationary, the medium was replaced with DME + 10% calf serum ± Suc-Con A. The medium was replaced again 23 hours later, just prior to mitosis. "(+-)" cells received Suc-Con A at the first medium change, none at the second; "(-+)" cells received no Suc-Con A at the first medium change, but Suc-Con A at the second; and so on. Cell counting was as described in Fig. 1. We believe that the 10-15% differences between the "(++)" and "(-+)" and "(+-)" and "(--)" curves are the result of a small amount of cell death, rather than of a growth inhibition effect.

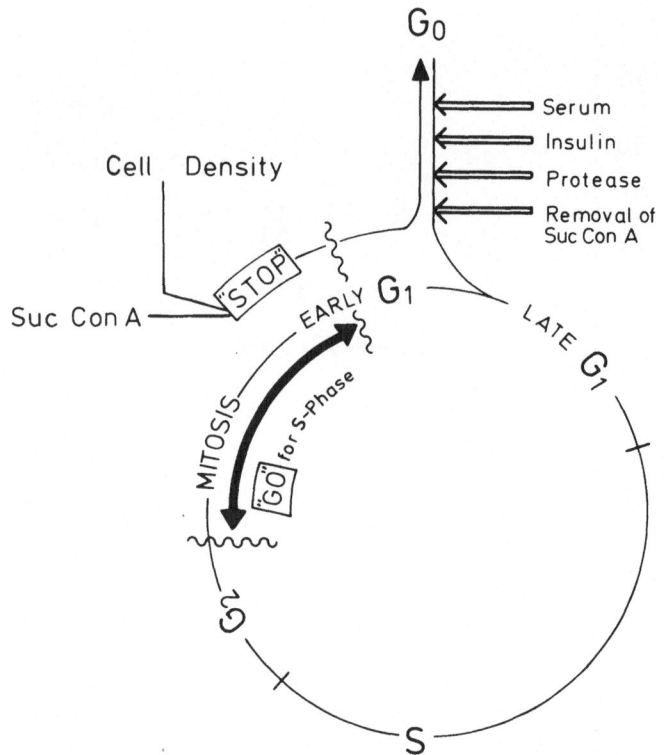

Fig. 4. Proposed interrelationships of various factors regulating the growth of 3T3 cells and the involvement of the mitotic cell surface configuration. The "GO" signal occurs early in mitosis, concomitant with the change to the agglutinable state; the "STOP" signal must be received before the cells revert back to the non-agglutinable state, sometimes early in G-1. ⟵⟶ indicates the period of the cell cycle during which the cells are agglutinable. Note that G_0 cells are past the point during which cells can receive the "STOP" signal.

3T3 cells may not be contradictory to the binding studies of Trowbridge and Hilborn (1974), since we believe that the binding that is important for growth inhibition occurs during mitosis and that more and perhaps different binding sites are exposed during this time. Studies are currently under way to examine this possibility. It should also be considered that binding characteristics normally studied for a short time in phosphate buffered saline may not be comparable with those in the tissue culture dish while cells are growing for several days in growth medium.

The finding that Suc-Con A causes growth inhibition by a specific interaction with cells during mitosis correlates well with the earlier finding that WGA and Con A bind to untransformed cells in mitosis but do not bind to interphase cells. In Fig. 4 we present a diagrammatic representation of some of the factors regulating cellular growth in our system, as well as some theoretical explanations concerning their interrelationship.

Untransformed 3T3 cells proceed through the cell cycle and are triggered into a new round of the cell cycle by a "GO" signal read during the mitosis or early G-1 phase of the previous cycle. It has been demonstrated that compounds which induce cell growth in 3T3 cells cause a very rapid drop in the intracellular levels of cyclic AMP (cAMP) (Burger, Bombik et al., 1972). An inhibition of this decrease in the intracellular cAMP level inhibits entrance into S-phase (DNA synthesis). A similar drop in the intracellular level of cAMP also occurs during mitosis, and it is possible that a drop in the level of cAMP plays a part in this "GO" signal.

When the cell density of a growing culture reaches a certain level, growth stops and the cells accumulate in G_0, a specific stage of G-1. We believe that a "STOP" signal exists that commits the cell to G_0 and results in the inhibition of cell growth. This "STOP" signal must occur after the "GO" signal but obviously before G_0. Studies using synchronized density-inhibited cells with 50% serum show that a drop in the intracellular level of cAMP occurs even in the mitosis immediately prior to the entrance of the cell into G_0 (Bombik and Burger, 1974). Thus an inhibition of the drop in cAMP levels is probably not the natural mechanism of growth restriction, although increased cAMP may be found in cells which have accumulated in G_0.

Our working hypothesis is that Suc-Con A mimicks the mechanism of density-dependent inhibition of growth and works through the same "STOP" signal to route the cells into G_0. Studies are underway, however, to establish the effect (if any) of Suc-Con A on changes in the levels of intracellular cAMP.

Cells that have stopped growing and are resting in G_0 can be induced to re-enter the growth cycle by a number of treatments, including serum stimulation, insulin treatment, mild proteolysis and in the case of Suc-Con A-inhibited cells by removal of Suc-Con A. Serum stimulation, insulin treatment, and mild proteolysis all cause the surface membrane to change very rapidly to the agglutinable state (Bombik, Bechtel and Burger, 1974). This membrane-mediated change is accompanied by a rapid drop in the intracellular cAMP levels (Burger, Bombik et al., 1972) and eventually leads to DNA replication and to one round of cell division.

If Suc-Con A is added concomitantly with a serum pulse, cell division is not inhibited even though the agglutinable surface configuration is known to be present. This suggests that not only is there a specific cell surface configuration required for Suc-Con A to exert its inhibitory effect, but that the particular surface requirements must occur during the mitotic stage of the cycle. We suggest therefore that cells are capable of receiving a "STOP" signal only in mitosis and early G-1, i.e., that part of the cell cycle when untransformed cells exhibit the lectin-detectable cell surface configuration. If growing cells proceed through this period in the cell cycle without receiving a "STOP" signal, they are committed both to a complete round of DNA replication and to cell division. Cells which do receive the "STOP" signal accumulate in G_0, and are also "past the point of no return". If G_0 cells are stimulated to re-enter the growth cycle, they are unable to receive a "STOP" signal either from Suc-Con A or from high cell density until they have proceeded through the entire cell cycle, entered mitosis and once again assumed an agglutinable surface configuration.

In conclusion, our present working hypothesis holds that growing untransformed cells upon entering mitosis undergo a lectin-detectable surface membrane change and might receive a "GO" signal which initiates the next round of cell division. Some time thereafter, but before the surface membrane returns to the non-agglutinable state, the cells receive a "STOP" signal that initiates the resting G_0 state. If the cells do not receive the "STOP" signal they are committed to another complete round of cell division.

REFERENCES

1. Aaronson, S., and Todaro, G. (1968). "Basis for acquisition of malignant potential by mouse cells cultivated *in vitro*." *Science* 162, 1024.

2. Bombik, B., Bechtel, T., and Burger, M. (1974). Unpublished observation.

3. Bombik, B., and Burger, M. (1974). Unpublished observation.

4. Burger, M. (1969). "A difference in the architecture of the surface membrane of normal and virally transformed cells." *Proc. Nat. Acad. Sci. U. S. A.* 62, 994.

5. Burger, M. (1970). "Proteolytic enzymes initiating cell division and escape from contact inhibition of growth." *Nature* 227, 170.

6. Burger, M., Bombik, B., Breckenridge, B., and Sheppard, J. (1972). "Growth control and cyclic alterations of cyclic AMP in the cell cycle." *Nature New Biol.* 239, 161.

7. Burger, M., and Martin, G. (1972). "Agglutination of cells transformed by Rous sarcoma virus by wheat germ agglutinin and

concanavalin A." *Nature New Biol.*, **237**, 9.

8. Burger, M., and Noonan, K. (1970). "Restoration of normal growth by covering of agglutinin sites on tumour cell surface." *Nature* **228**, 512.

9. Culp, L. and Black, P. (1972). "Contact-inhibited revertant cell lines isolated from simian virus 40-transformed cells." *J. Virol.* **9**, 611.

10. Eckhart, W., Dulbecco, R., and Burger, M. (1971). "Temperature dependent surface changes in cells infected or transformed by a thermosensitive mutant of polyoma virus." *Proc. Nat. Acad. Sci. U. S. A.* **68**, 283.

11. Fox, T., Sheppard, J., and Burger, M. (1971). "Cyclic membrane changes in animal cells: Transformed cells permanently display a surface architecture detected in normal cells only during mitosis." *Proc. Nat. Acad. Sci. U. S. A.* **68**, 244.

12. Gunther, G., Wang, J., Yahara, I., Cunningham, B., and Edelman, G. (1973). "Concanavalin A derivatives with altered biological activities." *Proc. Nat. Acad. Sci. U. S. A.* **70**, 1012.

13. Noonan, K., and Burger, M. (1973a). "Binding of ^{3}H-Concanavalin A to normal and transformed cells." *J. Biol. Chem.* **248**, 4286.

14. Noonan, K., and Burger, M. (1973b). "Induction of 3T3 cell division at the monolayer stage." *Exp. Cell Res.* **80**, 405.

15. Noonan, K., Renger, H., Basilica, C., and Burger, M. (1973). "Surface changes in temperature-sensitive simian virus 40 transformed cells." *Proc. Nat. Acad. Sci. U. S. A.* **70**, 347.

16. Ozanne, B. (1973). "Varients of simian virus 40-transformed 3T3 cells that are resistant to Concanavalin A." *J. Virol.* **12**, 79.

17. Pollack, R., and Burger, M. (1969). "Surface-specific characteristics of a contact-inhibited cell line containing the SV40 genome." *Proc. Nat. Acad. Sci. U. S. A.* **62**, 1074.

18. Sefton, B., and Rubin, H. (1970). "Release from density dependent growth inhibition by proteolytic enzymes." *Nature* **227**, 843.

19. Shoham, J., and Sachs, L. (1974). "Different cyclic changes in the surface membrane of normal and malignant transformed cells." *Exp. Cell Res.* **85**, 8.

20. Trowbridge, I., and Hilborn, D. (1974). "Effects of succinyl-Con A on the growth of normal and transformed cells." *Nature*

250, 304.

21. Weston, J., and Hendricks, K. (1972). "Reversible transformation by urea of contact inhibited fibroblasts." *Proc. Nat. Acad. Sci. U. S. A.* 69, 3727.

12

CELL CYCLE DEPENDENT AGGLUTINABILITY, DISTRIBUTION OF CONCANAVALIN A BINDING SITES AND SURFACE MORPHOLOGY OF NORMAL AND TRANSFORMED FIBROBLASTS.*

J. G. Collard, J. H. M. Temmink, and L. A. Smets
The Netherlands Cancer Institute
Amsterdam, The Netherlands

ABSTRACT

In studies on phenotypic reversion of transformed cells to normal growth patterns, we investigated the effect of dibutyryl cyclic AMP (dbc-AMP) and a protease inhibitor (TLCK) on growth of SV40-transformed mouse fibroblasts (3T3). The results did not support the hypothesis that transformed cells grown with dbc-AMP or TLCK are induced to contact-mediated growth control. The growth rate of SV-3T3 cells grown with the drugs was strongly reduced, due to accumulation of the cells in the G_2 phase of the cell cycle. In addition, decreased agglutinability with concanavalin A (Con A) of those SV-3T3 cells was not caused by a direct effect of the drugs on the cell surface, but by partial synchronization of the cells in the G_2 phase of the cycle. In synchronized cultures agglutinability of transformed cells reached a minimum in G_2 and was maximal in mitosis and G_1. Normal cells agglutinated only in mitosis. This suggested that agglutinability of cells is somehow cell cycle dependent.

Cytochemical investigations on normal and transformed 3T3 cells had shown that Con A-induced redistribution of binding sites on the surface of these cells is not correlated with agglutinability. The present work on replicas confirmed this, but indicated also that normal 3T3 cells have more extended lamellipodia with less Con A binding sites than SV-3T3 cells. Preliminary scanning electron

---

*The facilities for the scanning electron microscopy was available through the Department of Electron Microscopy of the University of Amsterdam.

microscope data showed cell cycle dependent changes in 3T3 cells and also showed that confluent 3T3 and SV-3T3 cells suspended for agglutination tests had a different surface morphology. These results may represent additional factors important for differences in cell agglutinability by Con A.

I. INTRODUCTION

The plasma membrane is thought to play an important role in growth regulation of cells. The plant lectin concanavalin A (Con A) is known to bind to the cell membrane and to agglutinate transformed cells more easily than normal cells (Inbar and Sachs, 1969). Increased agglutination with Con A correlates generally with tumorgenicity (Inbar et al., 1972) and loss of growth control (Burger, 1970; Burger, 1971). Con A binds to specific sugar residues on the outside of the cell membrane and can be made visible in the electron microscope by labeling with various marker molecules (Bernhard and Avrameas, 1971; Nicolson and Singer, 1971; Smith and Revel, 1972; Stobo and Rosenthal, 1972). Therefore we used Con A in previous studies on differences in the structure of the cell membrane between normal and SV40-transformed mouse fibroblasts in relation to differences in growth control *in vitro* (Collard and Temmink, 1974; Temmink and Collard, 1974; Temmink et al., 1974). These studies and similar investigations by others (Nicolson, 1971; Comoglio and Guglielmone, 1972; Bretton et al., 1972; Martinez-Palomo et al., 1972; de Petris et al., 1973; Garrido et al., 1974; Huet et al., 1974; Rowlatt et al., 1974) led to a number of apparently conflicting results regarding the correlation between changes in growth control and agglutinability on the one hand, and changes in amount or distribution of Con A binding sites on the cell membrane on the other hand.

In the literature it has been suggested that dibutyryl cyclic AMP (dbc-AMP) (Hsie and Puck, 1971; Johnson et al., 1971) and protease inhibitors (Schnebli and Burger, 1972) induce normal growth patterns in transformed cells and that transformed cells grown with these reagents have a low agglutination with plant lectins (Sheppard, 1971; Schnebli and Burger, 1971). For that reason we studied phenotypic reversion of transformed cells to normal growth patterns, as induced by dbc-AMP and protease inhibitors. We were able to confirm that the growth rate of SV40-transformed 3T3 cells (mouse fibroblast) is strongly reduced by dbc-AMP (Smets, 1972) or protease inhibitors (Collard and Smets, 1974). Our results indicate, however, that the reduction in growth rate is correlated with an accumulation of the cells in the G_2 phase of the cell cycle and not in G_1, as would have been expected if contact mediated growth had been induced (Smets, 1972; Collard and Smets, 1974).

This suggested that the position of cells in the cell cycle may be an important factor in agglutination. This had been confirmed by our observations on cell cycle dependent changes in agglutinability of Ebstein-Barr virus transformed lymphocytes (Smets, 1973). Therefore we tried to find similar changes in agglutination during the cell cycle in synchronized cultures of normal and SV40 virus-transformed fibroblasts. The results of this study and other investigations (Rubin and Everhart, 1973; Porter et al., 1973a,b) demonstrate that the overall surface morphology of some cells changes during the cell cycle. This supports the hypothesis that differences in agglutinability by Con A between normal and transformed cells do not only result from assumed differences in mobility of Con A binding sites, but also from differences in the gross surface morphology between these cells. Some scanning and transmission electron microscope experiments described in this report give additional circumstantial evidence for this notion and will be discussed in relation to the existing theories on cell agglutination by Con A and growth control in vitro.

II. MATERIALS AND METHODS

A. Tissue Culture

Mouse 3T3 fibroblasts and SV40 virus-transformed fibroblasts (SV-3T3) were obtained commercially from Flow Laboratories, U. S. A. Cells were grown in plastic Petri dishes (10 cm dia.) containing 10 ml of Dulbecco's modified Eagle's medium, supplemented with 10% fetal calf serum and antibiotics in a humidified CO_2 incubator at 37°C. In the experiments, cells were seeded at a density of 5×10^4 cells/ml of nutrient medium. After 24 hours of growth the culture medium was replaced by fresh medium with the different drugs as indicated in the experiments. Control cells were grown under the same conditions except that no drugs were added to the medium. For the electron microscopic studies, cells were grown on cover slips in plastic Petri dishes (5 cm dia.), filled with 5 ml of culture medium.

B. Agglutination Assays

Cells were removed from the Petri dishes with 5×10^{-5} M EDTA in Ca^{++} and Mg^{++} free phosphate buffered saline (PBS), washed once with PBS containing Ca^{++} and Mg^{++} and suspended in PBS to a final concentration of approx. 1×10^6 cells/ml. The agglutination assays were performed at room temperature with 1 ml of cell suspension in a small Erlenmeyer flask on a gyratory shaker. Con A (Calbiochem, A grade) was added in 0.1 ml of PBS to an amount necessary to obtain a final concentration of 25 µg/ml. Maximal agglutination in less than 20 min was scored as (+++++) without further differentiation and incomplete agglutination in 20 min was recorded on the usual scale (0 - ++++). In all experiments cell suspensions without Con A were used as the controls.

C. Measurement of the DNA Distribution

Cells were suspended with EDTA or trypsin, washed once in tyrode salt solution and fixed in suspension by adding an excess of 96% ethanol. The fixed cells were washed in Tris-HCl buffer (pH 7.3) and incubated for 20 min at 37°C with 0.1% ribonuclease in the same buffer. After washing with buffer the cells were resuspended in 0.1% crude pepsin in 0.2 N HCl and incubated for 20 min at 37°C. The cells were washed again and resuspended in Tris-HCl buffer containing 10 ppm ethidium bromide 15 min before DNA fluorescence was measured with a Phywe ICP 11 impulse cytophotometer. With this apparatus, the DNA content of at least 20,000 cells was measured, stored proportionally in a 120 channel analyzer, and plotted on a frequency distribution curve.

D. Description of Different Drugs

Dibutyryl adenosine 3':5'-cyclic monophosphate (dbc-AMP) was obtained from Calbiochem and theophylline from Merck. The protease inhibitor N-α-p-Tosyl-l-Lysine chloromethyl ketone HCl (TLCK, Sigma) is an active site titrant that reacts irreversibly with trypsin. p-tosyl-l-arginine methyl ester HCl (TAME, Sigma) is a substrate analogue and acts competitively on proteases and esterases. Egg-White Trypsin inhibitor (EWTI, Ovomucoid, B grade, Calbiochem) forms a poorly dissociating macromolecular complex with the proteases of the trypsin family (Schnebli and Burger, 1972).

E. Synchronization Procedures

Cells in mitosis were isolated by the shake-off method (Terasima and Tolmach, 1963) and early G_1 cells were obtained following plating of the mitotic cells for 1-2 hours in culture dishes. Populations of cells at the G_1/S boundary were obtained by synchronization in excess thymidine (5 mM) or fluoro-uridine (FUdr) (20 µg/ml). For this purpose, SV-3T3 cells from stationary cultures were inoculated at a density of 1.5×10^5 cells per cm^2 and grown for 18 hours in medium containing the inhibitor. In the case of 3T3 cells, the cultures were inoculated with cells from confluent monolayers (saturation density: $5 \times 10^4/cm^2$) at half saturation density and the serum concentration was raised to 20%. Early and late S cells were obtained from cultures synchronized with excess thymidine, by washing the cells twice with prewarmed regular medium and growing them in this medium for 3 and 7 hours respectively.

To prepare populations of pure G_2 cells, cultures were initiated as described for synchronization in S phase, with omission of the inhibitors. The cultures were irradiated with 1800 Roentgen of X-rays some 5 hours after inoculation and collected 24 hours thereafter. At that time, more than 90% of the cells had accumulated at the radiation-induced block in G_2. Aliquots from all preparations of synchronized cells were subjected to cytophotometry to calculate the average position of the cells in the cycle. Details of histogram analysis have been published elsewhere (Smets, 1973).

F. Transmission Electron Microscopy of Replicas

Cells grown on cover slips were rinsed, then treated *in situ* with Con A (50 µg/ml) and hemocyanin (HC) (500 µg/ml) for 15 min at 25°C, washed, and fixed for 30 min in 2.5% glutaraldehyde buffered with 0.1 M cacodylate buffer (pH 7.2). After dehydration in ethanol and amyl acetate the material was dried in a blast of warm air from a hair dryer. The compressing effect on the cells of this drying method is advantageous for transmission electron microscopy, where little depth of field is available. The preparation of the replicas was done according to the method described by Smith and Revel (1972).

G. Scanning Electron Microscopy

Cells on cover slips or suspended cells attached to a confluent monolayer on cover slips were washed and fixed for 30 min in 2.5% glutaraldehyde buffered with 0.1 M cacodylate buffer (pH 7.2). The suspended cells were prepared with 5×10^{-5} M EDTA in Ca^{++} and Mg^{++} free PBS, washed twice with PBS containing Ca^{++} and Mg^{++}, and prefixed after 15 min in 2.5% glutaraldehyde. After fixation for 1 h, the suspended cells were washed three times in PBS and layered on cover slips with a monolayer of confluent 3T3 cells. After 4 h at 37°C the cover slips with attached cells were fixed as described. After dehydration in ethanol and amyl acetate the cells were dried by the critical point method (Anderson, 1951) with CO_2. The dried cells on glass were mounted on stubs and covered with gold in a Balzer freeze etching apparatus. The material was studied in a Cambridge Stereoscan.

III. RESULTS

A. Effect of dbc-AMP and Protease Inhibitors on Normal and Transformed Cells

Previous experiments on phenotypic changes in SV-3T3 cells induced by dbc-AMP and protease inhibitors had revealed that restoration of normal growth patterns cannot be deduced safely from the phenotypic appearance in a light microscope only (Paul, 1972; Smets, 1972). Therefore we studied the effect of dbc-AMP and protease inhibitors on cell cycle kinetics and agglutinability of normal and SV 40-transformed 3T3 cells.

For that purpose, cells were grown during 48 hours with different drugs and subsequently prepared for cytophotometry and ag-

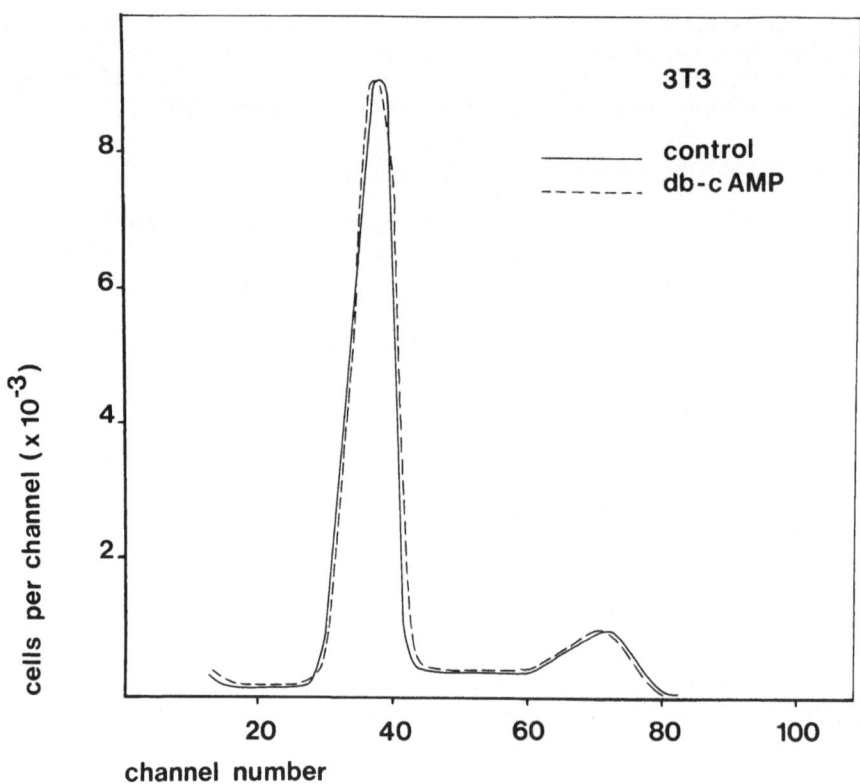

Fig. 1. Impulse cytophotometric distribution of 3T3 and SV-3T3 cells, grown during 48 hours with different drugs, on basis of DNA/cell content.

glutination tests. Fig. 1 and 2 represent the impulse cytophotometric diagrams of 3T3 and SV-3T3 cells grown during 48 hours in 10^{-3} M dbc-AMP and 10^{-3} M theophylline. Fig. 3 and 4 show the corresponding diagrams for cells grown during 48 hours with the protease inhibitor TLCK. It is apparent that transformed cells grown with these drugs accumulate in the G_2 phase of the cell cycle, whereas normal cells do not. Thus the reduced growth of transformed cells with these drugs is correlated with preferential retardation of the cells in the G_2 phase of the cell cycle and not in G_1 as would be expected if contact mediated growth had been induced.

When transformed cells, grown for 48 hours in dbc-AMP or in different protease inhibitors, were tested for agglutination, we found a correlation between amount of G_2 cells per culture and decrease in agglutinability (Table I). TLCK, dbc-AMP and X-ray

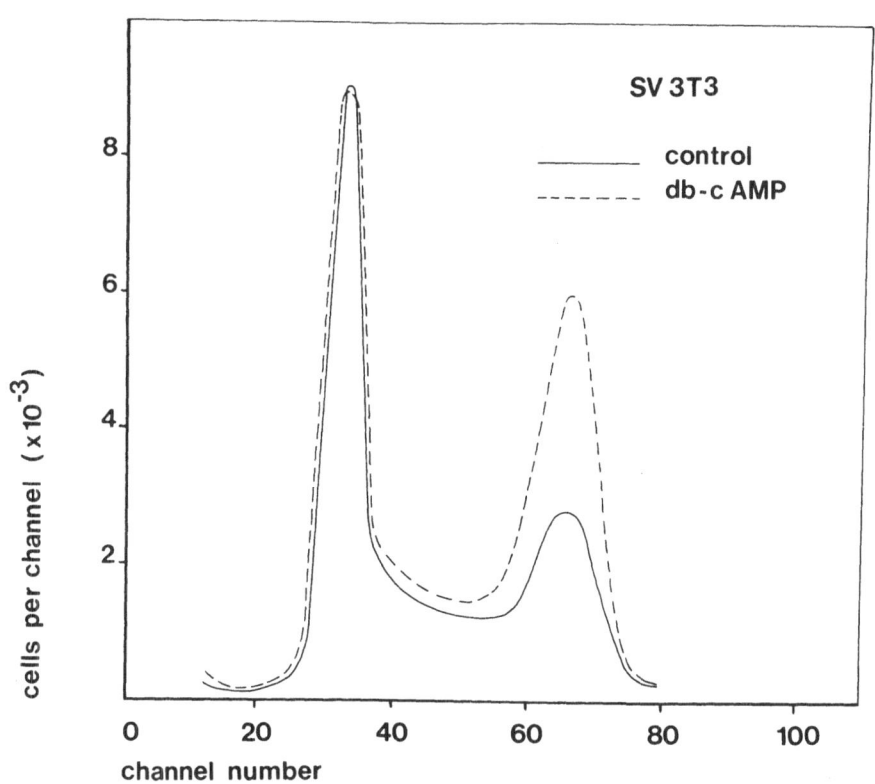

Fig. 2. Impulse cytophotometric distribution of 3T3 and SV-3T3 cells, grown during 48 hours with different drugs, on basis of DNA/cell content.

TABLE I

CHANGES IN AGGLUTINABILITY OF SV-3T3 CELLS AFTER TREATMENT WITH DIFFERENT PROTEASE INHIBITORS AND dbc-AMP FOR 48 HOURS OR WITH 1800 R OF X-RAYS 24 HOURS PRIOR TO TESTING.

Treatment	% cells in G_2	Agglutination after 20 min.
Control	30	+ + + +
TAME, 500 µg/ml.	28	+ + + +
EWTI, 500 µg/ml.	30	+ + + +
TLCK, 100 µg/ml.	63	+ + +
dbc-AMP 10^{-3} M.	46	+ +
X-rays	85	+ +

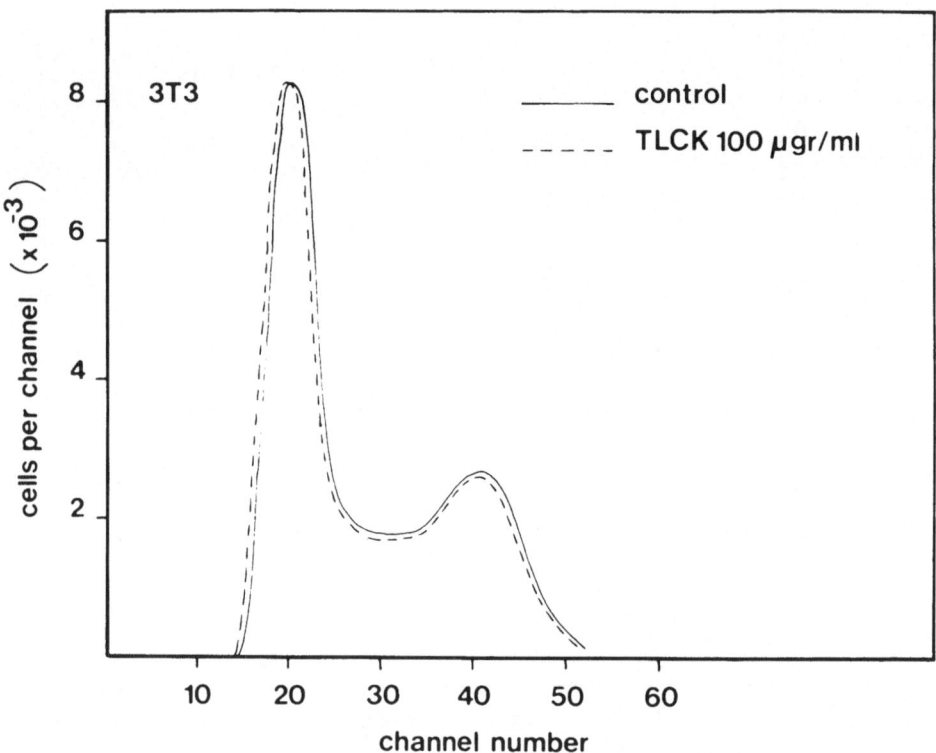

Fig. 3. Impulse cytophotometric distribution of 3T3 and SV-3T3 cells, grown during 48 hours with different drugs, on basis of DNA/cell content.

TABLE II

AGGLUTINATION WITH CON A OF 3T3 AND SV-3T3 CELLS IN VARIOUS PHASES OF THE CELL CYCLE.

Cycle phase	Agglutination after 20 mins at 25 μg/ml.	
	3T3 cells	SV-3T3 cells
Mitosis	+ + + + +	+ + + + +
Early G_1	+	+ + + +
G_1/S	0	+ + + +
Early S	0	+ + +
Late S	0	+(+)
G_2	0	+(+)
Asynchronous cells	0	+ + + +

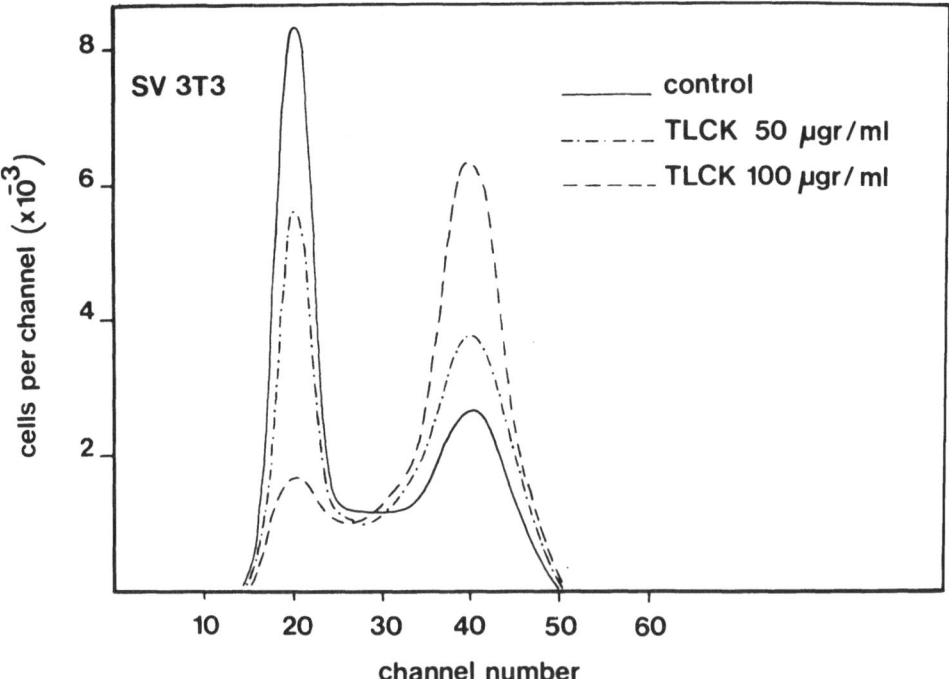

Fig. 4. Impulse cytophotometric distribution of 3T3 and SV-3T3 cells, grown during 48 hours with different drugs, on basis of DNA/cell content.

treatment caused the cells to accumulate in the G_2 phase of the cell cycle, whereas TAME and EWTI did not. Thus, the agglutination is reduced only by those drugs that induce an accumulation in the G_2 phase, and we suppose that this partial synchronization caused the decrease in agglutination with Con A.

B. Cell Cycle Dependent Agglutinability with Con A

The results in Table I suggested a correlation between the capacity of various treatments to reduce cytoagglutination and their effect on the distribution of cells over the cell cycle, viz. (partial) accumulation in the G_2 phase of the cell cycle.

Therefore, we started experiments to measure the agglutination of cells by Con A as a function of their position in the cell cycle. Previous results with Ebstein-Barr virus-transformed human lymphocytes suggested that the decrease in agglutinability started with the initiation of DNA synthesis and reached a minimum at the end of this process when cells entered G_2 (Smets, 1973). In those cells the transition from low agglutinability of G_2 cells to high agglutinability of mitotic cells occured rather abruptly between late prophase and metaphase.

As shown in Table II, agglutinability of SV-3T3 cells decreases when cells pass from G_1 to G_2 and is maximal in mitotic cells. This decrease was measured more quantitatively by comparing the mid-point agglutination of SV-3T3 cells in G_2 phase as compared with asynchronous cells. G_2 phase cells showed a 6-7 fold decrease in agglutinability. Mid-point agglutination of G_2 cells was reached at a Con A concentration of 70-80 µg/ml, whereas the mid-point agglutination of asynchronous cells was reached with 10-15 µg/ml of Con A. Normal 3T3 cells showed mid-point agglutination at approx. 1000 µg/ml of Con A. In order to exclude the possibility that the decreased agglutinability was caused by a direct effect of X-rays on the cells, cells were grown in medium containing FUdr subsequent to X-irradiation. In these cultures cells were blocked at the entry of the S phase and not in G_2, and their agglutination was as high as in non-irradiated, FUdr treated controls. Apparently, the effect of irradiation on agglutinability was linked directly to its synchronizing effect. Similar cell cycle dependent changes in agglutinability have also been found in the spontaneously transformed 3T3-F line. The results in Table II demonstrate that at the concentration used, normal 3T3 cells are only highly agglutinable when in mitosis.

C. Distribution of Con A Binding Sites on the Cell Membrane

Because of this cell cycle dependent agglutinability of 3T3 and SV-3T3 cells, we investigated next whether the distribution of Con A binding sites on the cell surface was also dependent on the cell cycle. This was done by investigating HC-labeled cells bound with Con A in replicas of normal and transformed 3T3 cells. It was hoped that this investigation might at the same time confirm results obtained in previous studies where normal and transformed 3T3 cells had been treated with Con A and several cytochemical markers. In those studies, we showed that on both cell types an irregular distribution of cell bound Con A was present on the cross sectioned plasma membrane (Collard and Temmink, 1974; Temmink and Collard, 1974; Temmink et al., 1974). These results were in good agreement with those of de Petris et al. (1973), but seemed to be in conflict with those of other investigators (Nicolson, 1971; Huet and Bernhard, 1974; Garrido et al., 1974).

When normal and transformed 3T3 cells were labeled with Con A and HC and the replicas of these cells were studied, the distribution of the labeled Con A on the membrane of 3T3 and SV-3T3 cells was irregular (Figs. 5A and B). Sometimes the irregularity of the distribution was less pronounced on SV-3T3 cells, because these

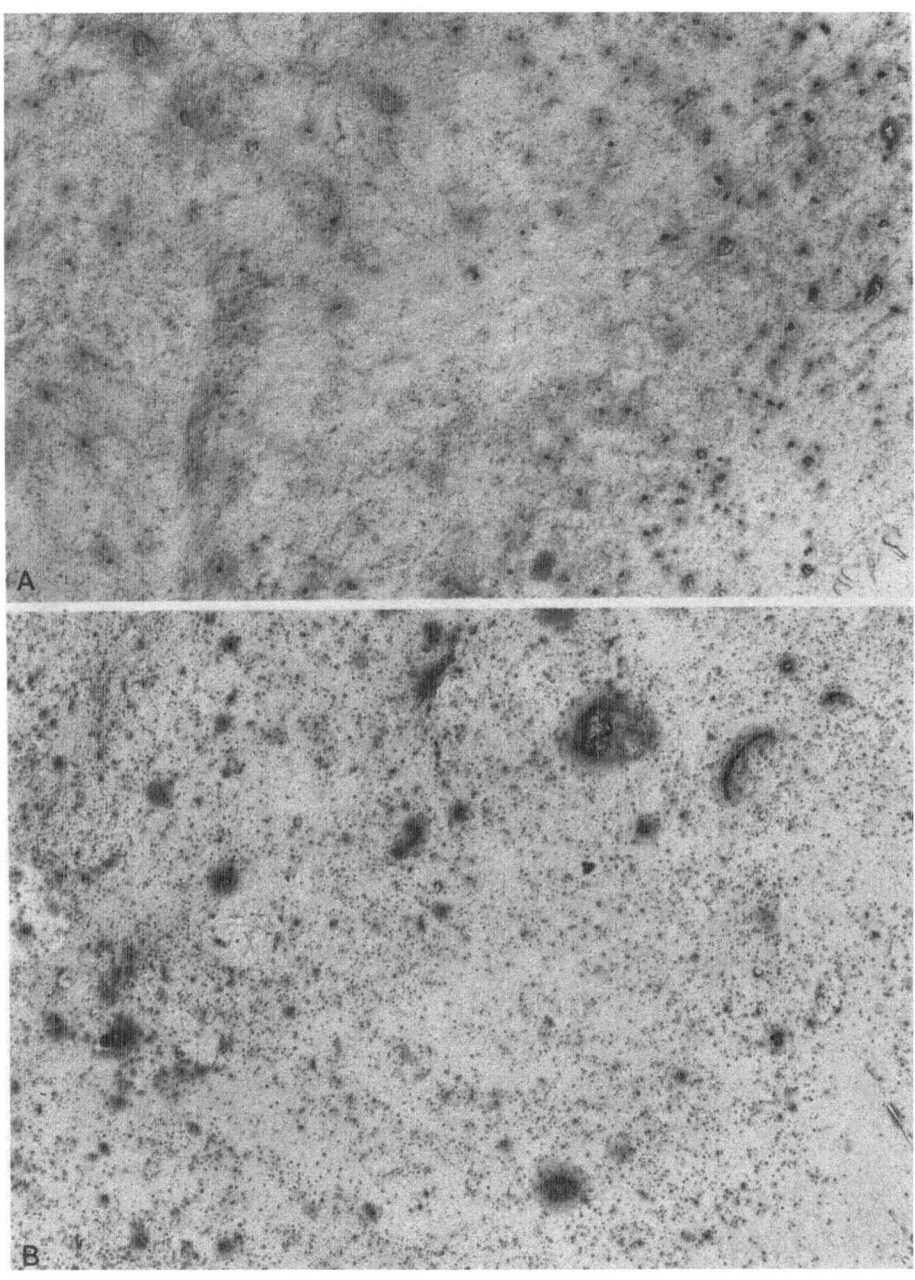

Fig. 5. Replica of HC-labeled mouse fibroblasts. Magn. 3,600 X.
A. 3T3 cell; B. SV-3T3 cell.

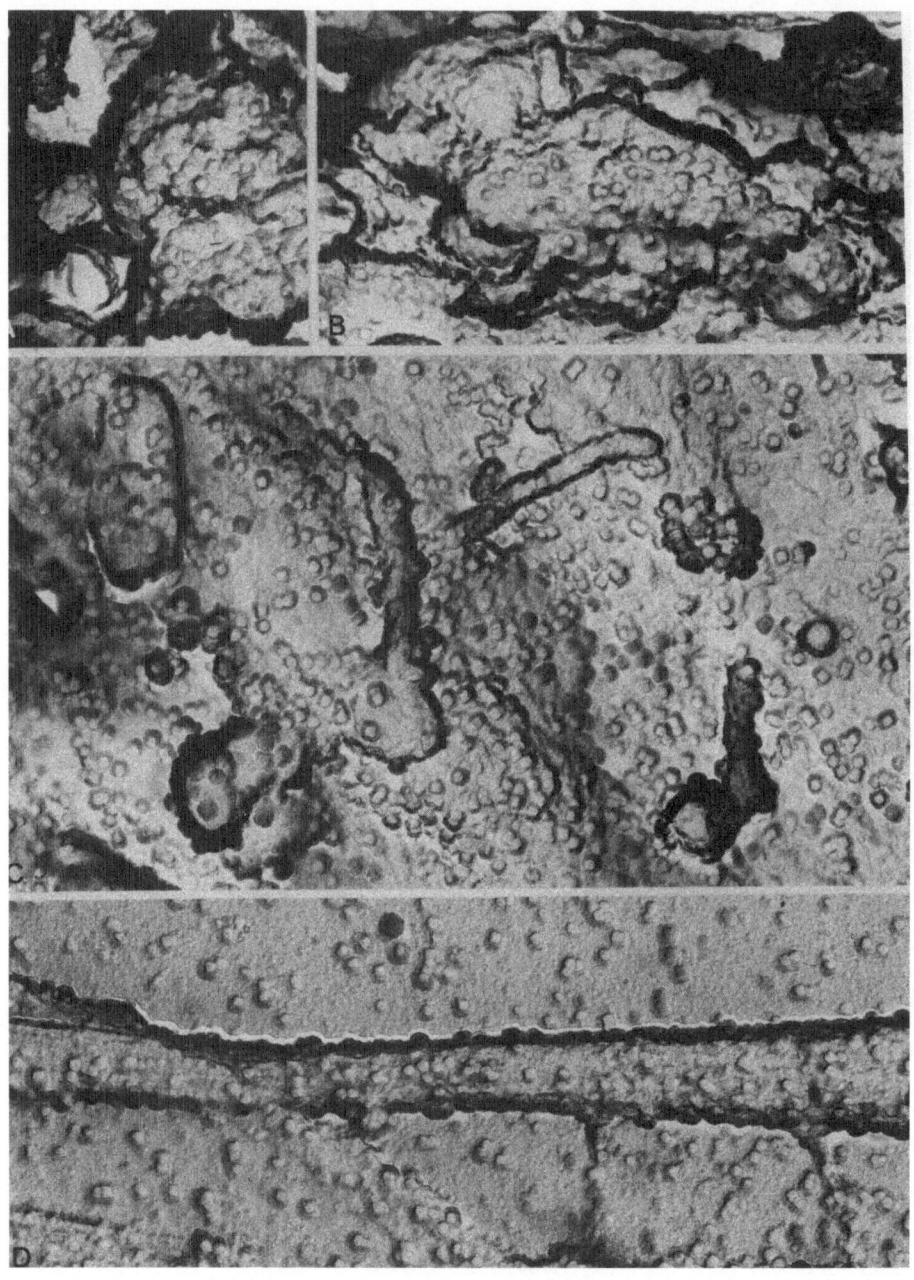

Fig. 6. Replica of parts of HC-labeled mouse fibroblasts. Magn. 45,000 X. A. Blebs on mitotic 3T3 cell; B. Blebs on mitotic SV-3T3 cell; C. Microvilli on SV-3T3 cell; D. Filopodium of 3T3 cell.

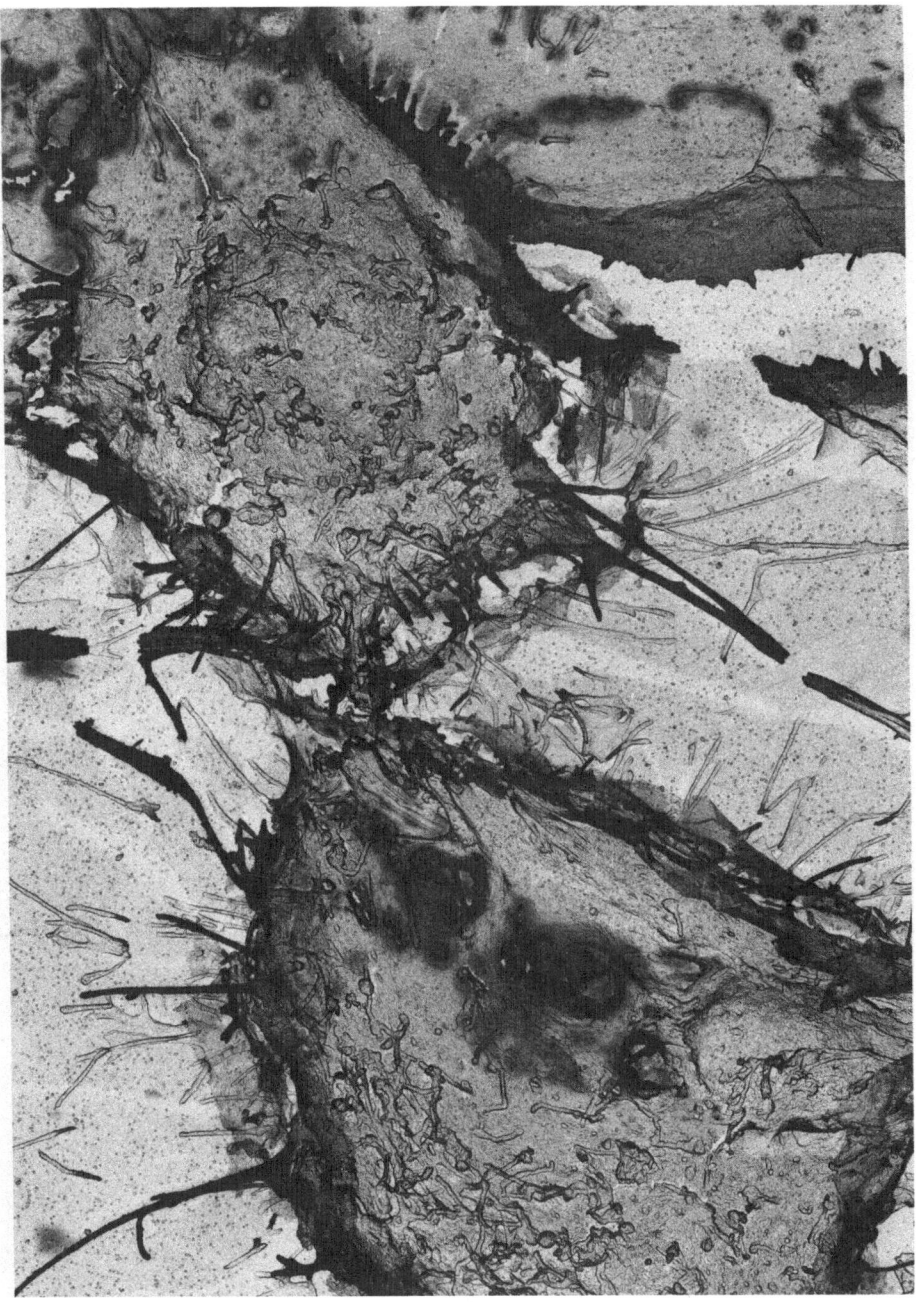

Fig. 7. Replica of HC-labeled 3T3 cell after cytokinesis. Magn. 3,400 X. HC molecules visible on some blebs, microvilli, and filopodia.

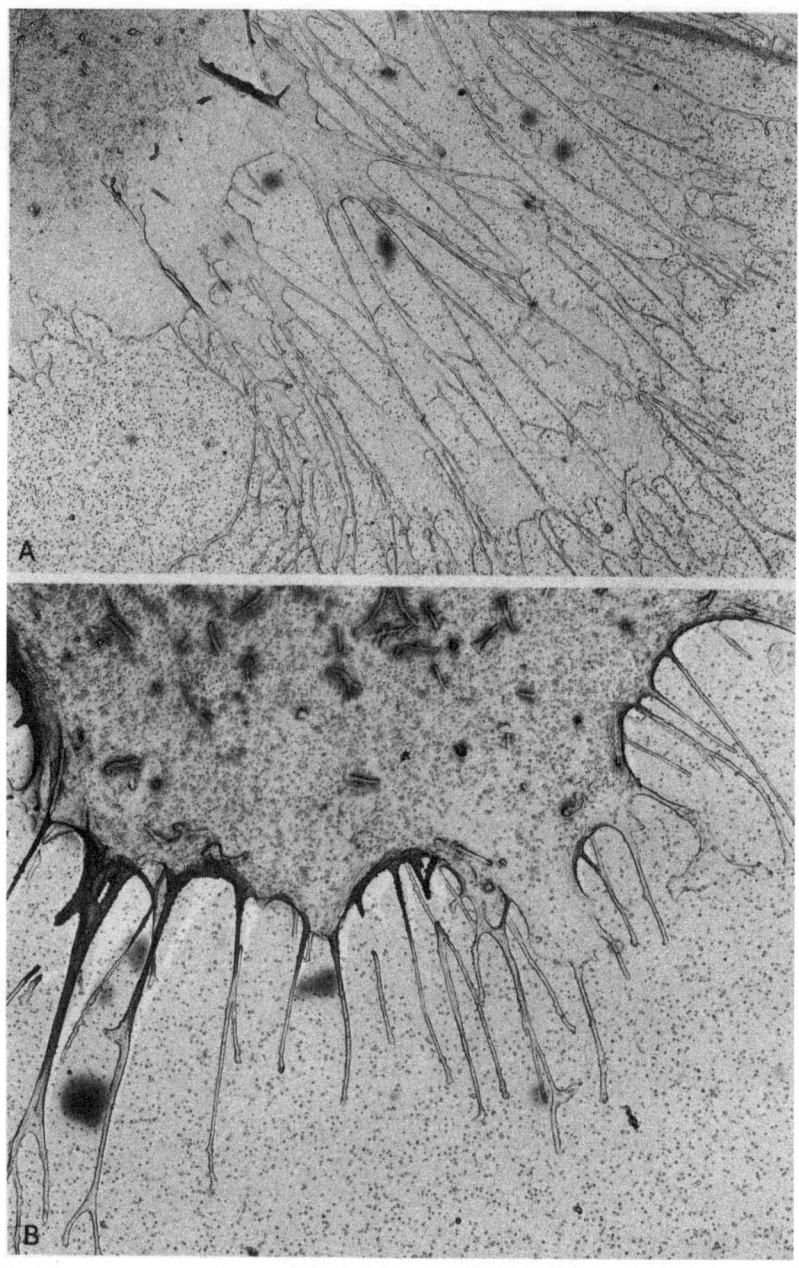

Fig. 8. Replica of lamellipodia of HC-labeled mouse fibroblasts. Magn. 3,600 X. A. 3T3 cell; B. SV-3T3 cell.

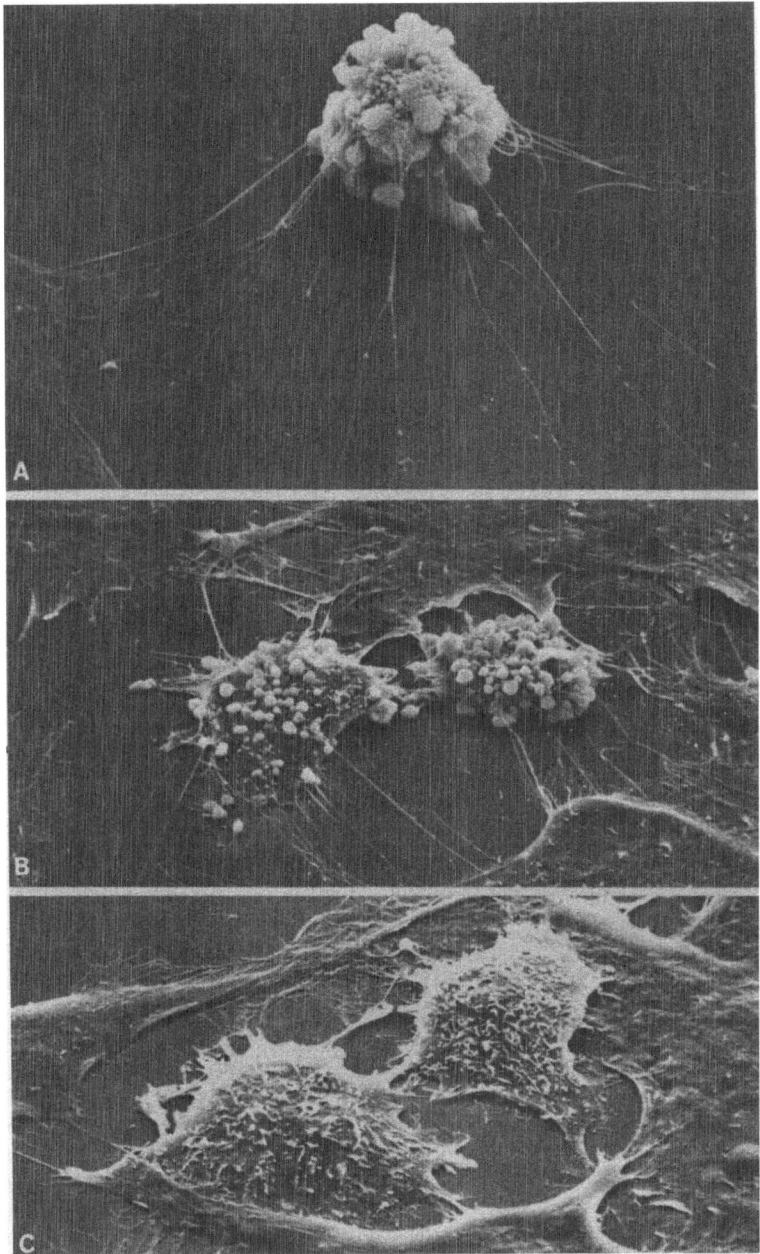

Fig. 9. Scanning electron micrographs of sub-confluent 3T3 fibroblasts *in situ*. Magn. 1,700 X. A. Cell in mitosis with filopodia attached to surface. B. Cells after cytokinesis with blebs and some microvilli. C. Cells in early G_1 with many microvilli.

cells generally had more HC-labeled Con A binding sites per unit surface area than 3T3 cells. The irregular distribution was found on most cells and over almost the complete cell surface, independent of the phase of the cell cycle. Thus, SV-3T3 cells in G_2 had the same distribution as 3T3 and SV-3T3 cells in other phases of the cell cycle. Blebs on cells in or shortly after mitosis (Figs. 6A and B, and 7) and microvilli on cells in early or middle G_1 (Fig. 6C) and even filapodia or retraction fibrils were covered with HC (Fig. 6D). However, there was one conspicuous difference between 3T3 and SV-3T3 cells. Normal 3T3 cells have very wide lamellipodia generally extending far from the nucleated center part of the cell body. These extensions were found to have very little HC-labeled Con A (Fig. 8A). In contrast, the lamellipodia of transformed cells were generally narrower and extended much less far from the cell body (Fig. 8B). These cellular extensions were often covered with more HC-labeled Con A than 3T3 lamellipodia, although with less HC than the rest of the cellular surface of SV-3T3 cells.

Thus the study of replicas of normal and transformed cells essentially confirmed our previously obtained data concerning the absence of differences in distribution of Con A binding sites between 3T3 and SV-3T3 cells. In addition, the study revealed differences in extension of the lamellipodia between normal and SV-3T3 cells and in detectable Con A binding sites on these lamellipodia. That these differences may have important implications for the explanation of the differences in agglutination between 3T3 and SV-3T3 cells will be discussed later.

D. Surface Morphology During the Cell Cycle

Our experiments had shown that the agglutinability of cells by Con A is cell cycle dependent and we confirmed in our study on replicas of HC-labeled cells that the differences in agglutination do not seem to be caused by differences in distribution of Con A binding sites between normal and transformed cells. Therefore it seemed worthwhile to investigate whether differences in agglutination could be correlated with cell cycle dependent morphological changes of the cells. 3T3 and SV-3T3 cells were prepared *in situ* for scanning electron microscopy and photographs were made of cells in different phases of the cell cycle.

Cells in S and G_2 were well spread over the cover slip and showed only very few surface projections. Upon entering mitosis dramatic morphological changes occured. The cells withdrew their lamellipodia and became rounded off (Fig. 9A). The surface became very irregular because of many blebs, and a number of filopodia retained contact with the glass surface. After mitotic division and cytokinesis (Fig. 9B) the two daughter cells gradually lost the surface blebs, but many microvilli were still visible (Fig. 9C). During G_1 the number of microvilli slowly decreased. Spreading of

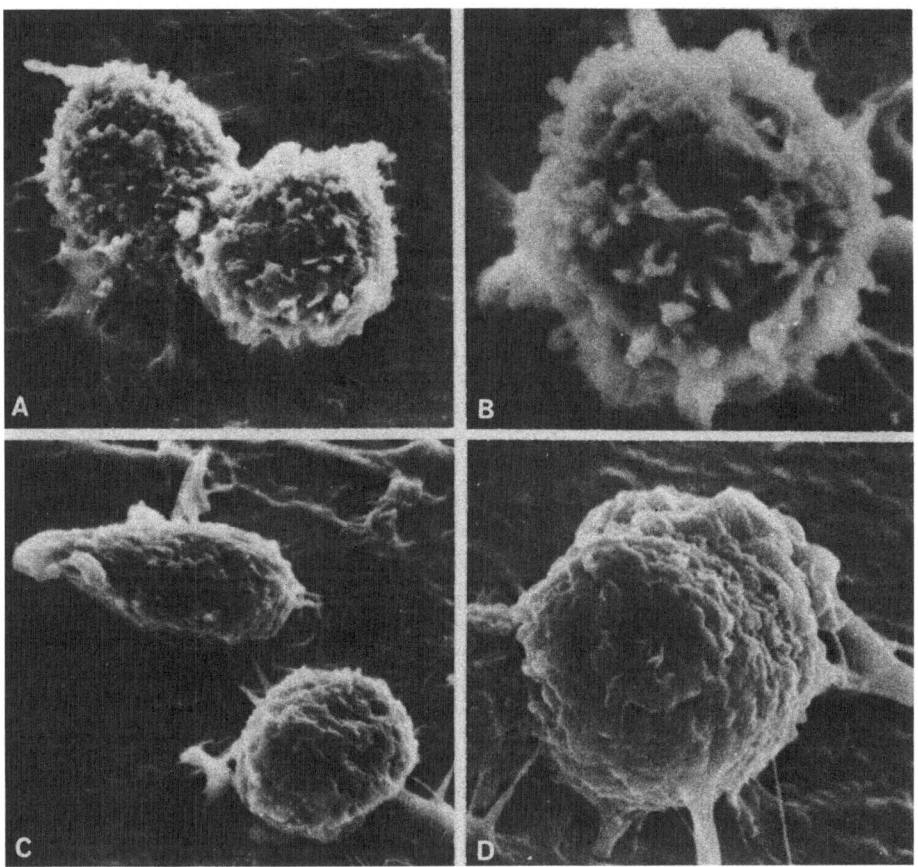

Fig. 10. Scanning electron micrographs of suspended fibroblasts as used in agglutination tests. The cells from confluent cultures were brought in suspension and, after prefixation, attached to a confluent layer of 3T3 cells. A. 3T3 cells, magn. 2,700 X; B. 3T3 cell, magn. 5,400 X; C. SV-3T3 cells, magn. 2,700 X; D. SV-3T3 cell, magn. 5,400 X.

the cells over the surface seemed to be guided by the old filopodia that had remained attached to the glass during mitosis.

The type of changes during the cell cycle was more or less the same for 3T3 and SV-3T3 cells. However the average degree of spreading was lower for SV-3T3 cells (confluent and superconfluent) than for 3T3 cells, and the morphological changes during the cell cycle were less pronounced in confluent cultures of transformed cells.

E. Surface Morphology of Suspended Cells

In order to correlate the surface morphology of normal and transformed cells with their difference in agglutinability by Con A, it is necessary to investigate the surface of these cells after they have been brought in suspension, as for agglutination assays. To that purpose we suspended normal and transformed cells with EDTA and used the same suspension for agglutination tests and for scanning electron microscopy of the surface.

Normal 3T3 cells, suspended with EDTA have a relatively rough surface with many protrusions and other irregularities (Fig. 10A and B). They also have a low agglutination with Con A (compare Table II). SV-3T3 cells, suspended with EDTA have a much more regular surface with less protrusions (Fig. 10C and D) and these cells have a high agglutination with Con A (compare Table II).

Thus there is a clear difference in morphological appearance between suspended normal and transformed cells, probably resulting from the phenotypic growth pattern in tissue culture.

IV. DISCUSSION

Several investigators have reported that drugs, such as dbc-AMP and protease inhibitors induce normal growth patterns in transformed cells (Hsie and Puck, 1971; Johnson et al., 1971; Schnebli, 1972) and in addition decrease the agglutinability of these cells by plant lectins (Sheppard, 1971; Schnebli and Burger, 1972). Our results do not support this concept of reversion to normal growth behaviour (Smets, 1972; Collard and Smets, 1974). SV 40-transformed fibroblasts grown with these drugs accumulate in the G_2 phase of the cell cycle and not in G_1, as would have been expected if contact mediated growth had been induced (Figs. 1 - 4). The decrease in agglutinability correlates with a synchronization in the G_2 phase of the cell cycle (Table I). Our results on the effect of dbc-AMP on SV-3T3 cells were confirmed by Paul (1972). More recently Schnebli (Schnebli and Haemmerli, 1974) supported our conclusions on the effect of the protease inhibitor TLCK and indicated that previous suggestions of phenotypic reversion induced by TLCK had been premature.

In the present report we also demonstrate a clear cell cycle dependent agglutinability in synchronized, transformed cells (Table II). Cells in late S and G_2 have a low agglutination with Con A as compared to asynchronous cells or cells in early phases of the cell cycle. The synchronization procedures do not influence the described changes in agglutination during the cell cycle because we found a similar decrease in agglutinability of G_2 cells from an asynchronous culture (Smets and de Ley).

Normal cells in mitosis are highly agglutinable with the concentration of Con A used in this series. We are presently investigating variations in agglutinability during the cell cycle of normal cells with high concentrations of Con A to see whether cell cycle dependent agglutination is a common phenomenon or only specific for transformed cells.

Our results suggest that dbc-AMP and TLCK reduce the agglutination with Con A by synchronizing SV-3T3 cells in the G_2 phase of the cell cycle. However it cannot be excluded that these drugs alter the cells in such a way that they become less agglutinable and at the same time are arrested in G_2. Reduced motility of cells has been observed by us with both drugs (Smets, 1972; Collard and Smets, 1974) and could well explain the simultaneous occurrence of both phenomena in these cells. This would also imply that with other cells or with a different experimental approach dbc-AMP and TLCK might decrease agglutinability without concomitant synchronization of cells in G_2 phase.

When it became clear that differences in number of accessible Con A binding sites between normal and transformed cells could not sufficiently explain differences in agglutinability (Arndt-Jovin and Berg, 1971; Ozanne and Sambrook, 1971; Temmink and Collard, 1974), it was suggested that differences in distribution of Con A binding sites might be responsible for differences in agglutinability (Nicolson, 1971), correlated with cell transformation. Experiments on normal and transformed hamster cells seemed to support this suggestion (Bretton *et al.*, 1972; Martinez-Palomo *et al.*, 1972; Rowlatt *et al.*, 1973; Huet and Barnhard, 1974; Garrido *et al.*, 1974) but results on rat cells were more ambiguous (Bretton *et al.*, 1972; Garrido *et al.*, 1974) and investigations on murine fibroblasts apparently contradicted the assumed correlation between increase in clustering and in agglutinability (Smith, 1972; de Petris *et al.*, 1973). Our own previous investigations on 3T3 and SV-3T3 cells with different cytochemical markers (Collard and Temmink, 1974; Temmink and Collard, 1974; Temmink *et al.*, 1974) were in good agreement with those of de Petris *et al.* (1973) and Smith and Revel (1972). Because application of the replica technique on 3T3 and SV-3T3 cells seemed to lead to different results (Rosenblith *et al.*, 1973; Ukena *et al.*, 1974), we tried to confirm our previous results by using in the present study the replica technique also and we were able to confirm them. Thus we agree with Garrido *et al.* (1974) that results obtained with one cell system and transforming agent cannot be generalized for other cell systems and transforming agents. However, we feel that a more far-reaching conclusion should be drawn. The conflicting results indicate that differences in redistribution of lectin binding sites are not sufficient cause for differences in agglutination (Loor, 1973), and that increased membrane fluidity (Singer and Nicolson, 1972), as measured by lectin induced clustering, does not always coincide with transformation.

Whether uninduced, large scale differences in distribution of Con A binding sites between normal and transformed cells play a role in the agglutination of these cells remains to be established. We found less HC-labeled Con A binding sites on the lamellipodia of 3T3 cells than on SV-3T3 cells and are inclined to think that this might be partially responsible for differences in agglutinability. The results of Ukena et al. (1974) on secondary organization are, however, at variance with our results. Although this may simply result from the very low detection rate that they were able to achieve, additional experiments are required to settle this matter.

Our experiments on the effect of the dbc-AMP and protease inhibitors had shown that these reagents affect the cell cycle of transformed cells. In addition, it had been demonstrated that agglutinability of 3T3 and SV-3T3 cells is dependent on their position in the cell cycle. Therefore it seemed worthwhile to investigate whether cell cycle dependent changes in surface morphology could be detected that might influence the agglutination behavior. In Chinese hamster ovary cells, cell cycle dependent changes in surface morphology had been demonstrated recently (Rubin and Everhart, 1973; Everhart and Rubin, 1974). We found similar changes in mouse fibroblasts, and were able to confirm the phenotypic difference between 3T3 and SV-3T3 cells, as described by Porter et al., (1973b). We thought that the morphological changes occuring during the cycle of normal and transformed cells might influence the agglutinability of these cells.

In order to obtain more evidence for the hypothesis that gross morphological differences (cell cycle dependent or otherwise) are important for differences in agglutinability, we started a number of new experiments. In the preliminary experiments described here, we were able to establish that morphological differences occur not only in attached cells spread on cover slips, but also remain detectable after detachment required for the agglutination tests. Additional experiments are needed to show that the morphological differences in detached cells do correlate with cell cycle dependent and transformation dependent phenotypic differences in cells growing on their substrate. Based on the information we have to date, we tentatively suggest that certain forms of transformation as well as the position in the cell cycle determine the morphology of growing cells and that this morphology is reflected in the detached cells used for agglutination tests. We suggest that these morphological differences between cells determine at least in part their difference in agglutination. More specifically, we are inclined to think that the lamellipodia of flattened cells appear as protrusions on the surface of cells following detachment from their substrate. Since these protrusions apparently contain less Con A binding sites and are more elaborate in normal cells, they might partly impair the agglutination response. In future experiments we hope to test this hypothesis.

REFERENCES

1. Anderson, T. F. (1951). "Techniques for the preservation of three-dimensional structure in preparing specimens for the electron microscope." *Trans. N. Y. Acad. Sci. Ser.* II, 13, 130.

2. Arndt-Jovin, D. J., and Berg, P. (1971). "Quantitative binding of ^{125}I-concanavalin A to normal and transformed cells." *J. Virol.* 8, 716.

3. Bernhard, W., and Avrameas, S. (1971). "Ultrastructural visualization of cellular carbohydrate components by means of concanavalin A." *Exptl. Cell Res.* 64, 232.

4. Bretton, R., Wicker, R., and Bernhard, W. (1972). "Ultrastructural localization of concanavalin A receptors in normal and SV40-transformed hamster and rat cells." *Int. J. Cancer.* 10, 397.

5. Burger, M. M. (1970). "Proteolytic enzymes initiating cell division and escape from contact inhibition of growth." *Nature* 227, 170.

6. Burger, M. M. (1971). "The significance of surface structure changes for growth control under crowded conditions. In: *Growth control in cell cultures* (Ed. G. E. Wolstenholme) p. 45. Churchill-Livingstone, London.

7. Collard, J. G., and Smets, L. A. (1974). "Effect of proteolytic inhibitors on growth and surface architecture of normal and transformed cells." *Exptl. Cell Res.* In press.

8. Collard, J. G., and Temmink, J. H. M. (1974). "Binding and cytochemical detection of cell-bound concanavalin A." *Exptl. Cell Res.* In press.

9. Comoglio, P. M., and Guglielmone, R. (1972). "Two-dimensional distribution of concanavalin A receptor molecules on fibroblast and lymphocyte plasma membranes." *FEBS Lett.* 27, 256.

10. De Petris, S., Raff, M. C., and Mallucci, L. (1973). "Ligand-induced redistribution of concanavalin A receptors on normal, trypsinized and transformed fibroblasts." *Nature New Biology* 244, 275.

11. Everhart, L. P., and Rubin, R. W. (1974). "Cyclic changes in the cell surface. II. The effect of cytochalasin B on the surface morphology of synchronized chinese hamster ovary cells." *J. Cell Biol.* 60, 442.

12. Garrido, J., Burglen, M-J., Samolyk, D., Wicker, R., and Bernhard, W. (1974). "Ultrastructural comparison between the distribution of concanavalin A and wheat germ agglutinin cell surface receptors of normal and transformed hamster and rat cell lines." *Cancer Res.* 34, 230.

13. Hsie, A. W., and Puck, T. T. (1971). "Morphological transformation of chinese hamster cells by dibutyryl adenosine cyclic 3':5'-monophosphate and testosterone." *Proc. Nat. Acad. Sci. U. S. A.* 68, 358.

14. Huet, Ch., and Bernhard, W. (1974). "Differences in the surface mobility between normal and SV40-, polyoma- and adenovirus-transformed hamster cells." *Int. J. Cancer* 13, 227.

15. Inbar, M., and Sachs, L. (1969). "Interaction of the carbohydrate-binding protein concanavalin A with normal and transformed cells." *Proc. Nat. Acad. Sci. U. S. A.* 63, 1418.

16. Inbar, M., Ben-Bassat, H., and Sachs, L. (1972). "Membrane changes associated with malignancy." *Nature New Biology* 236, 3.

17. Johnson, G. S., Friedman, R. M., and Pastan, I. (1971). "Restoration of several morphological characteristics of normal fibroblasts in sarcoma cells treated with adenosine-3':5'-cyclic monophosphate and its derivative." *Proc. Nat. Acad. Sci. U. S. A.* 68, 425.

18. Loor, F. (1973). "Lectin-induced lymphocyte agglutination." *Exptl. Cell Res.* 82, 415.

19. Martinez-Palomo, A., Wicker, R., and Bernhard, W. (1972). "Ultrastructural detection of concanavalin A surface receptors in normal and in polyoma-transformed cells." *Int. J. Cancer* 9, 676.

20. Nicolson, G. L. (1971). "Difference in topology of normal and tumour cell membranes as shown by different surface distributions of ferritin conjugated concanavalin A." *Nature New Biology* 233, 244.

21. Nicolson, G. L., and Singer, S. J. (1971). "Ferritin-conjugated plant agglutinins as specific saccharide stains for electron microscopy: Application to saccharides bound to cell membranes." *Proc. Nat. Acad. Sci. U. S. A.* 68, 942.

22. Ozanne, B., and Sambrook, J. (1971). "Binding of radioactively labeled concanavalin A and wheat germ agglutinin to normal and virus-transformed cells." *Nature New Biology* 232, 156.

23. Paul, D. (1972). "Effects of cyclic AMP on SV-3T3 cells in culture." *Nature New Biology* 240, 179.

24. Porter, K., Prescott, D., and Frye, J. (1973a). "Changes in surface morphology of chinese hamster ovary cells during the cell cycle." *J. Cell Biol.* 57, 815.

25. Porter, K. R., Todaro, G. J., and Fonte, V. (1973b). "A scanning electron microscope study of surface features of viral and spontaneous transformants of mouse Balb/3T3 cells." *J. Cell Biol.* 59, 633.

26. Rosenblith, J. Z., Ukena, T. E., Yin, H. H., Berlin, R. D., and Karnovsky, M. J. (1973). "A comparative evaluation of the distribution of concanavalin A binding sites on the surfaces of normal, virally-transformed, and protease-treated fibroblasts." *Proc. Nat. Acad. Sci. U. S. A.* 70, 1625.

27. Rowlatt, C., Wicker, R., and Bernhard, W. (1973). "Ultrastructural distribution of concanavalin A receptors on hamster embryo and adenovirus tumour cell cultures." *Int. J. Cancer* 11, 314.

28. Rubin, R. W., and Everhart, L. P. (1973). "The effect of cell-to-cell contact on the surface morphology of chinese hamster ovary cells." *J. Cell Biol.* 57, 837.

29. Schnebli, H. P., and Burger, M. M. (1972). "Selective inhibition of growth of transformed cells by protease inhibitors." *Proc. Nat. Acad. Sci. U. S. A.* 69, 3825.

30. Schnebli, H. P., and Haemmerli, G. (1974). "Protease inhibitors do not block transformed cells in the G1 phase of the cell cycle." *Nature* 248, 150.

31. Sheppard, J. R. (1971). "Restoration of contact-inhibited growth to transformed cells by dibutyryl adenosine 3':5'-cyclic monophosphate." *Proc. Nat. Acad. Sci. U. S. A.* 68, 1316.

32. Singer, S. J., and Nicolson, G. L. (1972). "The fluid mosaic model of the structure of cell membranes." *Science* 175, 720.

33. Smets, L. A. (1972). "Contact inhibition of transformed cells incompletely restored by dibutyryl cyclic-AMP." *Nature New Biology* 239, 123.

34. Smets, L. A. (1973). "Agglutination with Con A dependent on cell cycle." *Nature New Biology* 245, 113.

35. Smets, L. A., and de Ley, L. "Cell cycle dependent modulations

of the surface membrane of SV40-transformed 3T3 cells." Submitted for publication.

36. Smith, S. B., and Revel, J-P. (1972). "Mapping of concanavalin A binding sites on the surface of several cell types." *Dev. Biol.* <u>27</u>, 434.

37. Stobo, J. D., and Rosenthal, A. S. (1972). "Biologically active concanavalin A complexes suitable for light and electron microscopy." *Exptl. Cell Res.* <u>70</u>, 443.

38. Temmink, J. H. M., and Collard, J. G. (1974). "Correlation between cytochemical detection by horse-radish peroxidase and binding of concanavalin A on normal and transformed mouse fibroglasts." In: *Electron microscopy and cytochemistry*, p. 299. (Eds. E. Wisse, W. Th. Daems, I. Molenaar, and P. van Duyn) North-Holland, Amsterdam.

39. Temmink, J. H. M., Collard, J. G., Spits, H., and Roos, H. (1964). "A comparative study of four cytochemical detection methods of concanavalin A binding sites on the cell membrane." Submitted for publication.

40. Terasima, A., and Tolmach, L. J. (1963). "Growth and nucleic acid synthesis in synchronously dividing populations of HeLa cells." *Exptl. Cell Res.* <u>30</u>, 344.

41. Ukena, T. E., Borysenko, J. Z., Karnovsky, M. J., and Berlin, R. D. (1974). "Effects of colchicine, cytochalasin B, and 2-deoxyglucose on the topographical organization of surface-bound concanavalin A in normal and transformed fibroblasts." *J. Cell Biol.* <u>61</u>, 70.

13

CONCANAVALIN A AND OTHER LECTINS IN THE STUDY

OF TUMOR CELL SURFACE ORGANIZATION*

        Sten Friberg and Sten Hammarström
        Karolinska Institute and
        Wenner-Gren Institute
        Stockholm, Sweden

ABSTRACT

Cell surface structures of two mouse ascites tumors were studied using lectins. The tumors are sublines of the spontaneous mammary adenocarcinoma TA3 in strain A, differing in two main characteristics. Subline TA3-St grows only in syngeneic mice and has high expression of H-2 antigens. Subline TA3-Ha, in contrast, proliferates in all mouse strains, and has low amounts of exposed H-2 antigens. Concanavalin A (Con A), phytohemagglutinin (PHA) and *Helix pomatia* anti A hemagglutinin (HP) were used in agglutination tests, in binding experiments with ^{125}I-labelled lectins and also in fluorescence studies with FITC-labelled lectins.

Con A and PHA agglutinated the TA3-St cells but not the TA3-Ha cells. However, fluorescein-labelled Con A and PHA were bound to all cells (> 90%) of both sublines. Moreover both cell types contained an identical number of Con A receptors. The same result was obtained when the number of PHA receptors on the two sublines was compared.

HP agglutinated TA3-Ha cells but not TA3-St cells. However, in this case, the difference in agglutinability between the lines was due to the presence or absence of HP receptors. All TA3-Ha cells contained large numbers of HP receptors. In contrast the majority (> 90%) of the TA3-St cells lacked HP receptors.

---

*This investigation was supported by the Contract Grant No. NIH-NCI-E-69-2005 and the Swedish Science Research Council Grant No. B 3485-003.

TABLE I

SUMMARY OF SOME CHARACTERISTICS OF TA3-Ha AND TA3-St ASCITES TUMOR LINES

Characteristic	Tumor	
	TA3-Ha	TA3-St
Growth in allogeneic mice	Yes	No
H-2^a antigen expression	Low	High
Susceptibility to immune (anti-H-2^a) cytotoxicity *in vitro*	No	Yes
Electrophoretic mobility	Fast	Slow

Trypsin released the HP receptors from TA3-Ha cells, at the same time making these cells agglutinable by both Con A and PHA. We conclude that the Ta3-Ha cells have a trypsin sensitive surface glycoprotein (or glycoproteins) detectable by HP, which is absent on most TA3-St cells. It is possible that this glycoprotein interferes with the agglutinability of TA3-Ha cells by Con A and PHA; whether it is also responsible for the low expression of H-2 alloantigen on this cell remains to be seen.

I. INTRODUCTION

Most experimental and some spontaneous tumors have been shown to possess new antigens capable of provoking an immunological host reaction. If the immune response is strong, and the tumor cells susceptible to the rejection mechanisms, the tumor may be retarded in its growth, or even eliminated. However, if the tumor cell population has the capacity to develop immunoresistant sublines, the immune defense of the host would be rendered ineffective. The development of immunoresistant tumors may thus be of profound importance in the tumor-host balance.

In order to understand the phenomenon of immunoresistance it is essential to know in what way immunoresistant tumors differ from their immunosensitive counterparts. Detailed cell-biological comparisons were therefore made between two mouse sublines, TA3-Ha and TA3-St, which differed greatly in immunoresistance. Both tumors were established transplantation lines which had been converted to the ascites form of growth. The original tumor TA3 was a spontaneous mammary adenocarcinoma of strain A mice carrying histocompatibility antigens H-2^a (Hauschka, 1960).

Previous work has demonstrated that the two sublines differed in a number of properties (Friberg, 1972a and b). Some of these are listed in Table I. The TA3-St subline could grow progressively only in allogeneic (= immunosensitive) mice while the TA3-Ha subline caused lethal takes in all mouse strains investigated as well as in some rats and hamsters (= immunoresistant) (Friberg, 1972a). Alloantigen expression was markedly different in the two sublines. TA3-St cells expressed approximately 15 times as much H-2^a antigens on the surface as compared to TA3-Ha cells (Friberg, 1972b). In accord with the above findings, the TA3-St line was susceptible to anti-H-2^a antibodies plus complement and to immune lymphocytes with H-2^a specificity, while the TA3-Ha line was resistant. The two cell types also differed in electrophoretic mobility; the TA3-Ha cells had a higher anodal mobility than the TA3-St cells.

The purpose of this investigation was to compare the surface characteristics of the two sublines, using purified lectins as surface probes. The following three lectins were used: concanavalin A (Con A), *Phaseolus vulgaris* lectin (PHA) and *Helix pomatia* anti-A hemagglutinin (HP). Interaction between cells and lectins was measured by (1) agglutination experiments, (2) direct immunofluorescense utilizing fluorescein isothiocyanate - (FITC) - labelled lectins and (3) cell binding experiments using ^{125}I-labelled lectins.

II. METHODS

Con A and HP were highly purified preparations prepared by immunoadsorbent techniques according to Agrawal and Goldstein (1967) and Hammarström and Kabat (1969), respectively. PHA was a commercially available purified product (MR68), devoid of erythroagglutinin, from Wellcome Research Laboratories, Beckenham, Kent. The carbohydrate specificities of Con A and HP have been worked out in detail (Goldstein *et al.*, 1974; Hammarström, 1974). Con A binds to α-linked-D-mannosyl-, D-glucosyl- and *N*-acetyl-D-glucosaminoyl nonreducing endgroups as well as to internal 1,2 linked D-mannose residues (Goldstein *et al.*, 1974). HP interacts with the following nonreducing endgroups (in order of decreasing binding constants) α-*N*-acetyl-D-galactosaminoyl > α-*N*-acetyl-D-glucosaminoyl ≃ β-*N*-acetyl-D-galactosaminoyl > β-*N*-acetyl-D-glucosaminoyl > α-D-galactosyl (Hammarström, 1974). PHA-leucoagglutinin has a complex carbohydrate specificity. Cell agglutinin or precipitation of glycoprotein by PHA is not inhibited by simple sugars or methylglycosides (Hammarström *et al.*, 1974; Molnar *et al.*, 1973).

III. RESULTS

A. Agglutination of TA3-Ha and TA3-St Cells by Con A, PHA and HP

Table II summarizes the results of several experiments in which the ability of the three lectins to agglutinate TA3-Ha and TA3-St cells was investigated. TA3-St is easily agglutinated by both Con A and PHA. No agglutination with HP was demonstrated. Conversely,

TABLE II

AGGLUTINABILITY OF TA3-St AND TA3-Ha CELLS BY CON A, PHA AND HP BEFORE AND AFTER ENZYME TREATMENT

Lectin	Cell-line treatment	Minimal concentration of agglutinin (µg/ml) needed for agglutination					
		TA3-St none	TA3-Ha none	TA3-St + trypsin*	TA3-Ha + trypsin	TA3-St + NANase**	TA3-Ha + NANase
Con A		30	> 2000	30-60	8	30	> 2000
PHA		4	> 2000	4	8	8	250-2000
HP		> 2000	8	N.D.§	125	N.D.§	8

*50 µg trypsin/10^7 cells 37°C, 30 min.

**10 µg neuraminidase/10^7 cells 37°C, 30 min.

§N.D.: Not done

TA3-Ha was strongly agglutinated by HP but not at all with Con A or PHA. Thus, a highly significant difference between the two sublines in lectin-agglutinability was demonstrated.

Agglutinability of cells may be modified by enzyme treatment. As seen from Table II, TA3-Ha cells were rendered fully agglutinable by both Con A and PHA after treatment with trypsin. The same results were seen after treatment of TA3-Ha cells with other proteolytic enzymes, e.g., papain, pronase and subtilisin. Trypsin treatment of the TA3-St cells, on the other hand, had no significant effect on their agglutinability with Con A or PHA.

Since it could not be excluded that inagglutinability of TA3-Ha cells by Con A and PHA was due to the higher negative net charge of these cells (compare Table I), causing intercellular repulsion, agglutination experiments were also performed with neuraminidase-treated cells. As is seen from Table II, removal of most of the sialic acid from the surface did not significantly change the agglutinability of the TA3-Ha cells, nor of the TA3-St cells, with any of the lectins.

These findings indicate that receptors for Con A and PHA are present on both tumor cell sublines. Both receptors, furthermore, appear to be relatively trypsin resistant. However, on the TA3-Ha cells, Con A and PHA receptors are not fully exposed.

It was observed that a mixture of Con A and PHA caused strong agglutination of the TA3-Ha cells. Fixed TA3-Ha cells behaved like untreated cells, arguing against the possibility that lectin-induced topographical alterations of the cell membrane were responsible for the effect. The interaction of a mixture of Con A and PHA in causing agglutination of TA3-Ha cells was investigated by sensitizing TA3-Ha cells with Con A; after removing unbound Con A, cells were incubated with ^{125}I-PHA. Control unsensitized cells were also incubated with ^{125}I-PHA. Half of each of the cell suspensions was then exposed to 0.1 M α-methyl-D-mannoside, and the other halves to D-arabinose. The results are plotted graphically (Fig. 1), and demonstrate that (1) Con A does not prevent the binding of PHA, indicating separate binding sites; (2) D-arabinose is ineffective in removing complexed ^{125}I-PHA from Con A-sensitized TA3-Ha cells; (3) by contrast, α-methyl-D-mannoside can break the PHA-Con A complex even when the latter lectin is surface bound; (4) release of ^{125}I-PHA from the cell surface is resistant to the presence of either α-methyl-D-mannoside or D-arabinose.

These experiments lend further support to the notion that the receptors for Con A on the surface of TA3-Ha cells are less exposed than those on TA3-St cells, perhaps explaining why the TA3-Ha cells are not agglutinated by Con A.

It is of great interest that HP only agglutinates the TA3-Ha

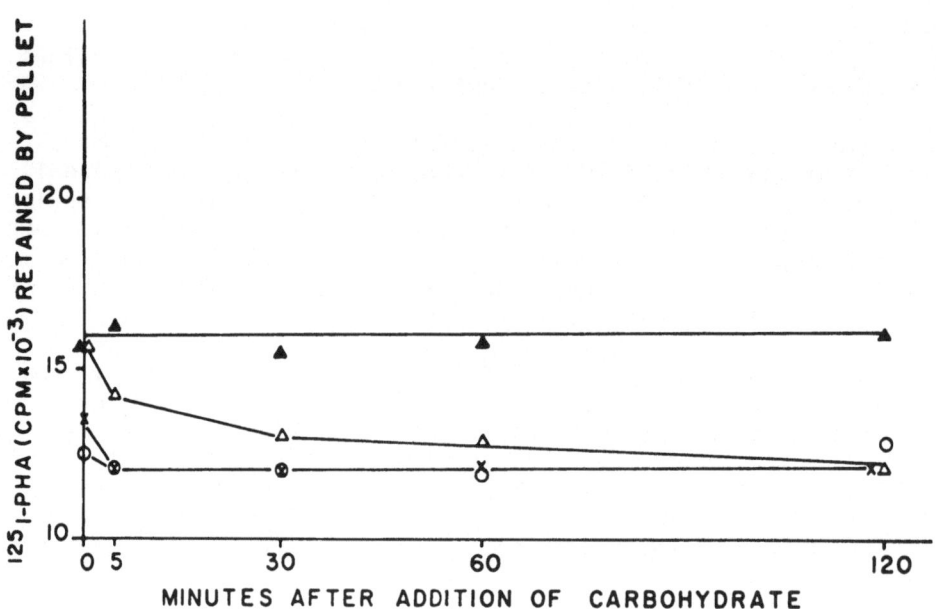

Fig. 1. Effect of Me-α-D-Manp or D-arabinose on release of ^{125}I-PHA from TA3-Ha cells not treated or pre-treated with Con A. Symbols: ▲——▲ = TA3-Ha cells pre-treated with Con A and ^{125}I-PHA, exposed to D-arabinose; △——△ = similarly treated cells exposed to Me-α-D-Manp; x——x = TA3-Ha control cells incubated with ^{125}I-PHA alone, exposed to D-arabinose; o——o = control cells exposed to Me-α-D-Manp. Ordinate: Time (min) after addition of carbohydrate. Abscissa: Radioactivity (c.p.m.) retained by cells (10^7/tube). Results are based on two experiments. See text for details.

cells and that the receptor molecule(s) for this lectin appears to be trypsin sensitive (Table II). These findings may indicate that HP in fact detects some surface material which interferes with or blocks the receptor molecules for PHA and Con A.

B. <u>Immunofluorescence studies on TA3-Ha and TA3-St cells with FITC-HP, FITC-Con A and FITC-PHA</u>

To obtain further information about the interaction between the lectins and the two TA3 cell lines, immunofluorescence experiments with FITC-labelled lectins were performed. Fig. 2 shows the results obtained with FITC-labelled HP tested against living TA3-Ha and TA3-St cells as well as control cells. The percentage of cells with positive membrane fluorescence in each suspension is plotted against the dilution of labelled lectin. As can be seen, > 90% of the TA3-Ha

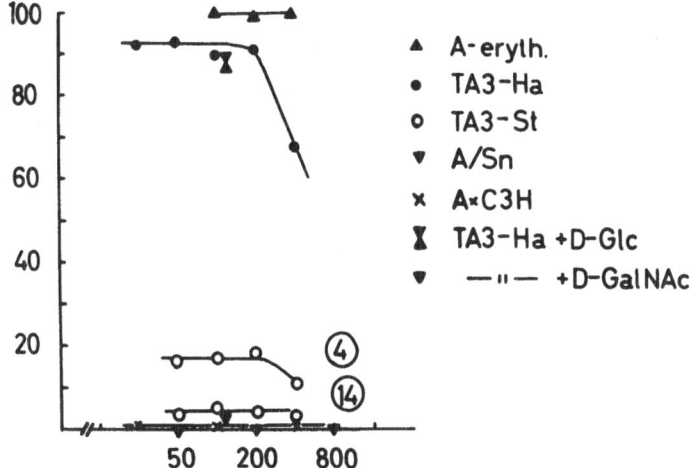

Fig. 2. Reactivity of FITC-HP with TA3-Ha and TA3-St cells and various controls. Abscissa: Dilution of FITC-HP, starting concentration 0.75 mg/ml. Ordinate: Percentage of positively stained cells.

cells are stained by FITC-HP. In contrast, only 5-10% of the TA3-St cells showed positive membrane fluorescence. Peritoneal host lymphoid cells of the mouse strains carrying the ascites tumor did not show positive fluorescence with FITC-HP. This latter finding most likely indicates that the small fraction of positively stained cells in the TA3-St population represents tumor cells. The percentage of positive TA3-St cells varied slightly depending on the time elapsed between inoculation and harvest (Fig. 1 shows TA3-St cells harvested after 4 and 14 days). The significance of this finding is unclear. Binding of FITC-HP was inhibited by D-GalNAc but not with D-Glc in accordance with the known specificity of the lectin (Hammarström and Kabat, 1969). Fig. 3a and b show two examples of positive membrane fluorescence of TA3-Ha cells and of one TA3-St cell.

Immunofluorescence studies with FITC-labelled Con A and PHA on TA3-Ha and TA3-St cells showed that all cells (> 90%) of both cell lines were stained. Moreover, the titration curve for labelled Con A was the same for both cell populations. This was also true for labelled PHA. Fig. 4a and b show positive membrane fluorescence with FITC-Con A on TA3-St and TA3-Ha cells, respectively. Note that only the TA3-St cells are agglutinated.

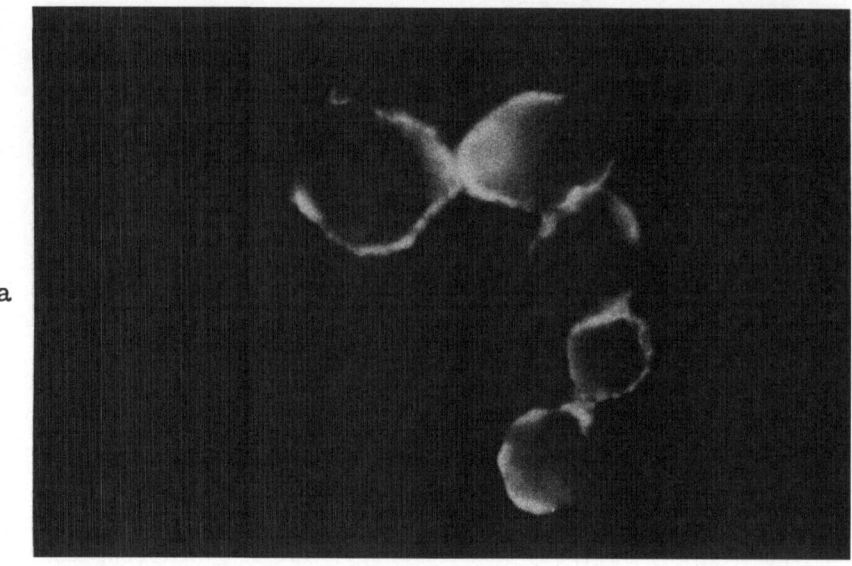

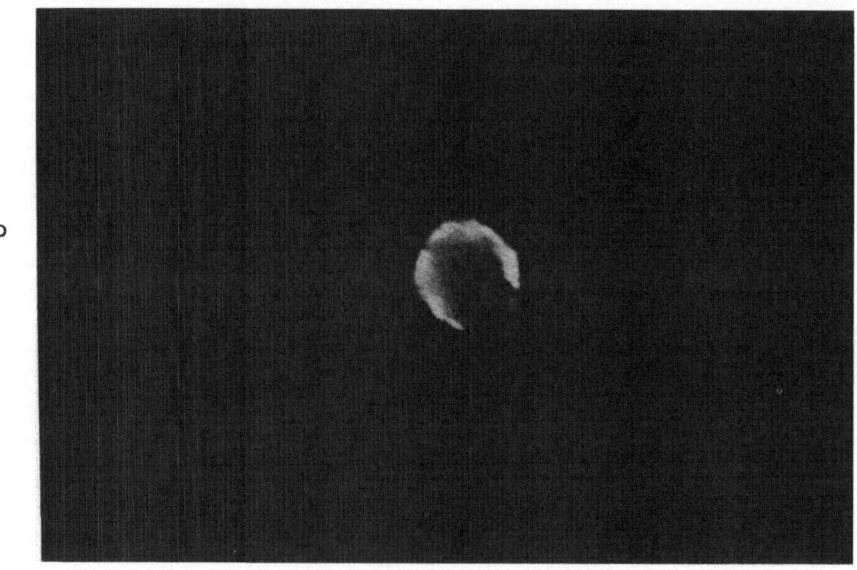

Fig. 3. Reactivity of TA3-Ha (3a) and TA3-St (3b) cells with FITC-HP. Note that the TA3-Ha cells are agglutinated. Both types of cells show an even distribution of stain. 5×10^5 cells were incubated with 5 µg FITC-labelled HP for 45 min at 4°C in the presence of 0.02% NaN_3 (non-capping conditions). x 490

a

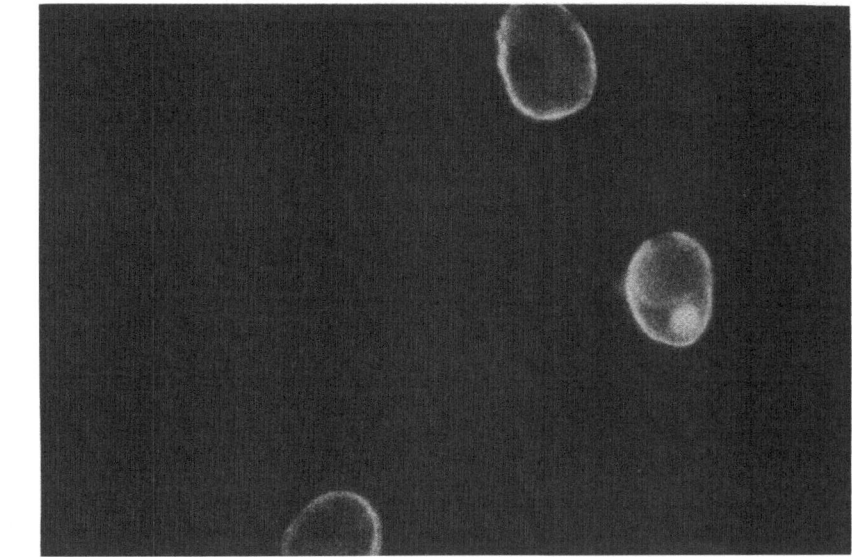

b

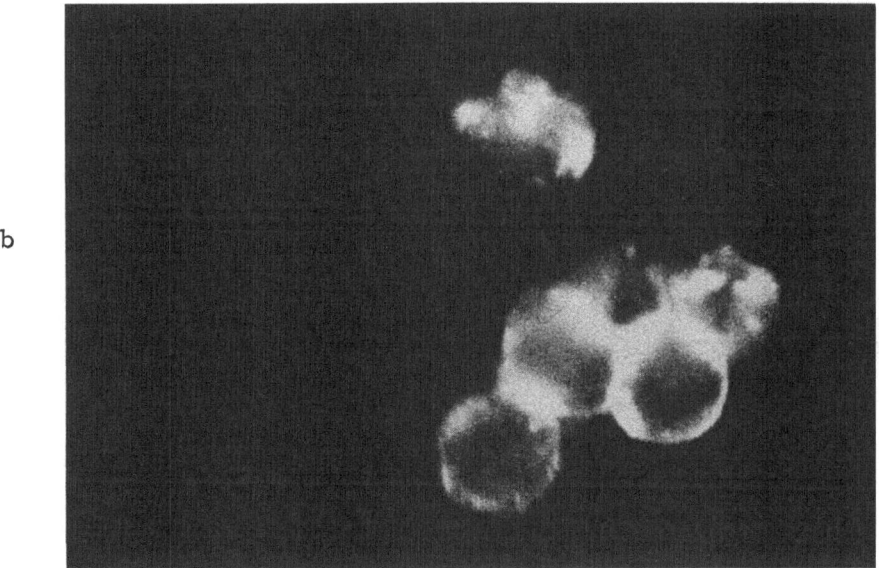

Fig. 4. Reactivity of TA3-Ha (2a) and TA3-St (2b) cells with FITC Con A. Both types of cells show an even distribution of stain but only the TA3-St cells are agglutinated. 10 µg of FITC-labelled Con A was added to 5×10^5 cells under non-capping conditions (for details, see legend to Fig. 3). x 490

TABLE III

CO-CAPPING EXPERIMENTS WITH HP AND CON A
ON TA3-Ha CELLS*

Staining sequence (first and second reagent)	Co-capping[§]
HP red - HP green	+
HP red - Con A green	-
Con A green - HP red	-
HP green - HP red	+

*The cells were incubated with the first reagent, FITC (green) or TRITC (red) labelled lectin, at 4°C for 30 min. The cells were then washed and kept at 37°C for 45 min to induce cap formation. They were then stained with the second reagent in the presence of 0.02% NaN_3 at 0°C. 40-50% of the cells showed caps.

[§]+, staining of the cap only with the second reagent. -, staining of the entire cell surface with the second reagent.

In order to study the relationship between the surface receptors for HP and Con A on TA3-Ha cells more closely, co-capping experiments with TRITC-labelled HP (red) and FITC-labelled Con A (green) were performed. The results of these experiments are summarized in Table III. In control experiments, TRITC-labelled HP and FITC-labelled HP were used. Only the cap produced by the first reagent was stained by the second reagent (= co-capping). In contrast, TA3-Ha cells, capped with TRITC-HP, showed positive fluorescence over the entire cell membrane with FITC-Con A. The reverse experiment, FITC-Con A added as the first reagent to produce a cap and TRITC-HP added as the second reagent, gave the same result as above. These experiments strongly indicate that the majority of the receptors for HP and for Con A are located on separate macromolecules, since the two receptors moved independently on the tumor cell surface.

C. <u>Binding of ^{125}I-labelled HP, -Con A and -PHA to TA3-Ha and TA3-St cells</u>

Information about the quantitative relationship between the lectin receptors on the two tumor cell sublines were obtained by cell binding experiments. Increasing amounts of ^{125}I-labelled Con

CELL SURFACE ARCHITECTURE

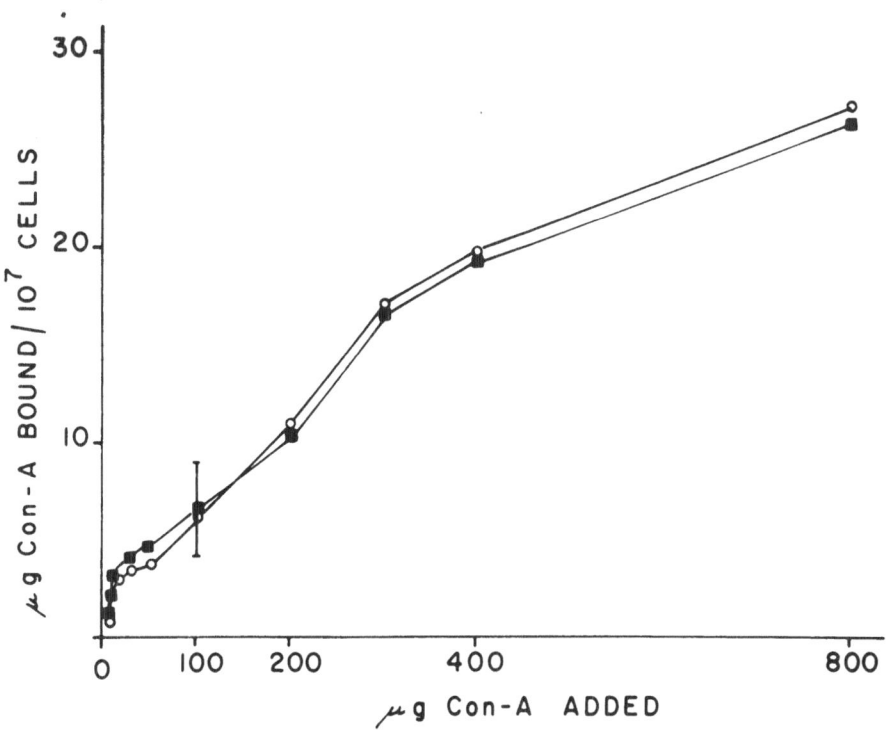

Fig. 5. Binding of ^{125}I-Con A to TA3-Ha (■———■) and to TA3-St (o———o) cells. Cells 1 x 10^7/tube) were incubated (30 min/20°C) with increasing amounts of ^{125}I-Con A. Ordinate: μg Con A added. Abscissa: μg Con A bound. Results are based on six experiments. Vertical bars indicate standard error (S.E.) of the mean. Where S.E. is not shown, its value is less than 10% of the mean.

A, PHA or HP, respectively, were added to a constant number of tumor cells. Figs. 5, 6, and 7 show the binding curves obtained with TA3-St and TA3-Ha cells and the three lectins respectively. Table IV summarizes the apparent binding constants in liters/mole for the interaction between lectin and cell and the minimal number of lectin receptors on the two cell types. The values were calculated from the Law of Mass Action. Two main findings emerge from these measurements: (1) TA3-Ha and TA3-St cells bound the same number of Con A molecules. The same was true for PHA. Moreover, the apparent binding constants for the interaction between Con A and TA3-Ha cells or TA3-St cells were identical. The same was found when the binding of PHA to the two sublines were compared. (2) TA3-Ha and TA3-St cells differed markedly in the number of HP-molecules bound per cell at saturation. Calculated on the whole population, TA3-Ha cells bound about 30 times more HP molecules per cell as compared to TA3-St cells. However, immunofluorescence studies with FITC-HP had al-

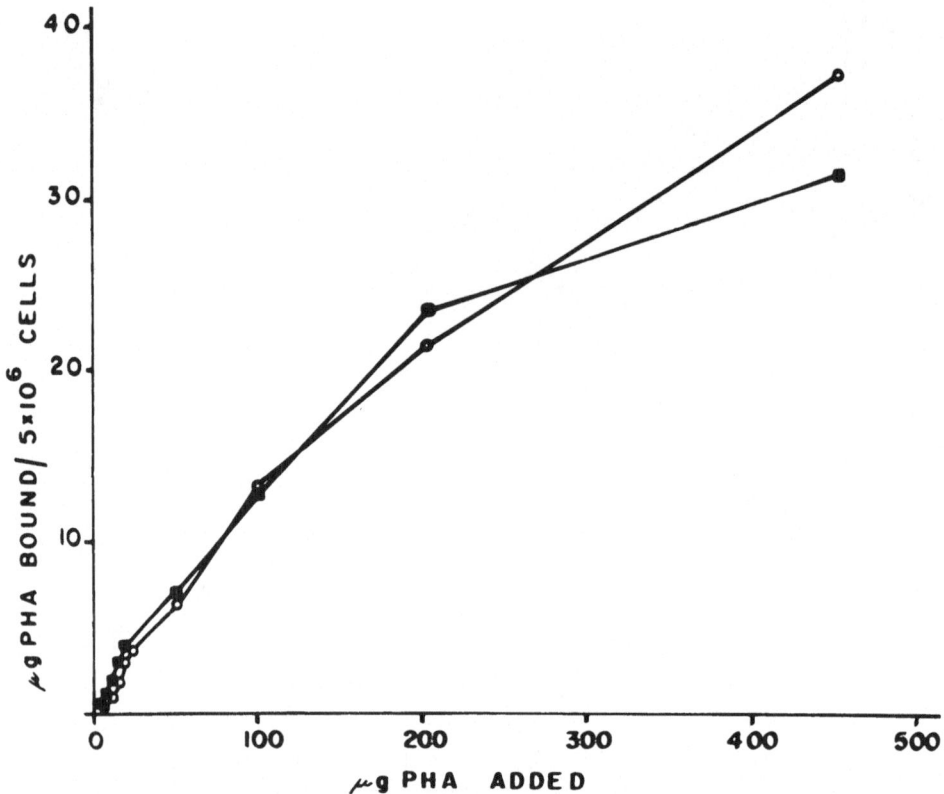

Fig. 6. Binding of ^{125}I-PHA by TA3-Ha and TA3-St cells. Symbols and ordinate parameters as in Fig. 5. Cell number was kept constant at 5 x 10^6/tube. Results are based on two experiments.

ready shown that only 5-10% of the TA3-St cells were stained. This indicates that a small fraction of these cells have about the same surface density of HP-receptors as the TA3-Ha cells, while the majority (90-95%) of the TA3-St cells lack HP-receptors. Also, as can be seen from Table IV, the apparent association constants for the interaction between HP and the two cell lines were essentially the same, indicating that carbohydrate endgroups on the receptors are very similar.

As mentioned earlier, the HP-receptors on TA3-Ha cells appear to be trypsin sensitive. To verify this, TA3-Ha cells were treated with low concentrations of purified trypsin (Worthington, Freehold, N. J. TPCK-treated) for various time periods and the amount of ^{125}I-labelled HP bound to the enzyme treated cells was determined. As can be seen from Table V, up to 90% of the HP receptors could be removed. In contrast, neuraminidase, in high concentrations, did

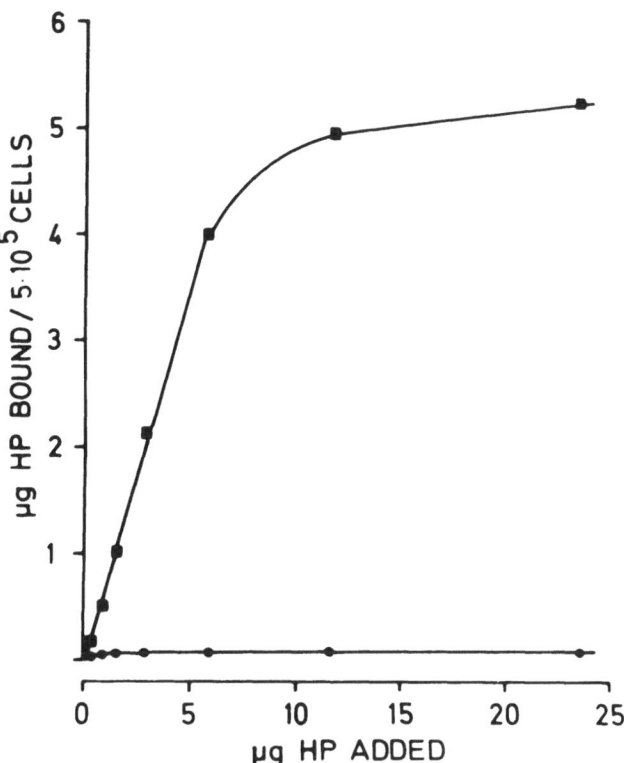

Fig. 7. Binding of ^{125}I-HP to TA3-Ha (■——■) and TA3-St (●——●) cells. Cells (5 x 10^5/tube) were incubated (30 min/20°C) with increasing amounts of ^{125}I-HP. Ordinate: μg HP added. Abscissa: μg HP bound/5 x 10^5 cells.

TABLE IV

APPARENT K-VALUES AND MINIMAL NUMBERS OF AGGLUTININ RECEPTORS ON TA3-Ha AND TA3-St CELLS FOR HP, CON A AND PHA

Agglutinin	TA3-Ha		TA3-St	
	n x 10^{-6}	K x 10^{-6}	N x 10^{-6}	K x 10^{-6}
HP	34	600	1.3	300
Con A (I)	2.8	26	2.0	39
(II)	30	0.16	36	0.13
PHA	22	1.5	21	1.6

TABLE V

HP-RECEPTORS ON TA3-Ha CELLS
EFFECT OF TRYPSIN AND NEURAMINIDASE

Exp.	Treatment	Time (min)	HP-receptors (% reduction)
1	Trypsin 4 µg/10^8 cells	10	58
	" "	20	71
	" "	40	91
2	Trypsin 1 µg/10^8 cells	20	55
3	Neuraminidase 100 µg/10^8 cells	60	2

not have any effect on the binding of labelled HP to the cells. These results confirm the agglutination experiments.

IV. DISCUSSION

Taken together, the results obtained with three lectins as cell-membrane probes demonstrate that the cell surface of the TA3-Ha and TA3-St sublines are markedly different. While all cells in both sublines have surface receptors for Con A and PHA which quantitatively and qualitatively appear to be the same, these receptors seem to be less accessible on the TA3-Ha cells, since untreated TA3-Ha cells were not agglutinated by Con A or PHA. The simplest explanation for the poor accessibility of these receptors on the TA3-Ha cells is that there is a substance (or substances) on the surface of these cells which interferes with the exposure of these receptors. This hypothetical substance then has the following characteristics: (1) it is sensitive to trypsin, since trypsin treatment of TA3-Ha cells renders these cells fully agglutinable by Con A and PHA; (2) it is a glycoprotein or glycolipid carrying the carbohydrate endgroups (α, β-D-GalNac; α, β-D-GNAc or α-D-Gal) which can bind the lectin HP, since trypsin sensitive HP receptors are present on TA3-Ha cells but not on the majority of the TA3-St cells; (3) this substance is not physically connected with the macromolecules carrying the Con A receptors, since co-capping experiments demonstrated that the HP and Con A receptor molecules moved independently on the cell surface.

These results are of interest in relation to the finding that TA3-Ha cell membrane contains non-exposed H-2^a alloantigens (Molnar et al., 1937; Friberg and Lilliehöök, 1973). Direct measurement of the capacity of intact TA3-Ha cells to absorb anti-H-2^a antiserum (both K- and D- end determinants) showed that these cells bound 10-

12 times less antibodies than TA3-St cells. However, ultrasonication or freeze-thawing of TA3-Ha cells essentially restored their antibody absorbing capacity to the level of the TA3-St cells. Again, it may be that the lectin HP detects some surface material which interferes with the exposure of H-2 alloantigens on the TA3-Ha cells. Since immunoresistence of tumors is at least partly related to the amount of H-2 antigens exposed on the cell surface, further studies on the trypsin sensitive, HP-positive surface material on TA3-Ha cells may be rewarding.

REFERENCES

1. Agrawal, B., and Goldstein, I. J. (1967). "Protein-carbohydrate interaction. VI. Isolation of concanavalin A by specific adsorption on cross-linked dextran gels." *Biochim. Biophys. Acta* 147, 262.

2. Friberg, S., Jr. (1972a). "Comparison of an immunoresistant and an immunosusceptible ascites subline from murine tumor TA3. I. Transplantability, morphology and some physico-chemical characteristics." *J. Nat. Cancer Inst.* 48, 1463.

3. Friberg, S., Jr. (1972b). "Comparison of an immunoresistant and an immunosusceptible ascites subline from murine tumor TA3. II. Immunosensitivity and antibody-binding capacity *in vitro*, and immunogenicity in allogeneic mice." *J. Nat. Cancer Inst.* 48, 1477.

4. Friberg, S., Jr., and Lillichöök, B. (1973). "Existence of non-exposed H-2 antigens in immunoresistant murine tumor." *Nature New Biology* 241, 112.

5. Goldstein, I. J., Reichert, C. M., and Misaki, A. (1974). "Interaction of concanavalin A with model substrates." *Ann. N. Y. Acad. Sci.* 234, 283.

6. Hammarström, S. (1974). "Structure, specificity, binding properties and some biological activities of a blood group A-reactive hemagglutinin from the snail *Helix pomatia*." *Ann. N. Y. Acad. Sci.* 234, 183.

7. Hammarström, S., Goldstein, I. J., and Weber, T. (1974). Unpublished results.

8. Hammarström, S., and Kabat, E. A. (1969). "Purification and characterization of a blood group A reactive hemagglutinin from the snail *Helix pomatia* and a study of its combining site." *Biochemistry* 8, 2696.

9. Hauschka, T. S. (1960). "Cell population studies on mouse ascites tumors." *Trans. N. Y. Acad. Sci.* <u>63</u>, 640.

10. Molnar, J., Klein, G., and Friberg, S., Jr. (1973). "Subcellular localization of murine histocompatibility antigens in tumor cells." *Transplantation* <u>16</u>, 93.

14

MODIFICATION OF THE BIOLOGICAL ACTIVITIES OF

CONCANAVALIN A BY ANTI-CONCANAVALIN A

> Myron A. Leon
> St. Luke's Hospital
> Cleveland, Ohio, U. S. A.

ABSTRACT

Concanavalin A (Con A) bound to cell membrane glycoproteins, may be dissociated from the membrane receptors by competitive ligands such as α-methyl-D-mannoside. Addition of antibody to Con A to the system forms complexes of antibody and Con A which are still bound to the membrane receptors. Such complexes are not dissociable from the membrane by α-methyl-D-mannoside. Presence of the complexes on the membrane is monitored by radioactive label or by passive lysis of the cells with guinea pig complement.

Antibody to Con A can completely suppress Con A mediated stimulation of lymphocytes as measured by incorporation of tritiated thymidine. However production of lymphokines involved in migration inhibition, enhancement of plaque forming cells or blastogenesis are differentially suppressed by antibody to Con A.

I. INTRODUCTION

In this paper an attempt will be made to delineate some of the areas where modification of the biological properties of concanavalin A (Con A) by antibody specific for Con A may offer significant possibilities for research.

When a multivalent lectin such as Con A reacts with carbohydrate groups on a cell membrane, the reactions may fall into two general classes. In the first class of reactions a molecule of Con A binds to carbohydrate groups present on a single glycoprotein or glycolipid resulting in intramolecular cross-linking. In the second class of reactions a molecule of Con A binds to carbohydrate groups present on adjacent molecules resulting in intermolecular cross-linking. The reactions of the second class may be further subdivided according to whether the adjacent molecules cross-linked by the Con A are the same or different molecules (Fig. 1).

The relative proportions of the various types of linkages formed would depend on the geometry of the carbohydrate groups on the membrane surface, their association constants for Con A and whether or not Con A is in excess. The number of valences of a given molecule of Con A involved in binding to membrane components (Fig. 1) is purely diagrammatic and depends on similar considerations. We have demonstrated previously that polysaccharides or glycoproteins can readily attach to Con A-erythrocyte complexes (Leon and Young, 1970). Whether such attachment is to free valences of Con A or whether there is competition by the added polysaccharides resulting in partial dissociation of Con A-membrane complexes is not known.

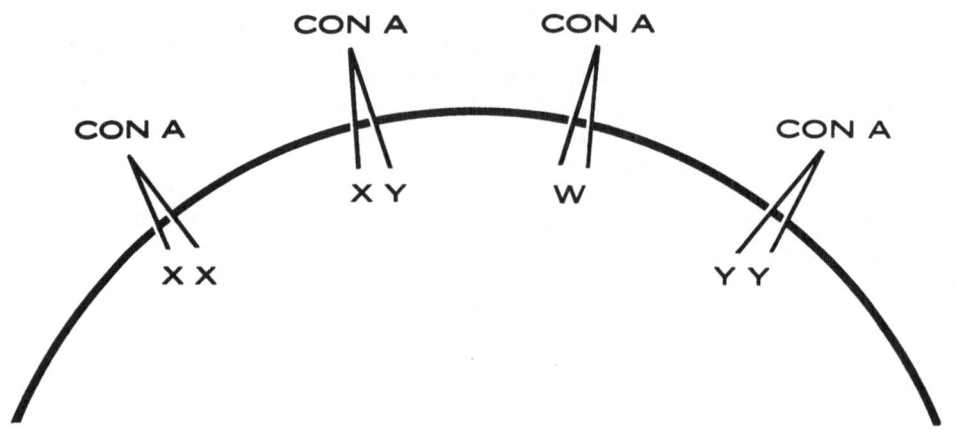

SCHEME FOR NEAREST NEIGHBOR ANALYSIS

Fig. 1. Schematic representation of binding of Con A to glycoproteins W, X, or Y on a cell membrane; binding to two carbohydrate residues in the same glycoprotein (W), to carbohydrate residues on identical adjacent glycoproteins (XX, YY) or to carbohydrate residues on dissimilar adjacent glycoproteins (XY).

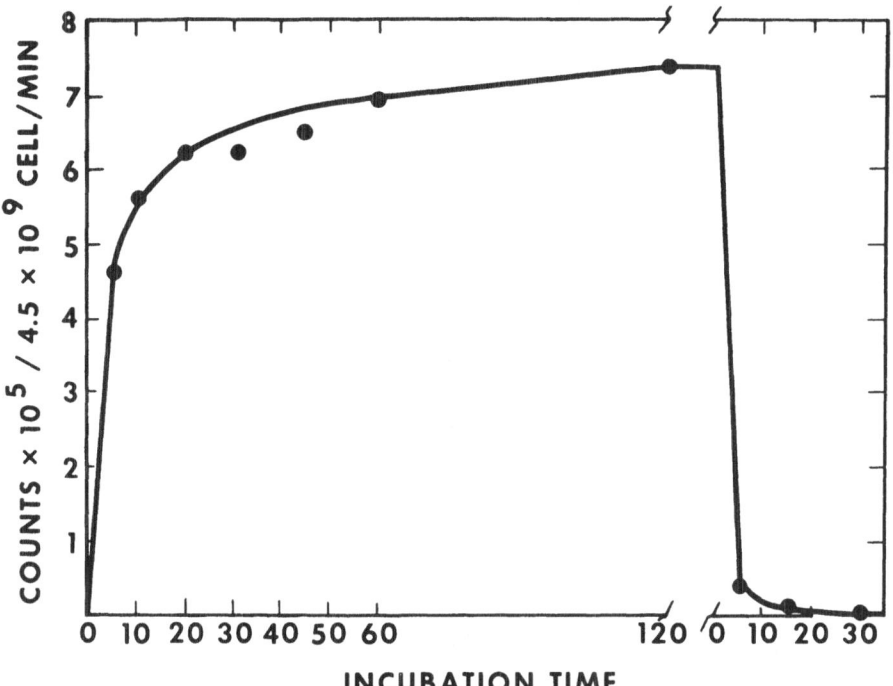

Fig. 2. Binding of ^{125}I-labeled Con A to sheep erythrocytes at 37°C followed by dissociation in the presence of 0.15 M α-methyl-D-mannoside.

These considerations form the basis of an approach to "nearest neighbor analysis" of surface membrane glycoproteins and glycolipids recently proposed by this laboratory (Leon et al., 1974). The approach requires the development of methods for removing the complexes of membrane components and Con A from the cell membrane without dissociating the Con A-carbohydrate complex. Analysis of such mixtures for the proportions of complexes of the type X-Con A-X, Y-Con A-Y, and X-Con A-Y would give information as to the frequency of finding molecules X and Y as neighbors when the cell membrane is in a particular state.

II. METHODS, RESULTS AND DISCUSSIONS

The problem of maintaining the integrity of the Con A-membrane component complexes during extraction from the membrane and subsequent analysis may be approached from two directions: The first approach, which will be presented here, involves increasing the stability of Con A-carbohydrate bonds. The second approach requires

(a) Con A + Erythrocyte (E) → Con A-E

(b) Anti-Con A + Con A-E → Anti-Con A-Con A-E

(c) Anti-Con A-Con A-E + Complement → Lysis

Fig. 3. Scheme for detection of Con A on erythrocytes by reaction with rabbit anti-Con A and subsequent lysis by guinea pig complement.

selection of methods which disrupt the membrane but do not affect the Con A-carbohydrate bond (Leon, 1974).

The model chosen for these initial studies was the sheep erythrocyte-Con A system. As shown in Fig. 2 Con A readily reacts with sheep erythrocytes. Addition of the competitive ligand α-methyl-D-mannoside (Goldstein et al., 1965) (Fig. 2) rapidly dissociates Con A from the erythrocyte membrane. The method chosen to stabilize the Con A-erythrocyte membrane complexes was addition of a specific cross-linking agent, namely, antibody to Con A. The sequence of reactions is shown in Fig. 3a and b.

As shown in Table I, a single extraction with α-methyl-D-mannoside removed approximately 60% of the Con A from the control erythrocytes. Addition of increasing amounts of anti-Con A to the cells, prior to extraction with α-methyl-D-mannoside, progressively stabilized the Con A-membrane component complexes. However, the amount of such complexes stabilized by anti-Con A appeared to reach a pla-

TABLE I

EFFECT OF ANTI-CON A ON DISSOCIATION OF CON A

Anti-Con A	Extract	Cells*
1/360	3,530	14,130
1/720	3,520	14,020
1/1440	4,030	13,760
1/2880	5,150	12,010
1/5760	8,280	8,940
Saline	10,710	6,820

*Sheep erythrocytes incubated with ^{125}I-Con A for 30 min at 37°C centrifuged, washed, extracted with 0.05 M α-methyl-D-mannoside in 0.125 M saline for 15 min at 37°C and recentrifuged.

TABLE II

EFFECT OF IONIC STRENGTH ON PASSIVE LYSIS OF SHEEP ERYTHROCYTES SENSITIZED WITH CON A AND ANTI-CON A*

Ionic Strength (M)	0.150	0.137	0.120	0.117	0.113	0.100	0.090	0.083
% Lysis	26	53	64	83	87	78	60	33

*Sheep erythrocytes 1×10^9 incubated with 5γ Con A 30 min at 37°C, centrifuged, washed, reincubated with anti-Con A 30 min at 37°C, centrifuged and washed.

teau corresponding to approximately 80% of the Con A. The 20% of the bound Con A which could still be extracted by mannoside might represent Con A which for spatial reasons was unavailable for reaction with anti-Con A. Alternatively, some Con A molecules on the membrane may have been cross-linked by antibody more extensively than others, due to appropriate exposure of multiple antigenic determinants on these Con A molecules. This consideration may apply particularly to reaction of anti-Con A with Con A tetramers as compared with reaction with any Con A dimers present.

To discriminate between Con A-membrane complexes unreacted with antibody and complexes reacted with antibody, the passive lysis system shown in (c) of Fig. 3 was utilized. Initial experiments demonstrated that (1) despite the fact that Con A reacts with components of complement (Leon, 1967) Con A coated erythrocytes do not hemolyze in the presence of complement unless sensitized with antibody; (2) the extent of passive lysis is proportional to concentration of Con A on the cells at non-agglutinating doses of Con A; (3) the extent of passive lysis is proportional to the concentration of anti-Con A added to the system over a restricted range of concentrations of anti-Con A; (4) the system is quite sensitive to the ionic strength of the medium with maximal lysis obtained at $\mu = 0.113$ (Table II).

Dilutions of anti-Con A were then incubated with constant amounts of Con A-erythrocyte complexes, centrifuged, washed and extracted with either α-methyl-D-mannoside or saline. Following the extraction, all the complexes were washed again with saline and the extent of lysis with complement was measured. As shown in Table III, for all dilutions of anti-Con A tested, the extent of lysis of cells extracted with mannoside was significantly lower than the extent of lysis of cells extracted with saline. This decrease in lysis indicates that some complement-fixing anti-Con A-Con A complexes can be removed from the membrane by extraction with mannoside while other such complexes are more resistant to extraction. The more

TABLE III

EFFECT OF α-METHYL-D-MANNOSIDE ON PASSIVE LYSIS
BY SENSITIZED SHEEP ERYTHROCYTES*

Anti-Con A Dilution	Extraction Buffer	
	α-methyl-D-mannoside	Saline
	% Lysis	% Lysis
1/50	12	71
1/75	12	71
1/100	12	68
1/200	7	57
1/400	4	53

*Sheep erythrocytes 1×10^9 incubated with 5 μg Con A 30 min at 37°C, centrifuged, washed and reincubated with various dilutions of anti-Con A for 30 min at 37°C. After centrifugation and washing the cells were extracted for 15 min at 37°C with 0.05 M α-methyl-D-mannoside in 0.12 M NaCl or with normal saline. After centrifugation and washing the cells were lysed with guinea pig complement.

resistant complexes are presumably those which contain Con A that has reacted with more than one antibody molecule, i.e., those complexes containing Con A extensively cross-linked by antibody.

An interesting possibility for exploration is the nature of the lesions produced by passive lysis in the anti-Con A-Con A system. Cell membranes with significant heterogeneity in Con A-binding sites might produce lesions of quite different characteristics depending on the nature of the Con A-binding site.

Con A stimulates resting lymphocytes to secrete a variety of biologically active materials called lymphokines (David, 1973) as well as to undergo blastogenesis (Douglas et al., 1969; Powell and Leon, 1970). Stimulation may be followed by measuring lymphokine production or incorporation of labeled precursors such as leucine, uridine or thymidine into protein, RNA or DNA respectively. Addition of anti-Con A one hour after incubation of Con A with human lymphocyte cultures suppresses incorporation of tritiated thymidine (Jones, 1973; Lindahl-Kiessling et al., 1973; Sell and Sheppard, 1974) as shown in Fig. 4. Later additions of anti-Con A show progressively less suppression (Table IV). However, despite the suppression of tritiated thymidine incorporation by anti-Con A the culture can readily be shown to be quite different in behavior from a virginal culture.

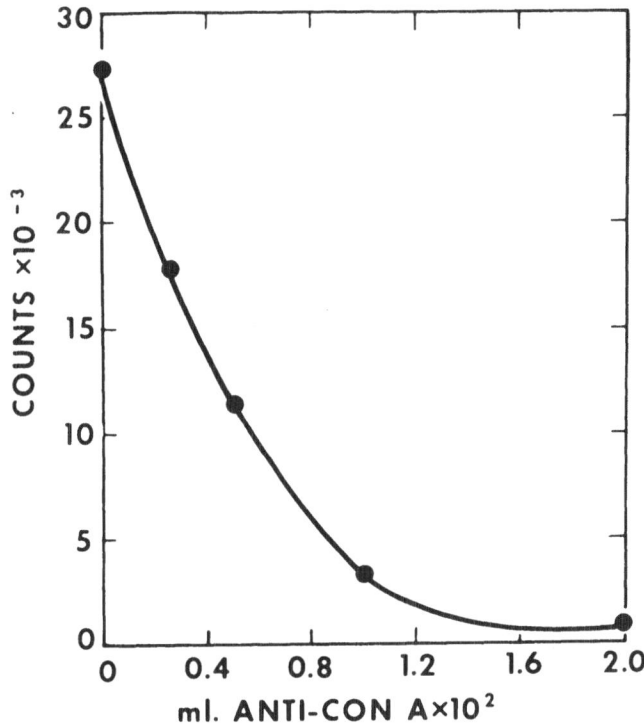

Fig. 4. Suppression of tritiated thymidine incorporation (at 72 hrs) by addition of rabbit anti-Con A to human lymphocytes 60 min after stimulating the lymphocytes with 10 µg Con A/1.5 x 10^6 cells.

For example, addition of a second mitogen such as the lentil phytohemagglutinin (LcH) to a suppressed culture of human lymphocytes results in levels of thymidine incorporation which far exceed that produced by addition of LcH alone (Table V). Furthermore the suppressed cultures, initiated with 0.2γ or 1γ of Con A, show higher levels of thymidine incorporation with LcH than do the cultures which did not receive anti-Con A. Attempts to demonstrate that the enhancement is due to lymphokine production by removing supernatants from suppressed cultures at various times and adding them to lymphocytes prior to stimulating with LcH have not been consistently successful. However, direct measurement of production of the lymphokine migration inhibitory factor (MIF) by Con A in the presence and absence of anti-Con A shows that MIF production is not suppressed (Table VI).

Clearly anti-Con A offers potential as a reagent for studying

TABLE IV

SUPPRESSION OF CON A-MEDIATED HUMAN LYMPHOCYTE STIMULATION

Reagent	Time	Thymidine Incorporation*
	Hrs	cpm
Anti-Con A	1	1,803
Anti-Con A	4	8,090
Anti-Con A	8	17,949
Anti-Con A	24	61,187
Normal Serum	1	53,754
Normal Serum	4	61,272
Normal Serum	8	56,848
Normal Serum	24	61,338

*Measured by pulsing cultures between 67 and 72 hrs.

TABLE V

EFFECT OF ANTI-CON A ON SEQUENTIAL STIMULATION* BY CON A AND LcH[a]

Initial Reactant	Subsequent Reactants			
	Saline	Anti-Con A[†]	LcH[‡]	Anti-Con A[†]+LcH[‡]
	cpm	cpm	cpm	cpm
Saline	579	514	6,213	5,653
Con A 0.2γ	577	380	23,623	30,250
Con A 1.0γ	4,314	866	26,759	36,236
Con A 5.0γ	55,210	744	69,724	42,170

*Measured by incorporation of tritiated thymidine between 67 and 72 hrs

[a]Lentil phytohemagglutinin (LcH)

[†]Added 1 hr after initial reactant

[‡]Added 2 hrs after initial reactant

TABLE VI

EFFECT OF ANTI-CON A ON MIF[a] PRODUCTION

Reagent	Migration Inhibition*
Con A	74%
Con A + Anti-Con A[†]	71%
Anti-Con A	6%

[a]Lymphokine migration inhibitory factor (MIF)

*Measured with BALB/c mouse peritoneal cells

[†]Added 1 hr after Con A

membrane structure and for dissecting the multiple biochemical reactions provoked by reaction of cells with Con A. In many instances it is more convenient to use anti-Con A than mannoside to study Con A-induced biochemical reactions, since mannoside may have effects on cellular biochemistry unrelated to effects of Con A. What applies to the Con A-anti-Con A system presumably also applies to other lectin-anti-lectin systems as initial studies with Lch-anti-Lch indicate (Stein et al., 1972).

REFERENCES

1. David, J. R. (1973). "Lymphocyte mediators and cellular hypersensitivity." *New Eng. J. Med.* **288**, 143.

2. Douglas, S. D., Kamin, R. M., Davis, W. C., and Fudenberg, H. H. (1969). In "Biochemical and morphologic aspects of phytomitogens: Jack Bean, wax bean, pokeweed, and phytohemagglutinin." *Proc. 3rd Leukocyte Culture Conf.*, W. O. Rieke, Ed., Appleton-Century-Crofts, N. Y., p. 607.

3. Goldstein, I. J., Hollerman, C. E., and Smith, E. E. (1965). "Protein-carbohydrate interaction. II. Inhibition studies on the interaction of concanavalin A with polysaccharides." *Biochemistry* **4**, 876.

4. Jones, G. (1973). "Lymphocyte activation. 3. The prolonged requirement for mitogen in phytohaemagglutinin and concanavalin A-stimulated cultures." *J. Immunol.* **110**, 1262.

5. Leon, M. A. (1967). "Concanavalin A reaction with human normal immunoglobulin G. and myeloma immunoglobulin G." *Science* **158**, 1325.

6. Leon, M. A., and Young, N. M. (1970). "Concanavalin A: a reagent for sensitization of erythrocytes with glycoproteins and polysaccharides." *J. Immunol.* **104**, 1556.

7. Leon, M. A. (1974). Unpublished observation.

8. Leon, M. A., Jocius, I. J., and Yokohari, R. *Ann. N. Y. Acad. Sci.* In press.

9. Lindahl-Kiessling, K., Mattsson, A., Skoog, V., and Weber, T. (1973). In *Proc. of the 7th Leukocyte Culture Con.*, F. Daguillard, Ed., Academic Press, N. Y.

10. Powell, A. E., and Leon, M. A. (1970). "Reversible interaction of human lymphocytes with the mitogen concanavalin A." *Exptl. Cell. Res.* **62**, 315.

11. Sell, S., and Sheppard, H. W., Jr. (1974). "Studies on rabbit lymphocytes *in vitro* XVIII. Kinetics of reversible Con A stimulation and restimulation of blast transformation after blocking with anti-Con A." *Exptl. Cell Res.* 84, 153.

12. Stein, M. D., Sage, H. J., and Leon, M. A. (1972). "Studies on a phytohemaglutinin from the common lentil. VI. Stimulation of human peripheral lymphocytes in culture by *Lens culinaris* hemagglutinin A." *Exptl. Cell Res.* 75, 475.

15

CONCANAVALIN A AS A PROBE FOR STUDYING THE MECHANISM OF METABOLIC STIMULATION OF LEUKOCYTES*

D. Romeo, G. Zabucchi, M. Jug, N. Miani,
M. R. Soranzo
University of Trieste
Trieste, Italy

ABSTRACT

The disruption of the molecular organization of the plasma membrane of leukocytes by phagocytosable particles, or by agents such as surfactants, antibodies, phospholipase C, fatty acids and chemotactic factors, leads to a stimulation of the phagocyte oxidative metabolism. Concanavalin A (Con A) has been used as a tool to study the mechanism of this metabolic regulation.

The binding of Con A to the surface of polymorphonuclear leukocytes (PMNL) or macrophages produces a rapid enhancement of oxygen uptake and glucose oxidation through the hexose monophosphate pathway (HMP). This is explained by an activation of the granular NADPH oxidase, the key enzyme in the metabolic stimulation. The effect of Con A is not due to endocytosed lectin, since Con A covalently coupled to large sepharose beads still acts as stimulant.

The metabolic changes caused by Con A are reversible. If, after the onset of stimulation, sugars with high affinity for Con A are added to the leukocyte suspension, the activity of granular NADPH oxidase and the rate of respiration and glucose oxidation return to their resting values.

The metabolic burst, while partially supressed by treatment of PMNL with iodoacetate, sodium fluoride and cytochalasin B, is slightly increased by colchicine.

---

*This work was aided by a grant from the Italian National Research Council (No. 73.00441.04).

Con A induces a selective release of granular enzymes (β-glucuronidase, peroxidase, alkaline phosphatase) from PMNL, whereas no leakage of cytoplasmic enzymes is observed. The enzyme release is inhibited by iodoacetate and by drugs known to increase cell levels of cyclic AMP.

Based on a current view of the mode of interaction between Con A and cell surfaces, a model of the metabolic disruption of leukocytes is presented.

I. INTRODUCTION

The process of phagocytosis is initiated with the attachment of a bacterium or other particulate matter to the surface of the cell, followed by a complex series of events in which the microorganism is surrounded by the plasma membrane, sequestered into a vacuole and eventually digested. In the mononuclear and in the polymorphonuclear leukocyte (PMNL) (Rossi et al., 1972; Sbarra and Karnovsky, 1959) this process is accompanied by a number of metabolic modifications. In particular, there is an enhancement of oxygen uptake, of H_2O_2 production and of glucose oxidation via the hexose monophosphate pathway (HMP). Stimulation of the oxidative metabolism is independent of particle ingestion (Selvaraj et al., 1967; Simberkoff and Elsbach, 1971), requires the intactness of the leukocyte membrane (Selvaraj and Sbarra, 1966) and starts a few seconds after the addition of particles to the phagocytes (Rossi and Patriarca, 1972; Patriarca et al., 1971). Thus it seems to be timed by the stage of collisions or adhesion of the particle to the cell surface and it is not related to post-engulfment events.

In 1968 Rossi and Zatti suggested that the triggering mechanism for the stimulation of the oxidative metabolism should be provided by an early and rapid change of the molecular organization of the plasma membrane. This hypothesis is substantiated by the discovery that a disruption of the leukocyte surface membrane with detergents, anti-leukocyte antibodies, phospholipase C, fatty acids, chemotactic factors or non-phagocytosable immune complexes (Rossi et al., 1968) mimics the metabolic stimulation observed during phagocytosis. Further, by using a fluorescent probe of membrane structure we have shown an apparent conformational transition of plasma membrane components immediately following the interaction between polystyrene beads and PMNL (Romeo et al., 1970).

In this view, the phagocytosis-induced metabolic disturbance provides another example of the capacity exhibited by the plasma membrane in efficiently conveying signals from the cell environment to the cytoplasm. This transducing capacity is observed in embryonic development and organization (Curtis, 1967), social behavior of neoplastic cells (Aaronson and Todaro, 1968; Burger; 1970), fertilization

(Nazakawa et al., 1970), lymphocyte stimualtion by antigens or mitogens (Greaves and Bauminger, 1972; Ling, 1968) and regulation of cell metabolism by a variety of hormones (Cuatrecasas, 1969; Hecht et al., 1972; Pastan et al., 1966).

The nature of the alteration of membrane properties which leads to the metabolic stimulation of phagocytes is still obscure. In search for an agent exhibiting a rather well understood mode of interaction with the surface of cells, we have decided to study the effects of Concanavalin A (Con A) on the metabolism of polymorphonuclear and mononuclear phagocytes. These cells have been shown to have Con A binding sites on their surface membrane (Berlin, 1972; Smith and Hollers, 1970). As already reported (Romeo, 1974; Romeo et al., 1973, 1974a,b), leukocytes exposed to either immobilized or soluble Con A become activated and also show an increased respiration and $1-^{14}C$-glucose oxidation. This increase is accounted for by an activation of a granular NADPH oxidizing system, which is responsible for the stimulation of oxidative metabolism concomitant with phagocytosis (Patriarca et al, 1971; Romeo, 1974; Romeo et al., 1973a,b; Rossi et al., 1972, 1974). Con A induces also a selective release of granular enzymes from PMNL, less pronounced but similar to that promoted by phagocytosis.

II. MATERIALS AND METHODS

A. Cells

Guinea-pig PMNL were isolated from peritoneal exudates induced with saline (Rossi and Zatti, 1964). Contaminating erythrocytes were lysed by a short treatment of cell pellets with 0.2% NaCl. Cells were suspended in calcium-free Krebs-Ringer phosphate buffer, pH 7.4 (KRP).

B. Concanavalin A

Con A (Sigma grade III) was conjugated with fluorescein isothiocyanate by either the method of Smith and Hollers (1970) or of Rinderknecht (1962). Coupling of the lectin to sepharose 2B (Pharmacia) was performed essentially according to Porath et al. (1967). Fluorescein-Con A was included in the mixture and elution of unbound lectin was monitored by fluorometry. Before any experimentation, leakage of Con A from the beads was checked by the same technique. α-methyl-D-mannopyranoside (α-MM) was obtained from Koch-Light and lactose from Fluka; mannan was isolated from baker's yeast and extensively dialysed against isotonic NaCl.

C. Oxidative Metabolism

Oxygen uptake was measured with a Clark-type oxygen electrode and glucose oxidation by counting $^{14}CO_2$ produced from labelled glucose, as previously reported (Patriarca et al., 1971; Romeo et al., 1973). In a few experiments, beef liver catalase (Sigma) was added to respiring cells in the electrode vessel in order to demonstrate an accumulation of H_2O_2.

D. Enzyme Activities

Granules sedimenting at 20,000 g were isolated from PMNL incubated for 5-10 min at 37°C in the absence or in the presence of Con A (25-40 µg/ml) as previously reported (Patriarca et al., 1971). The rate of NADPH oxidation by these granules was assayed in the presence of 1 mM KCN. To measure enzyme release, PMNL were incubated for 15 min at 37°C in the absence or in the presence of Con A (50-75 µg/ml). Cells, separated from media at 800 g for 10 min, were disrupted by sonication. Peroxidase activity was measured spectrophotometrically at 37°C by recording the peroxidation of guaiacol (Romeo et al., 1973) (1 unit = 1 µmole tetraguaiacol/min). β-Glucuronidase and alkaline phosphatase activities were assayed at 37°C by following the enzymatic splitting of phenolphthalein-β-glucuronide (Gianetto and de Duve, 1955) (1 unit = 1 µg phenolphthalein/hr) and p-nitrophenyl phosphate (Michell et al., 1970) (1 unit = 1 µmole p-nitrophenol/10 min), respectively. Lactate dehydrogenase and glucose-6-phosphate dehydrogenase were assayed according to Bergmayer et al. (1965) and according to Kornberg and Horecker (1955), respectively (1 unit = 1 µmole NAD^+ or $NADP^+$/ min).

E. Analytical Procedures

Adenosine 3':5'-monophosphate (c-AMP) was determined by the method of Fisch et al. (1972), after extraction of the nucleotide from PMNL according to the microtechnique of Cooper et al. (1973).

III. RESULTS

A. Oxidative Metabolism

1. *Immobilized Con A.* When sepharose 2B (average diameter 190 µm) covalently coupled to Con A is added to a PMNL suspension, cells become immediately bound to the non-phagocytosable beads and are activated as shown by increased $1-^{14}C$-glucose oxidation (Fig. 1). The simultaneous presence of a sugar (α-MM, 100 mM) with high affinity for Con A (Goldstein et al., 1965) prevents both binding and metabolic stimulation of leukocytes. Sepharose activated with CNBr but not coupled to the lectin fails to bind cells and to exhibit any stimulatory effect.

2. *Soluble Con A.* When treated with soluble Con A, PMNL form small aggregates. Fluorescein-tagged Con A is particularly concen-

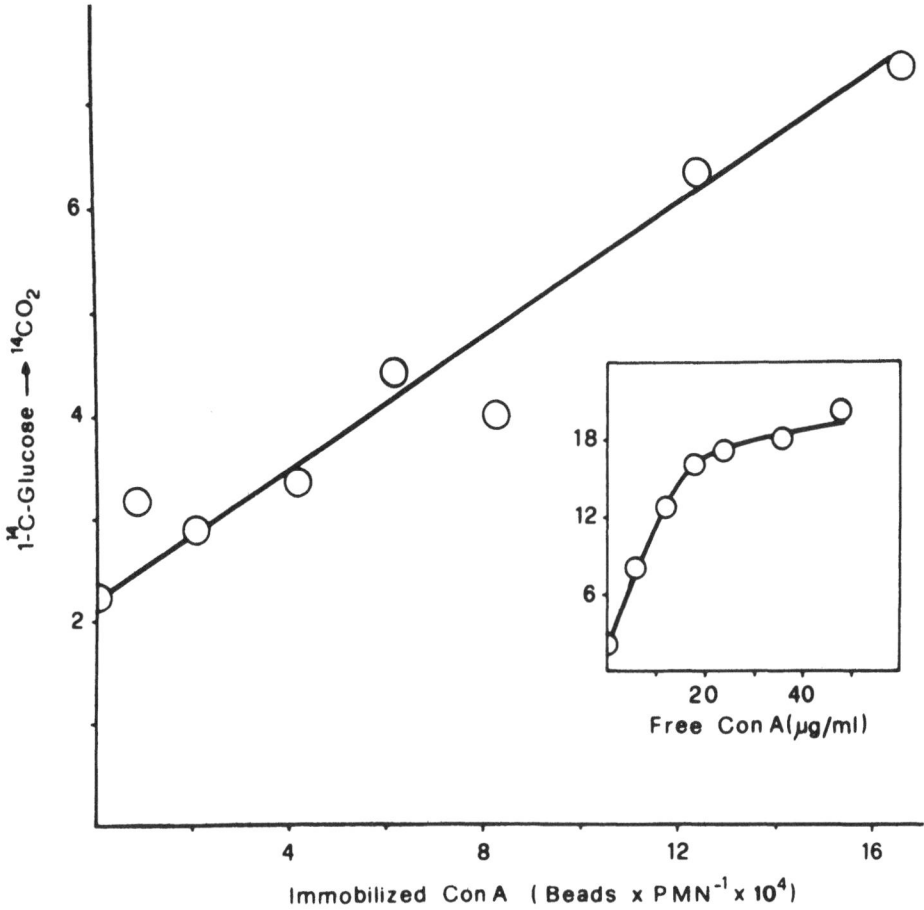

Fig. 1. Stimulation of glucose oxidation in PMNL by non-phagocytosable sepharose-bound Con A or by free Con A.

Ordinate: Conversion of labelled glucose to labelled CO_2 ($^{14}CO_2$ cpm $\times 10^{-4}$/µCi/10 min/1 $\times 10^7$ cells)

trated in the areas of contact between cells. In isolated cells the labelled lectin is almost exclusively found at the cell surface (10 min incubation at 37°C, 30 µg Con A/ml).

As shown in the insert of Fig. 1, maximum stimulation of 1-^{14}C-glucose oxidation by Con A is reached at about 25 µg/ml. The increase in $^{14}CO_2$ production is insensitive to KCN (1 mM) and antimycin A (2.5 µg/ml), thus excluding an activation of mitochondrial metabolism while strongly indicating a specific activation of the HMP pathway (Rossi et al., 1972).

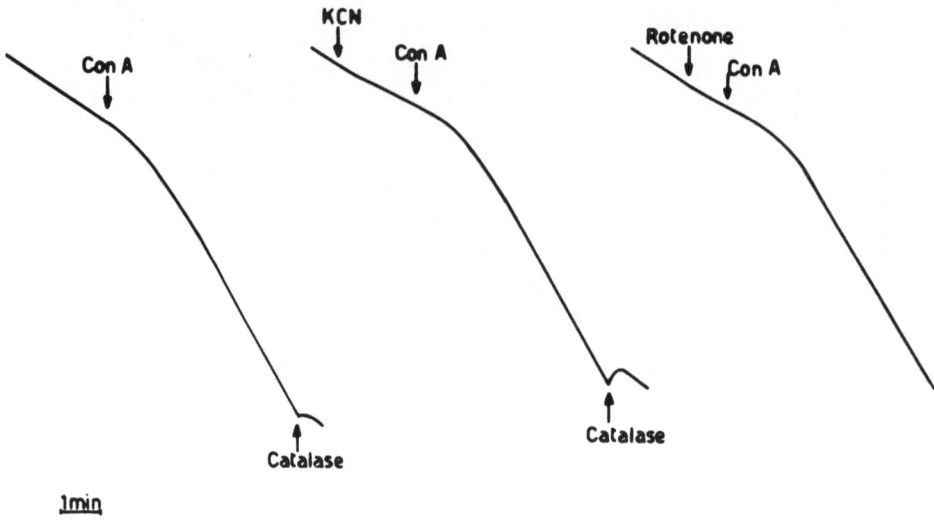

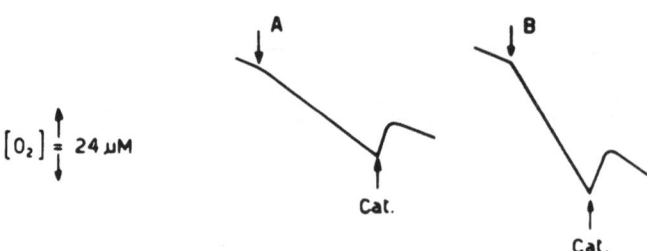

Fig. 2. Stimulatory effect of Con A on respiration of PMNL and on the related NADPH oxidation by granules.

Graphs shown are polarographic tracings of oxygen consumption (top) by whole cells (1.5×10^7) and by granules oxidizing NADPH (bottom). A = granules from 4×10^6 resting cells; B = granules from 4×10^6 cells treated with Con A. Con A = 30 µg/ml; KCN = 1 mM; rotenone = 5 µg/ml. Oxidation of NADPH as described by Patriarca et al. (1971). Exogenous liver catalase added to show production of H_2O_2.

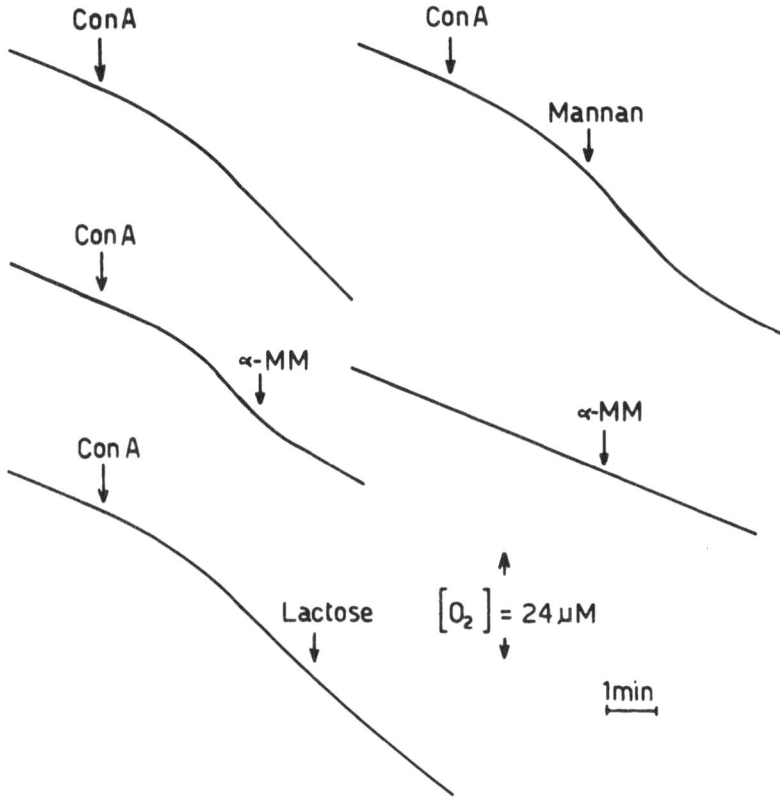

Fig. 3. Reversibility of Con A-induced respiratory stimulation in PMNL by specific sugars.

PMNL = 2×10^7; Con A = 35 µg/ml; α-MM = 50 mM; lactose = 50mM; yeast mannan corresponding to 12 mM mannose.

The kinetics of metabolic stimulation of PMNL by Con A is very fast. As shown in Fig. 2, a few seconds after the addition of the lectin to the cell suspension the rate of respiration is already enhanced. The burst of oxygen uptake is not dependent on mitochondrial activity, since it is insensitive to KCN and rotenone. Respiring cells release a small amount of H_2O_2, as is indicated by the oxygen produced upon addition of an excess of exogenous catalase. In the presence of endogenous peroxidase and catalase-inhibiting cyanide, a greater amount of H_2O_2 is released from and very likely accumulated in the cell.

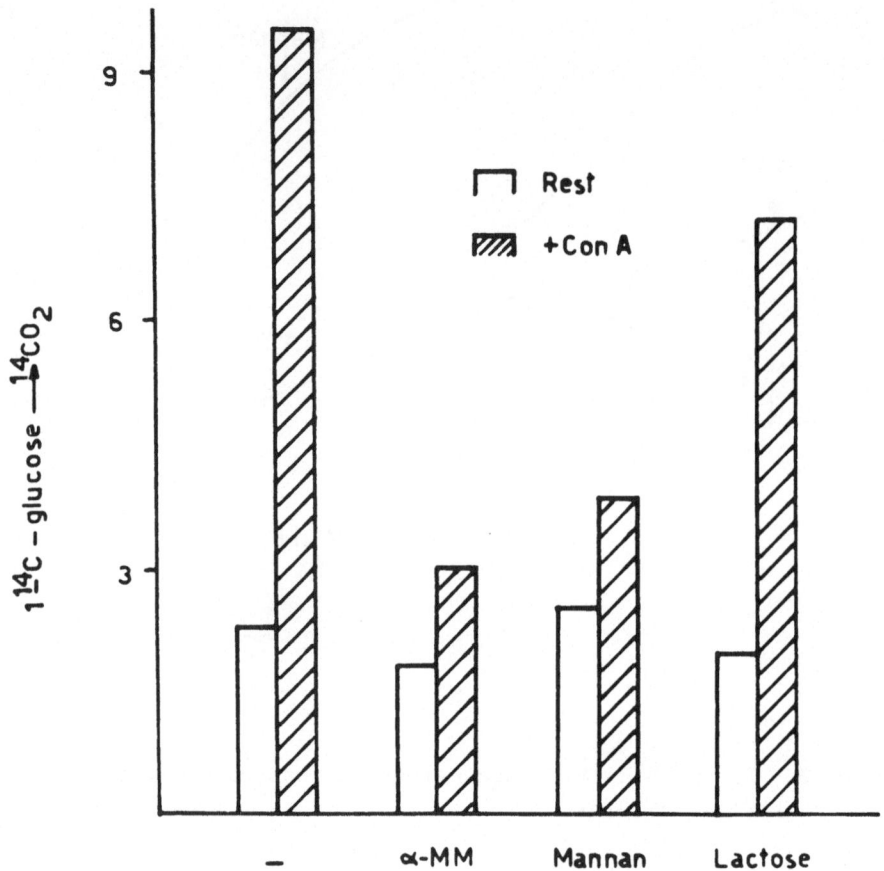

Fig. 4. Reversibility of Con A-induced stimulation of glucose oxidation in PMNL by specific sugars.

1 x 10[7] PMNL were incubated in the absence or in the presence of Con A (15 μg/ml); after 10 min, either α-MM (60 mM) or yeast mannan (corresponding to 25 mM mannose) or lactose (50 mM) were added to both resting and treated cells, followed by 1-^{14}C-glucose. The $^{14}CO_2$ produced in 10 min incubation was trapped in KOH and counted by scintillation spectroscopy (Romeo et al., 1973a).

Ordinate: $^{14}CO_2$ cpm x 10^{-4}/μCi/10 min/1 x 10^7 cells.

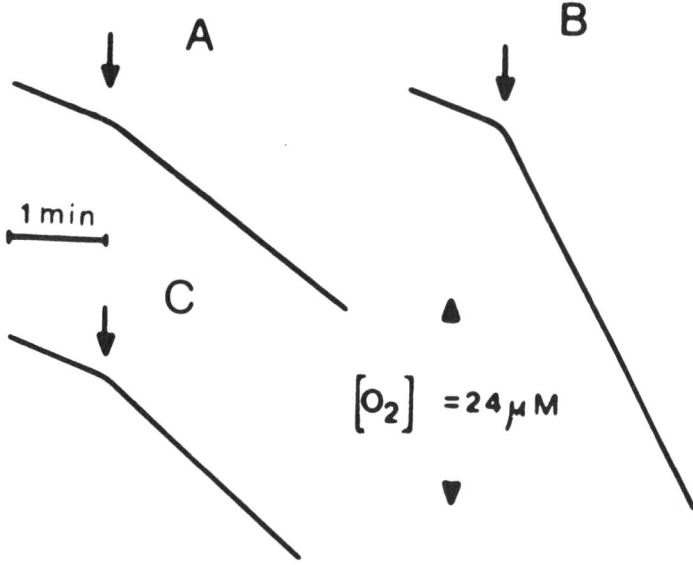

Fig. 5. Reversibility of activation of the NADPH oxidizing system. Rate of oxidation of NADPH assayed with granules of 4×10^6 PMNL (Patriarca et al., 1971). A = resting cells; B = cells treated with Con A (35 µg/ml, 6 min); C = cells treated with Con A, and then with α-MM (60 mM, 5 min).

The treatment of intact PMNL with Con A induces an activation of KCN-insensitive oxidation of NADPH by isolated granules (Fig. 2). The oxidation of this reduced nucleotide is accompanied by an almost stoichiometric formation of H_2O_2. If isolated granules are exposed to Con A, there is no change in the rate of NADPH oxidation. A similar pattern of stimulation, as described in this section, is observed with macrophages (Romeo et al., 1973, 1974).

3. <u>Effect of various drugs on the metabolic stimulation.</u> Preincubation of PMNL with 20 mM NaF or 2.5×10^{-4} M iodoacetate exerts a 30% and 70% suppression of the respiratory burst respectively. Cytochalasin B (5-10 µg/ml) has a slight inhibitory effect (about 30%). Colchicine, at a concentration of 10^{-6} to 10^{-5} M causes a further increase (30-60%) in the increment of both oxygen uptake and glucose oxidation.

4. <u>Reversibility.</u> Con A can be removed from its binding sites on cell surfaces by competition with specific sugars (Goldstein et

al., 1965). As shown in Fig. 3, if carbohydrates with high affinity for Con A, such as α-MM or mannan, are added to stimulated cells, within 1-2 min the rate of PMNL respiration is damped down to a preburst value. No effect is seen with lactose which binds poorly to Con A. Respiration of resting cells is not affected by the addition of the carbohydrates.

A post-treatment of PMNL with α-MM or mannan following exposure to Con A reduces the increment in $1-^{14}C$-glucose oxidation by about 85% (Fig. 4). On the other hand, the effect of lactose is very small. As determined in parallel experiments, the effect of α-MM on the rate of glucose oxidation is not due to inhibition of glucose-6-phosphate dehydrogenase.

If PMNL are incubated first with Con A alone and then supplemented with competing sugars, such as α-MM, the rate of NADPH oxidation by granules isolated from them is comparable with that of granules of resting cells (Fig. 5). Incubation of "activated" granules with α-MM does not affect the rate of NADPH oxidation. Thus, reversibility is achieved only when specific sugars act on intact cells.

B. Enzyme Release

Extracellular release of lysosomal enzymes from PMNL incubated with Con A was monitored by assaying the enzyme activities remaining in the cell pellet after removal of the media. This procedure was adopted because these enzymes are labile in KRP at 37°C and appreciable amounts of their activity are not recovered from the medium (Holmes *et al.*, 1969; Wright and Malawista, 1972). The presence of Con A further decreases their stability.

As shown in Table I, cytoplasmic enzymes, such as lactate dehydrogenase and glucose-6-phosphate dehydrogenase, do not leak from PMNL incubated with Con A. Conversely, the lectin induces a small but significant decrease of activity of granular enzymes, such as peroxidase, β-glucuronidase and alkaline phosphatase. This decrease is 30-60% of that observed in parallel experiments during phagocytosis of *B. mycoides* (bacteria: cell ratio = 150:1).

The decrease of β-glucuronidase and peroxidase activity of PMNL treated with Con A is reduced by various drugs known to prevent the release of granular enzymes from leukocytes (Henson, 1972; Zurier *et al.*, 1973). The addition to the cells 20 min before Con A of 5×10^{-5} M iodoacetate, 4×10^{-3} M PGE_1, 2×10^{-3} M aminophylline, or 5×10^{-4} M aminophylline + 5×10^{-4} M dbc-AMP inhibits the loss of activity of lysosomal enzymes by 40-100%.

If the treatment with Con A is performed under an atmosphere of nitrogen, there is no significant change in the release of β-glucu-

TABLE I

ACTIVITY OF CYTOPLASMIC AND GRANULAR ENZYMES IN PMN
LEUKOCYTES AFTER INCUBATION WITH CONCANAVALIN A

		Resting cells	Cells + Con A		Total Activity
Lactate DH	(5)	1.01 ±0.22	1.05 ±0.26	$P>0.5$	1.10 ±0.25
Glc-6-P DH	(3)	0.29 ±0.04	0.30 ±0.04	$P>0.2$	0.30 ±0.04
Peroxidase	(10)	1.03 ±0.17	0.92 ±0.17	$P<0.005$	1.07 ±0.18
β-Glucuronidase	(14)	23.1 ± 1.9	20.8 ±2.0	$P<0.005$	24.3 ±2.1
Alkaline phosphatase	(11)	1.27 ± 0.19	1.17 ±0.18	$P<0.025$	1.34 ±0.21

Mean values (enzyme units/10^7 cells) ± S.E.; number of experiments is shown in parentheses; Student's "t" test made on paired data.

Guinea pig PMNL in KRP (about 2×10^7 cells/ml) were incubated at 37°C for 15 min without or with Con A (50-75 µg/ml); cells were separated from media by centrifugation at 800 g for 10 min and lysed by sonication.

ronidase or peroxidase from PMNL. This suggests that in Con A-stimulated cells the loss of activity of lysosomal enzymes is not caused by the enhanced generation of oxidizing compounds (e.g., H_2O_2).

Con A covalently coupled to sepharose is also able to induce a decrease of β-glucuronidase and peroxidase content in PMNL.

C. Con A and c-AMP

Two, 5 and 10 min after the addition of Con A to PMNL there is no change in the cell levels of c-AMP.

IV. DISCUSSION

This paper shows that Con A is able to mimic phagocytosis-associated biochemical events, such as stimulation of the oxidative metabolism and the selective release of granular enzymes into extracellular fluid. We assume that the metabolic effects observed are mediated by lectin bound to the cell surface and not to Con A which has been pinocytosed (Oliver et al., 1974) or possibly discharged into the cytoplasm from endocytic vacuoles. Evidence supporting this assumption is as follows: (1) When, very shortly after the addition of Con A to PMNL, maximal activation of cell respiration is observed, the fluorescein-tagged lectin is found at the cell surface. (2) The metabolic stimulation is also triggered by Con A covalently coupled to non-phagocytosable beads. (3) The enhancement of oxidative metabolism is reversed by carbohydrates competing with cell surface binding sites for Con A. One could argue that Con A might exert its effects also from within the endocytic vacuoles derived from invaginated plasma membrane. However, among the sugars which are rapidly reversing the metabolic activation there is also mannan, which, even if endocytosed, would be sequestered in vesicles different from those containing Con A. (4) Treatment of isolated granules with Con A fails to activate the NADPH oxidizing system.

The indication that Con A acts on the cell surface to produce specific biochemical changes in phagocytes may be utilized to formulate a three-element system: (1) a Con A receptor; (b) a transduction mechanism, which conveys a signal; (3) a biochemical target which is affected. The receptor consists very likely of glycoprotein subunits containing internal $2\text{-}O\text{-}$substituted-α-D-mannopyranosyl units (Goldstein, 1975). One of the biochemical targets is represented by the NADPH oxidizing system, which is localized in the primary granules (Patriarca et al., 1973). The transduction or amplification system should be thought of in terms of current models of membrane structure. The most widely accepted model (Singer and Nicolson, 1972) postulates that the membrane matrix is made up of phospholipids arranged in a bilayer into which protein subunits are intercalated. Both lipids and proteins undergo a random lateral diffusion in the plane of the membrane. The topographical distribution of the glycoprotein subunits can, however, be disturbed by various agents. Smith and Revel (1972) have shown that Con A-receptor complexes appear in patches on the surface of PMNL. This association of mobile protein subunits might lead to the elaboration and transmission of the signal necessary for the activation of the NADPH oxidizing system. Further, it might promote the fusion of a few granules with an area situated just beneath or adjacent to the patch on the cytoplasmic face of the membrane, followed by discharge of the granule content into the extracellular fluid. It is unlikely that microfilaments mediate the respiratory burst induced by Con A in PMNL, since cytochalasin B exhibits only a slight inhibitory effect on this process. Besides interacting with the peripheral cytoplasmic filamentous network, this drug also affects plasma membrane functions (Zigmond and Hirsch, 1972). The potentiation of the metabolic burst by colchicine is consistent with the assumption that this drug abolishes the attachment of membrane proteins to cytoplasmic structures (perhaps microtubules), thus fostering their clustering by exogenous cross-linking agents (Berlin et al., 1974).

A similar association of mobile glycoprotein subunits might be induced by the other agents which promote a metabolic stimulation of PMNL and macrophages. Antibodies directed against surface antigens are known to cause "patching" and "capping" of membrane components in lymphocytes (Taylor et al., 1971). Treatment of erythrocyte ghosts with phospholipase C causes an intense aggregation of negatively charged intrinsic membrane molecules (Nicolson, 1973). Negatively or positively charged particles approaching the phagocyte surface might also favor the association of the prominently negatively charged membrane glycoproteins by repulsion or attraction respectively. That an "on-off" switch, presumably represented by the glycoprotein association-dissociation system, might operate at the level of the plasma membrane of leukocytes, is also suggested by the reversibility of the Con A-mediated metabolic stimulation. Specific carbohydrates, such as α-MM or mannan, by virtue of their high affinity for Con A, displace the cross-linking lectin from its cell binding sites, thus permitting the mobile glycoprotein units to resume their random distribution.

Our experiments have not yet indicated the nature of the signal generated during these modifications of molecular organization of the plasma membrane. Cyclic AMP does not seem to operate as a second messenger since its concentration does not change in phagocytes upon stimulation. The rapid increase in fluorescence intensity of 1-anilino-8-naphthalene sulphonate (ANS), used as a label of the PMNL surface, following a challenge with polystyrene beads (Romeo et al., 1970) suggests that in the very early stage of phagocytosis there might be a redistribution of charges across the plasma membrane. Variations in ANS fluorescence have been reported to reflect variations in the distribution of membrane charges (Azzi, 1969; Fortes and Hoffman, 1971) and in ion association with membranes (Gomperts et al., 1970). Indeed, as suggested by Gingell (1973), the aggregation of charged mobile protein subunits should lead to a local increase in membrane permeability to ions. We are presently investigating the possibility that the metabolic stimulation of phagocytes is preceded by rapid fluxes of ions across the membrane, in an analogous manner to the generation of potentials in excitable membranes.

REFERENCES

1. Aaronson, S. A., and Todaro, G. J. (1968). "Basis for the acquisition of malignant potential by mouse cells cultivated in vitro." Science 162, 1024.

2. Azzi, A. (1969). "Redistribution of the electrical charge of

the mitochondrial membrane during energy conservation." *Biochem. Biophys. Res. Commun.* 37, 254.

3. Bergmayer, H. U., Bernt, E., and Hesso, B. (1965). "Lactic dehydrogenase." In *Methods of enzymatic analysis* (H. U. Bergmayer, ed.), Academic Press, New York.

4. Berlin, R. D. (1972). "Effect of Concanavalin A on phagocytosis." *Nature New Biology* 235, 44.

5. Berlin, R. D., Oliver, J. M., Ukena, T. E., and Yin, H. H. (1974). "Control of cell surface topography." *Nature New Biology* 247, 45.

6. Burger, M. M. (1970). "A difference in the architecture of the surface membrane of normal and virally transformed cells." *Proc. Nat. Acad. Sci. U. S. A.* 62, 994.

7. Cooper, R. H., Ashcroft, S. J., and Randle, P. J. (1973). "Concentrations of adenosine 3':5' - cyclic monophosphate in mouse pancreatic islets measured by a protein-binding radioassay." *Biochem. J.* 134, 599.

8. Cuatrecasas, P. (1969). "Interaction of insulin with the cell membrane: the primary action of insulin." *Proc. Nat. Acad. Sci. U. S. A.* 63, 450.

9. Curtis, A. S. G. (1967). *The cell surface: its molecular role in morphogenesis.* Pergamon Press, Oxford.

10. Fisch, H. U., Pliška, U., and Schwyzer, R. (1972). "A covalent protein-Sepharose complex for the specific adsorption and assay of adenosine 3':5' - monophosphate." *Eur. J. Biochem.* 30, 1.

11. Fortes, P. A. G., and Hoffman, J. F. (1971). "Interactions of the fluorescent anion 1-anilino-8-naphthalene sulfonate with membrane charges in human red cell ghosts." *J. Membrane Biol.* 5, 154.

12. Gianetto, R., and de Duve, C. (1955). "Comparative study of the binding of acid phosphatase, β-glucuronidase and cathespin by rat-liver particles." *Biochem. J.* 59, 433.

13. Gingell, D. (1973). "Membrane permeability change by aggregation of mobile glycoprotein units." *J. Theor. Biol.* 38, 677.

14. Goldstein, I. J. (1975). Personal communication.

15. Goldstein, I. J., Hollerman, C. E., and Smith, E. E. (1965). "Protein-carbohydrate interaction. II. Inhibition studies on

the interactions of Concanavalin A with polysaccharides." *Biochemistry* 4, 876.

16. Gomperts, B., Lanteline, F., and Stock, R. (1970). "Ion association reactions with biological membranes studied with the fluorescent dye 1-anilino-8-naphthalene sulfonate." *J. Membrane Biol.* 3, 241.

17. Greaves, M. F., and Bauminger, S. (1972). "Activation of T and B cells by insoluble phytomitogens." *Nature New Biol.* 235, 67.

18. Hecht, Y. P., Dellacha, J. M., Santome, J. A., Paladini, A. C., Hurwitz, E., and Sela, M. (1972). Personal Communication.

19. Henson, P. M. (1972). "Pathologic mechanisms in neutrophil-mediated injury." *Am. J. Pathol.* 68, 593.

20. Holmes, B., Sater, J., Rodey, G., Park, B., and Good, R. (1969). "Changes in intracellular distribution of lysosomal enzymes during phagocytosis by human polymorphonuclear leukocytes." *Clin. Invest.* 48, 39a, 125 abs.

21. Kornberg, A., and Horecker, B. L. (1955). "Glucose-6-phosphate dehydrogenase." In *Methods in enzymology* (S. P. Colowick and N. O. Kaplan, eds.) vol. I.

22. Ling, N. R. (1968). *Lymphocyte stimulation*. North Holland Publishing Co., Amsterdam.

23. Michell, R. H., Karnovsky, M. J., and Karnowsky, M. L. (1970). "The distribution of some granule-associated enzymes in guinea-pig polymorphonuclear leukocytes." *Biochem. J.* 116, 207.

24. Nazakawa, T., Asani, K., Shoger, R., Fujiwara, A., and Yasumasu, I. (1970). "Ca^{++} uptake, H^+ ejection and respiration in sea-urchin eggs on fertilization." *Exp. Cell Res.* 63, 143.

25. Nicolson, G. L. (1973). "Anionic sites of human erythrocyte membranes. I. Effects of trypsin, phospholipase C, and pH on the topography of bound positively charged colloidal particles." *J. Cell Biol.* 57, 373.

26. Oliver, J. M., Ukena, T. E., and Berlin, R. D. (1974). "Effects of phagocytosis and colchicine on the distribution of lectin-binding sites on cell surfaces." *Proc. Nat. Acad. Sci. U. S. A.* 71, 394.

27. Pastan, I., Roth, J., and Macchia, V. (1966). "Binding of hormone to tissue: the first step in polypeptide hormone action." *Proc. Nat. Acad. Sci. U. S. A.* 56, 1802.

28. Patriarca, P., Cramer, R., Moncalvo, S., Rossi, F., and Romeo, D. (1971). "Enzymatic basis of metabolic stimulation in leukocytes during phagocytosis: the role of activated NADPH oxidase." *Arch. Biochem. Biophys.* 145, 255.

29. Patriarca, P., Cramer, R., Dri, P., Fant, L., Basford, R. E., and Rossi, F. (1973). "NADPH oxidizing activity in rabbit polymorphonuclear leukocytes: localization in azurophilic granules." *Biochem. Biophys. Res. Commun.* 53, 830.

30. Porath, J., Axen, R., and Ernback, S. (1967). "Chemical coupling of proteins to Agarose." *Nature New Biology* 215, 1491.

31. Rinderknecht, H. (1962). "Ultra-rapid fluorescent labelling of proteins." *Nature New Biology* 193, 167.

32. Romeo, D. (1974). "Modulation of phagocyte metabolism by perturbation of their surface with Concanavalin A." In *Comparative biochemistry and physiology of transport* (K. Bloch, L. Bolis, and S. E. Luria, eds.), in press.

33. Romeo, D., Cramer, R., and Rossi, F. (1970). "Use of 1-anilino-8-naphthalene sulfonate to study structural transitions in cell membrane of PMN leukocytes." *Biochem. Biophys. Res. Commun.* 41, 582.

34. Romeo, D., Zabucchi, G., and Rossi, F. (1973a). "Reversible metabolic stimulation of polymorphonuclear leukocytes and macrophages by Concanavalin A." *Nature New Biology* 243, 111.

35. Romeo, D., Zabucchi, G., Marzi, T., and Rossi, F. (1973a). "Kinetic and enzymatic features of metabolic stimulation of alveolar and peritoneal macrophages challenged with bacteria." *Exp. Cell Res.* 78, 423.

36. Romeo, D., Cramer, R., Marzi, T., Soranzo, M. R., Zabucchi, G., and Rossi, F. (1973). "Peroxidase activity of alveolar and peritoneal macrophages." *J. Reticuloend. Soc.* 13, 399.

37. Romeo, D., Zabucchi, G., Jug, M., and Rossi, F. (1974a). "Alteration of macrophage surface in the course of immunological activation: decay of metabolic response to Concanavalin A." *Cell. Immunol.*, in press.

38. Romeo, D., Jug, J., Zabucchi, G., and Rossi, F. (1974b). "Perturbation of leukocytes metabolism by nonphagocytosable Concanavalin A-coupled beads." *FEBS Letters*, in press.

39. Rossi, F., and Zatti, M. (1964). "Changes in the metabolic pattern of polymorphonuclear leukocytes during phagocytosis."

Brit. J. Exp. Pathol. 45, 548.

40. Rossi, F., and Zatti, M. (1968). "Mechanism of the respiratory stimulation in saponine treated leukocytes. The KCN insensitive oxidation of NADPH." Biochim. Biophys. Acta 153, 296.

41. Rossi, F., Romeo, D., and Patriarca, P. (1972). "Mechanism of phagocytosis associated oxidative metabolism in polymorphonuclear leukocytes and macrophages." RES: J. Reticuloend. Soc. 12, 127.

42. Rossi, F., Patriarca, P., and Romeo, D. (1974). "Regulation of oxidative metabolism and functions of phagocytes." In Future trends in inflammation (D. A. Willoughby, ed.)

43. Sbarra, A. J., and Karnovsky, M. L. (1959). "The biochemical basis of phagocytosis." J. Biol. Chem. 237, 1355.

44. Selvaraj, R. J., and Sbarra, A. J. (1966). "Relationship of glycolytic and oxidative metabolism to particle entry and destruction in phagocytosing cells." Nature New Biology 211, 1272.

45. Selvaraj, R. J., McRipley, R. J., and Sbarra, A. J. (1967). "The metabolic activities of leukocytes from limphoproliferative and myeloproliferative disorders during phagocytosis." Cancer Res. 27, 2287.

46. Simberkoff, M. S., and Elsback, P. (1971). "The interaction in vitro between polymorphonuclear leukocytes and mycoplasma." J. Exp. Med. 134, 1417.

47. Singer, S. J., and Nicolson, G. L. (1972). "The fluid mosaic model of the structure of cell membranes." Science 175, 720.

48. Smith, C. W. and Hollers, J. C. (1970). "The pattern of binding of fluorescein-labeled Concanavalin A to motil lymphocyte." J. Reticuloend. Soc. 8, 458.

49. Smith, S. B., and Revel, J. (1972). "Mapping of Concanavalin A binding sites on the surface of several cell types." Develop. Biol. 27, 434.

50. Taylor, R. B., Duffus, P. H., Raff, M. C., and de Petris, S. (1971). "Redistribution and pinocytosis of lymphocyte surface immunoglobulins molecules induced by anti-immunoglobulin antibody." Nature New Biology 233, 225.

51. Wright, D. G., and Malawista, S. E. (1972). "The mobilization and extracellular release of granular enzymes from human leuko-

cytes during phagocytosis." *J. Cell Biol.* 53, 788.

52. Zigmond, S. H., and Hirsch, J. G. (1972). "Cytochalasin B: inhibition of D-2-deoxyglucose transport into leukocytes and fibroblasts." *Science* 176, 1432.

53. Zurier, R. B., Hoffstein, S., and Weissmann, G. (1973). "Mechanisms of lysosomal enzyme release from human leukocytes. I. Effect of cyclic nucleotides and colchicine." *J. Cell Biol.* 58, 27.

16

ENHANCED CYTOTOXICITY IN MICE OF COMBINATIONS OF

CONCANAVALIN A AND SELECTED ANTITUMOR DRUGS*

S. G. Bradley, N. M. Marecki, J. S. Bond, A. E. Munson, and D. T. John
Virginia Commonwealth University
Richmond, Virginia, U. S. A.

ABSTRACT

Concanavalin A (Con A) injected intraperitoneally at a dose of 50 mg per kg was not lethal for male BALB/c mice. Six hours after administration of 5 mg Con A/kg, the proportion of circulating granulocytes had increased from 23% to 74% of the white cell population; by 24 hr, the proportion of granulocytes had decreased to 56%. Administration of 5 mg Con A/kg 24 hr before 200 mg of 5-[3,3-bis(2-chloroethyl)-triazeno]-imidazole-4-carboxamide per kg, or 100 mg of 5-fluorouracil per kg resulted in a significant enhancement of lethality. Simultaneous administration of 5 mg Con A/kg and 10 mg of daunomycin per kg also resulted in enhanced lethality. Administration of 5 mg Con A/kg 24 hr before 40 mg of 1,3-bis(2-chloroethyl)-1-nitrosourea per kg, 200 mg of 1-(2-chloroethyl)-3-cyclohexyl-1-nitrosourea per kg, 1000 mg of cytosine arabinoside per kg, 0.1 mg of mithramycin per kg, 2 mg of pactamycin per kg or 1 mg of vincristine per kg did not result in enhanced lethality. Lipid A prepared from *Escherichia coli* 0127:B8 Boivin lipopolysaccharide has been complexed to Con A. The lipid A-Con A complex (5mg/kg) was no more, or less effective in enhancing the lethality of 5-fluorouracil than 2.5 mg Con A/kg. The lipid A-Con A complex (40 mg/kg), given simultaneously with drug, enhanced lethality for mice given 0.1 mg mithramycin per kg or 1 mg vincristine per kg. In this regard, the lipid A-Con A complex had activity comparable to the complex formed between lipid A and bovine serum albumin. Conceivably, Con A can be used to enhance the sus-

---

*This investigation was supported in part by Public Health Service grant CA-13715 from the National Cancer Institute.

ceptibility of neoplastic cells to phase-specific antitumor drugs, especially those acting on deoxyribonucleic acid synthesis.

I. INTRODUCTION

Many antitumor drugs show an enhanced toxicity *in vivo* when administered in combination with bacterial endotoxin, a component of the cell wall of gram-negative bacteria (Marecki and Bradley, 1973; Rose and Bradley, 1971). Of 27 drugs known to be active in human cancer (Zubrod, 1974), we have examined 19 for their ability to interact *in vivo* with endotoxin (lipopolysaccharide [LPS]). At least 10 of these antitumor drugs in combination with LPS kill mice synergistically (Rose, 1973). Because cancer patients are particularly susceptible to infection by gram-negative bacteria, we have proposed that some of the adverse reactions observed in cancer patients undergoing chemotherapy may reflect synergistic toxicities involving LPS and antitumor drugs.

Endotoxin is a mitogen (Peavy *et al.*, 1973) as is concanavalin A (Con A). If the mitogenic activity of LPS is responsible for the enhanced susceptibility to certain antitumor drugs, particularly inhibitors of deoxyribonucleic acid (DNA) synthesis, Con A should be able to substitute for LPS. It should be noted, however, that Con A and LPS activate different types of cells. Con A selectively stimulates thymus-derived lymphocytes (T cells) *in vitro* (Stobo *et al.*, 1972) and not bone marrow-derived lymphocytes (B cells) unless the mitogen is presented to the cells in a locally concentrated, insoluble form (Anderson *et al.*, 1972b). Endotoxin preferentially activates B cells (Anderson *et al.*, 1972a).

We have been interested in determining whether the mitogenic effects of Con A have possible use in the treatment of cancer and infectious diseases. Most studies on biochemical and physiological responses to Con A have been done *in vitro* and in cell culture systems. Relatively few studies have measured *in vivo* effects of Con A on tumors or host responses (Eagan *et al.*, 1974). The experiments presented in this report have been designed to determine (a) whether Con A causes synergistic lethality with selected antitumor drugs and (b) whether Con A retards the growth of a solid tumor (Lewis lung carcinoma) *in vivo* and prolongs survival of mice. In addition, we have measured the effects of Con A administration on the percent of circulating granulocytes, adenosine 3',5'-cyclic monophosphate (cyclic AMP) concentrations in plasma and liver and survival of mice infected with a protozoan parasite (*Plasmodium*).

II. MATERIALS AND METHODS

A. Experimental Animals:

Male BALB/c mice weighing 20-25 g were used in all experiments, except those involving the Lewis lung carcinoma. Mice were obtained from Battelle Memorial Institute (Columbus, Ohio), Laboratory Supply Co., Inc. (Indianapolis, Ind.), and ARS/Sprague-Dawley (Madison, Wis.) and allowed to become conditioned to their surroundings at least one week prior to use. Mice were given free access to water and Purina Lab-Chow (Ralston Purina Company, St. Louis, Mo.). Antitumor agents and mitogens were prepared so that the desired dose was contained in 0.01 ml per g mouse weight. Con A (grade IV) (Sigma Chemical Co., St. Louis, Mo.), 1,3-bis(2-chloroethyl)-1-nitrosourea (BCNU) (NSC-409962), 5-[3,3-bis(2-chloroethyl)-1-triazeno]-imidazole-4-carboxamide (TIC-mustard) (NSC-82196), 1-(2-chloroethyl)-3-cyclohexyl-1-nitrosourea (CCNU) (NSC-63878), cytosine arabinoside (NSC-63878), daunomycin (NSC-82151), 5-fluorouracil (NSC-19893), mithramycin (NSC-24559), pactamycin (NSC-52947), and vincristine (NSC-67574) were dissolved or suspended in sterile distilled water, adjusted to pH 7, and used immediately. *Escherichia coli* 026:B6 LPS (Boivin method) (Difco, Detroit, Mich.) was suspended in 0.15 N NaCl. All injections were given by the intraperitoneal (ip) route.

B. Preparation of Lipid A Complexes:

Lipid A was extracted from *E. coli* 0127:B8 LPS (Boivin method) (Difco) by a procedure adapted from the method of Galanos *et al.* (1969). The absence of 2-keto-3-deoxyoctonate (KDO) as confirmed by the thiobarbiturate assay of Waravdekar and Saslaw (1959) was used to establish that hydrolysis was complete.

Complexes of lipid A and bovine serum albumin (BSA) (Sigma), or Con A were prepared according to methods described by Galanos *et al.* (1972). Immediately before use, the complexes were suspended in distilled water and dispersed with sonication.

C. Peripheral Leukocyte Counts:

Peripheral blood was obtained from the tail veins of mice at 6 and 24 hr after administration of various substances. Blood smears were made and stained with Wright's stain, and differential counts were performed. One hundred leukocytes were counted per slide and classed as monocytes or granulocytes.

D. Protozoan Infection:

Plasmodium berghei NYU-2 was maintained by blood passage in mice. The inoculum for each experiment was prepared by bleeding several infected mice 2 weeks after infection. Heparinized blood was pooled and the percentage of parasitized cells was determined by examining Giemsa-stained smears. No attempt was made to differentiate between individually or multiply infected erythrocytes or between different stages of development of individual parasites. Erythrocyte counts of the pooled blood were made in a hemocytometer and the inoculum mixed with buffered sterile saline so that each 0.01 ml contained 10^6 parasitized red blood cells. The mice were injected ip with 0.01 ml/g mouse weight (2×10^7 infected erythrocytes/mouse). One inoculum was used for all groups in the same experiment.

E. Cyclic AMP Concentration in Plasma and Liver:

Cyclic AMP concentrations in plasma and liver were assayed by the protein binding assay described by Gilman (1970). The binding protein was prepared from beef heart. For the preparation of samples (Johnson, 1972), groups of 6 mice were decapitated and blood was collected in a tube containing ethylenediamine-tetraacetic acid (final concentration 6 mM). The liver of a mouse was removed and immediately homogenized with cold 0.36 N perchloric acid (PCA) (1:5 w/v). After centrifugation (6,000 x g for 5 min), the plasma was removed and PCA added to it (final concentration, 0.3 N). One ml samples of the PCA-extracts were chromatographed on a Dowex-50W (hydrogen form) (Sigma) column (0.6 x 3 cm). The 4th through 7th ml of effluent contained 85% of the cyclic AMP as determined by radioisotopic dilution, using [^{3}H]-cyclic AMP. This fraction was lyophilized, dissolved in 0.5 ml distilled water and assayed for cyclic AMP. In each instance, a portion of the lyophilized sample was treated with 3',5'-cyclic nucleotide phosphodiesterase before assaying, to assess and make corrections for compounds in the sample other than cyclic AMP that interfere with the protein binding assay. Assays were done in duplicate. Each plasma value represented determinations for 3 groups (6 mice/group); each liver value represented determinations for 3 mice.

F. Lewis Lung Tumor:

The Lewis lung tumor was maintained in C_{57} BL/6 male mice (Flow Research Animals, Dublin, Va.) by subcutaneous implant of ca. 25 mg tumor fragments. For experimental use, 14-18 day tumors were excised, minced, and cleared of debris and necrotic tissue. The tumor tissue was then placed in 1% trypsin in Eagle's MEM with 1 unit of penicillin/ml and 1 µg of streptomycin/ml. After 60 min incubation at 22 C, fetal calf serum was added (final concentration 2%), the

TABLE I

LETHAL RESPONSE OF MICE TO 5-FLUOROURACIL ADMINISTERED AFTER CONCANAVALIN A

5-Fluorouracil 100 mg/kg	Concanavalin A 5 mg/kg	No. Mice	%Dead by Day 14
+	-	60	2
-	+	50	0
0 hr	+	10	0
+3 hr	+	10	70
+6 hr	+	20	60
+18 hr	+	20	90
+24 hr	+	20	100

Male BALB/c mice were injected intraperitoneally.

cells washed, enumerated, and suspended to 5×10^6 cells/ml. BDF_1 male mice were injected intramuscularly with 10^6 cells/mouse. The size of the primary tumor was calculated by the method described by Mayo et al. (1972), using the formula: $w = (a + b^2)/2$, where a = length in mm, b = width in mm and w = weight in mg. Test agents were administered daily by the ip route for 10 consecutive days following tumor inoculation. The size of the primary tumor was determined on days 12, 19, and 30, and mean survival times were recorded.

III. RESULTS

A. Interaction of Concanavalin A with Antitumor Drugs:

Administration of 5 mg of Con A/kg simultaneously with 100 mg of 5-FU/kg, was not lethal to mice (Table I). However, when 5-FU was given 3 to 24 hr after Con A, 70 to 100 percent of the mice died. It should be noted that 50 mg of Con A/kg alone was not lethal for mice and that the LD_{50} of 5-FU was 143 mg/kg. Similar experiments with 5-FU and LPS established that simultaneous administration of these agents did not result in potentiation of lethality, but when LPS was administered 24 hr prior to 5-FU, the lethality for mice was significantly enhanced (Marecki and Bradley, 1973).

TABLE II

LETHAL RESPONSE OF MICE TO TIC-MUSTARD ADMINISTERED AFTER CONCANAVALIN A

TIC-mustard 200 mg/kg	Concanavalin A 5 mg/kg	No. Mice	% Dead by Day 4
+	−	20	0
−	+	50	0
−24 hr	+	20	0
0 hr	+	20	0
+24 hr	+	20	55

Male BALB/c mice were injected intraperitoneally.

TABLE III

LETHAL RESPONSE OF MICE TO DAUNOMYCIN ADMINISTERED BEFORE OR WITH CONCANAVALIN A

Daunomycin 10 mg/kg	Concanavalin A 5 mg/kg	No. Mice	% Dead by Day 4
+	−	60	10
−	+	50	0
+24 hr	+	10	0
0 hr	+	10	60
−6 hr	+	10	40
−18 hr	+	10	50
−24 hr	+	10	50
−30 hr	+	10	40
−48 hr	+	10	40

Male BALB/c mice were injected intraperitoneally.

TABLE IV

FAILURE OF SELECTED ANTITUMOR AGENTS TO INTERACT
SYNERGISTICALLY WITH CONCANAVALIN A

Antitumor Agent	Concanavalin A	No. Mice	%Dead by Day 14
BCNU, 40 mg/kg	-	10	0
	5 mg/kg	10	10
CCNU, 200 mg/kg	-	10	0
	5 mg/kg	10	10
Cytosine arabinoside 1000 mg/kg	-	10	0
	5 mg/kg	10	0
Mithramycin, 0.1 mg/kg	-	10	10
	5 mg/kg	10	0
Pactamycin, 2 mg/kg	-	10	0
	5 mg/kg	10	0

All antitumor agents were administered 24 hr after concanavalin A.

The alkylating agent TIC-mustard also interacted synergistically with LPS only when administered 24 hr after LPS (Bradley, unpublished observations). This same sequence of administration resulted in an enhanced lethal response for mice when Con A was substituted for LPS (Table II). The LD_{50} of TIC-mustard by the intraperitoneal route was 275 mg/kg.

Daunomycin, which also inhibits DNA synthesis, interacted in a synergistic manner with LPS (Rose, 1973). However, unlike 5-FU and TIC-mustard, the LPS-daunomycin synergy occurred when the drug was administered simultaneously with or prior to LPS, rather than after LPS. Con A interacted with daunomycin in a similar manner. The simultaneous or delayed administration of 5 mg of Con A/kg with 10 mg of daunomycin/kg resulted in 40 to 60 percent lethality at 4 days after daunomycin (Table III). The LD_{50} of daunomycin alone was 15 mg/kg, as determined on day 4.

Con A and LPS did not produce synergistic effects with all

TABLE V

LETHAL RESPONSE OF MICE TO 5-FLUOROURACIL ADMINISTERED
AFTER LIPID A-PROTEIN COMPLEXES

Test Agent	5-Fluorouracil 100 mg/kg	No. Mice	%Dead by Day 14
None	+	60	2
BSA, 10 mg/kg	-	10	0
Con A, 50 mg/kg	-	10	0
Lipid A[a], 40 mg/kg	24 hr	10	0
LPS[b], 2 mg/kg	24 hr	40	90
Con A, 2.5 mg/kg	24 hr	20	60
Lipid A-BSA, 40 mg/kg	-	10	0
Lipid A-BSA, 40 mg/kg	24 hr	40	68
Lipid A-Con A, 5 mg/kg	-	10	0
Lipid A-Con A, 5 mg/kg	24 hr	20	45
Lipid A-Con A, 80 mg/kg	-	10	0
Lipid A-Con A, 80 mg/kg	24 hr	20	95

[a]Lipid A extracted from *E. coli* 0127:B8 LPS.

[b]*E. coli* 026:B6 LPS. A dose of 2 mg/kg was non-lethal to male BALB/c mice.

inhibitors of DNA synthesis. Non-lethal doses of the alkylating agents BCNU and CCNU did not interact in a synergistic manner with LPS or Con A regardless of the sequence of administration (Table IV). Cytosine arabinoside, another inhibitor of DNA synthesis, also did not interact with LPS regardless of the sequence of administration. Likewise, Con A did not potentiate the lethal response of mice to cytosine arabinoside (Table IV).

Con A was not able to substitute for LPS in all known antitumor drug-LPS synergies. The antibiotics mithramycin and pactamycin interacted synergistically with LPS when administered simultaneously, but not with Con A administered simultaneously, or when either drug followed Con A by 24 hr (Table IV).

TABLE VI

LETHAL RESPONSE OF MICE TO MITHRAMYCIN ADMINISTERED WITH LIPID A-PROTEIN COMPLEXES

Test Agent[a]	Mithramycin 0.1 mg/kg	No. Mice	%Dead by Day 4
None	+	10	0
LPS[b], 2 mg/kg	+	20	65
Con A, 5 mg/kg	+	10	0
Lipid A-BSA, 40 mg/kg	−	10	0
	+	10	60
Lipid A-Con A, 40 mg/kg	−	10	0
	+	10	70

[a]All agents were administered simultaneously.

[b]$E.\ coli$ 026:B6 LPS.

Complexes of lipid A with BSA or Con A were prepared and tested for the capability to enhance the lethality of 5-FU, mithramycin or vincristine. Mice administered 5-FU 24 hr after lipid A-BSA were killed synergistically (Table V). This lethal response was not significantly ($p > 0.05$) different from that produced by 2 mg of LPS/kg given 24 hr prior to 100 mg of 5-FU/kg. The lethal response to the administration of 5 mg of lipid A-Con A/kg 24 hr after 5-FU was not significantly ($p > 0.05$) different from the administration of 2.5 mg of Con A/kg 24 hr prior to 5-FU (Table V). The response to 5-FU with 80 mg of lipid A-Con A/kg was similar to that resulting from 5 mg of Con A/kg (Table I), or 2 mg of LPS/kg (Table V) given 24 hr prior to 5-FU. These results indicated that the lipid A-protein complexes, when given together with 5-FU, were no more effective in potentiating the lethal response in mice than was a small dose of a known mitogenic agent.

Con A did not potentiate the lethal response of mithramycin when administered 24 hr prior to (Table IV) or simultaneously with its administration (Table VI). However, when 0.1 mg of mithramycin/kg was administered simultaneously with 40 mg of lipid A-BSA/kg, or 40 mg of lipid A-Con A/kg, the potentiation of lethality was similar to that resulting from the simultaneous administration of mithramycin and 2 mg of LPS/kg.

TABLE VII

LETHAL RESPONSE OF MICE TO VINCRISTINE ADMINISTERED
WITH LIPID A-PROTEIN COMPLEXES

Test Agent[a]	Vincristine 1 mg/kg	No. Mice	%Dead by Day 3
None	+	40	0
LPS[b], 2 mg/kg	+	40	90
Con A, 5 mg/kg	+	10	0
Lipid A, 80 mg/kg	−	10	10
Lipid A-BSA, 40 mg/kg	+	20	60
80 mg/kg	+	20	80
Lipid A-Con A, 40 mg/kg	+	20	50
80 mg/kg	+	20	70

[a]All agents were administered simultaneously.

[b]*E. coli* 026:B6 LPS.

The simultaneous administration of 1 mg of vincristine/kg with 2 mg of LPS/kg resulted in 90 percent lethality (Marecki and Bradley, 1973). Five mg of Con A/kg did not interact synergistically with vincristine (Table VII). However, both lipid A-BSA and lipid A-Con A potentiated the lethal response of mice given 40 or 80 mg/kg of either complex simultaneously with 1 mg of vincristine/kg (Table VII). The administration of 80 mg of lipid A/kg did not result in an enhanced lethal response.

B. <u>Effects of Mitogens on Circulating Granulocytes</u>:

The peripheral leukocyte population of normal male BALB/c mice was found to be composed of 23 percent granulocytes and 77 percent monocytes (Table VIII). Six hr after the administrations of 2 mg of *E. coli* 026:B6 LPS/kg the two cell types occurred in approximately equal numbers, and by 24 hr, 81 percent of the peripheral leukocytes were granulocytic. The administration of 5 mg of Con A/kg brought about a marked early change in the leukocyte population. By 6 hr, the granulocytes reached 74 percent. However, this elevation did not persist, and by 24 hr there were approximately equal numbers of monocytes and granulocytes. The proportion of the circulating leukocytes

TABLE VIII

CHANGE IN CIRCULATING GRANULOCYTES AFTER LPS, CONCANAVALIN A, LIPID A-BSA, OR LIPID A-CON A

Treatment	Time of Sample[a]	%Monocytes	%Granulocytes
None		77	23
LPS[b], 2 mg/kg	+6 hr	48	52
	+24 hr	19	81
Con A, 5 mg/kg	+6 hr	26	74
	+24 hr	44	56
Lipid A-BSA, 40 mg/kg	+6 hr	35	65
	+24 hr	19	81
Lipid A-Con A, 40 mg/kg	+6 hr	40	60
	+24 hr	23	77

[a] Blood samples were collected from the tail veins of male BALB/c mice.

[b] *E. coli* 026:B6 LPS.

that were granulocytes 6 hr after administration of 40 mg of lipid A-BSA/kg or 40 mg of lipid A-Con A/kg was intermediate between the response provoked by 2 mg of LPS/kg and 5 mg of Con A/kg (Table VIII). The proportion of granulocytes was higher 24 hr after administration of 40 mg of a lipid A-protein complex/kg than after 6 hr.

C. Effect of Mitogens on the Lewis Lung Carcinoma:

Mice inoculated with Lewis lung carcinoma cells, and treated with LPS, lipid A, Con A or lipid A-Con A daily for 10 days, gained 2.1 to 2.7 g, whereas the untreated mice gained ca. 1 g. Con A alone and the lipid A-Con A complex significantly reduced the tumor weight 12 and 19 days after inoculation, whereas LPS and lipid A did not. Thirty days after tumor inoculation, all of the lipid A-treated mice and LPS-treated mice were dead, 86% of the lipid A-Con A-treated mice were dead, and 71% of the untreated mice were dead, but 63% of the Con A-treated mice were still alive. The mean survival time of Con A-treated mice, however, was not significantly greater than that of the untreated mice (Table IX).

TABLE IX

EFFECT OF LIPID A, CONCANAVALIN A, LPS, AND LIPID A-CON A
ON TUMOR GROWTH RATE AND SURVIVAL TIME OF MICE
INOCULATED WITH LEWIS LUNG CARCINOMA

Treatment[a]	Tumor Weights (mg)[b] Days After Tumor Inoculation		Mean Survival Time Days[b]
	Day 12	Day 19	
None	913 ± 118	4002 ± 330	27.3 ± 1.1
LPS[c], 1 mg/kg	924 ± 186	4109 ± 229	22.7 ± 1.0
Lipid A, 5 mg/kg	1035 ± 83	5062 ± 454	24.2 ± 0.7
Con A, 5 mg/kg	581 ± 70	2833 ± 209	31.3 ± 2.1
Lipid A-Con A, 10 mg/kg	197 ± 130	2918 ± 251	25.4 ± 1.8

[a]Male BDF$_1$ mice were injected intramuscularly with 10^6 tumor cells. Drugs were administered ip daily for 10 consecutive days. There were 8 animals in each group.

[b]Mean ± S.E.M.

[c]*E. coli* 0127:B8

D. Effect of Mitogens on Plasmodium Infection:

The course of many bacterial infections can be altered by the administration of LPS prior to the infecting agent. This has also been demonstrated in infections with *P. berghei* in mice (MacGregor et al., 1969). The administration of 2 mg of LPS/kg 24 hr prior to 2×10^7 infected RBC/mouse prolonged survival of infected mice (Fig. 1). Five mg of Con A/kg altered the lethal response in a manner similar to that of LPS. The altered response to infections was attributed in part to the increased numbers of circulating granulocytes induced by LPS and Con A.

E. Effects of Mitogens on Cyclic AMP Concentrations in Plasma and and Liver:

The concentrations of cyclic AMP in livers of mice injected with 5 mg Con A/kg or 8 mg LPS/kg were essentially the same as those in livers of control mice injected with saline (Table X). The cyclic AMP concentration in liver or plasma of control mice did not vary

TABLE X

EFFECT OF CONCANAVALIN A AND LPS ON CYCLIC AMP CONCENTRATIONS IN PLASMA AND LIVER

Treatment[a]	Time After Injection (hr)	Cyclic AMP in:	
		Plasma % of Control	Liver % of Control
Con A (5 mg/kg)	0	109	98
	1	80	96
	3	97	87
	6	105	100
	18	95	98
LPS[b] (8 mg/kg)	1	100	116
	2	108	116
	3	170	92
	6	167	96
	18	171	113

[a] All compounds were injected intraperitoneally into BALB/c mice. Control animals were injected with 0.9% NaCl. The concentration of cyclic AMP in plasma of control mice was 53 ± 8 pmoles/ml; in liver it was 947 ± 44 pmoles/g.

[b] *E. coli* 026:B6 LPS.

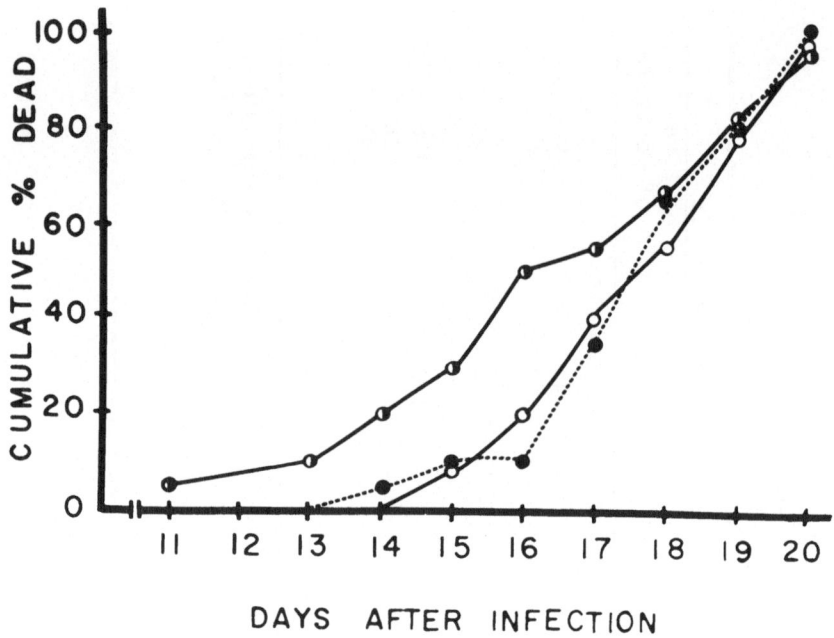

Fig. 1. Effect of mitogens upon mortality in male BALB/c mice infected with *Plasmodium berghei*. Mice were administered 5 mg of Con A/kg (●----●) or 2 mg of *E. coli* 026:B6 LPS/kg (0———0) 24 hr before 2×10^7 infected erythrocytes per mouse. One group of infected mice received no mitogen (●———●).

significantly with time after injection. An average value for cyclic AMP in the livers of control mice was 947 pmoles/g; in the plasma, it was 53 pmoles/ml. Plasma concentrations were affected differently by the mitogens; Con A had little or no effect on the plasma levels, whereas there was a consistent increase 3, 6, and 18 hr after LPS administration (Table X).

F. Absence of Endotoxin in the Concanavalin A Preparation:

Because it has been suggested that some of the effects attributed to Con A are due to contamination by bacterial endotoxin, a 50 µg sample of Con A was tested for capability to gel the commercial *Limulus* amoebocyte lysate (Reinhold and Fine, 1972). This amount of Con A did not cause gelation, although 10^{-5} to 10^{-6} µg of *E. coli* 026:B6 LPS did. Moreover, Con A was tested for its capability to kill mice sensitized to LPS by 0.5 mg of mithramycin/kg. Con A (50 mg/kg) given simultaneously with mithramycin did not enhance lethality. This *in vivo* assay was capable of detecting 10 ng LPS/kg. The Con A preparation used in this study contained less than 10^{-5}% LPS.

IV. DISCUSSION

It has been amply demonstrated that antitumor drugs capable of inhibiting protein or ribonucleic acid (RNA) synthesis are potent potentiators of endotoxicity (Rose, 1973). Recently Marecki and Bradley (1973) demonstrated that LPS sensitized mice to 5-FU, an inhibitor of DNA synthesis. Eaves and Bruce (1972) have shown that the sensitivity of hematopoietic cells of mice to 5-FU is increased by prior treatment with bacterial LPS. Eaves and Bruce concluded that proliferating hematopoietic stem cells are more sensitive to 5-FU than resting marrow cells. Accordingly, we proposed that mitogens such as LPS or Con A may sensitize mice to the lethal effects of a variety of antitumor agents that inhibit DNA synthesis. Indeed, both LPS and Con A sensitize mice to the lethal effects of 5-FU, TIC-mustard and daunomycin. Neither LPS nor Con A sensitize mice to the lethal effects of CCNU, BCNU or cytosine arabinoside. Mithramycin, pactamycin and vincristine sensitize mice to the lethal effects of LPS but Con A in combination with these three agents, does not result in enhanced lethality. These results indicate that inhibitors of protein or RNA synthesis interact *in vivo* with LPS by a mechanism(s) different from inhibitors of DNA synthesis or function. The mitogens LPS and Con A increase the proportion of granulocytes in the circulation of mice, and concurrently, these animals are rendered more vulnerable to 5-FU or TIC-mustard. It should be noted that LPS and Con A sensitize mice to daunomycin most markedly when the mitogen and drug are administered simultaneously, whereas the mitogens need to precede 5-FU or TIC-mustard administration. The effect of time of administration is consistent with the proposition that it is a metabolite of daunomycin and not daunomycin *per se* that is the toxic agent. It is significant that CCNU, BCNU and cytosine arabinoside, antitumor agents that affect DNA, do not interact *in vivo* with Con A to enhance lethality. It cannot be concluded, therefore, that mitogens sensitize mice to all agents that affect DNA.

Our results confirm that lipid A is the active moiety of LPS in precipitating lethal synergies with mithramycin or vincristine. Either bovine serum albumin or Con A is a suitable carrier for the lipid A. Lipid A does not enhance nor suppress the capability of Con A to sensitize mice to 5-FU. Because LPS and Con A have different spectra of mitogenic action, the lipid A-Con A complex may be expected to possess novel properties as a mitogen, immunoadjuvant, interferon inducer or cytotoxic agent. These characteristics have not yet been determined.

Con A alone can conceivably be used to enhance the susceptibility of neoplastic cells to phase-specific antitumor drugs, especially those acting on DNA synthesis. This potential has not yet been assessed; however, Con A alone is able to retard the growth of the Lewis lung tumor and the Moloney virus-induced lymphoma (Inbar et al., 1972). Moreover, Con A is able to prolong the survival of mice challenged with lethal doses of *Plasmodium*. Because infectious diseases are a serious complication of neoplastic disease, Con A may have dual beneficial effects, i.e., as an agent that stimulates host defenses to infections and as an agent that increases the therapeutic index of selected antitumor drugs.

REFERENCES

1. Anderson, J., Sjöberg, O., and Möller, G. (1972a). "Introduction of immunoglobulin and antibody synthesis *in vitro* by lipopolysaccharides." *Eur. J. Immunol.* 2, 349.

2. Anderson, J., Edelman, G. M., Möller, G., and Sjöberg, O. (1972b). "Activation of B lymphocytes by locally concentrated concanavalin A." *Eur. J. Immunol.* 2, 233.

3. Eagan, H. S., Reeder, W. J., and Ekstedt, R. D. (1974). "Effect of concanavalin A *in vivo* in suppressing the antibody response in mice." *J. Immunol.* 112, 63.

4. Eaves, A. C., and Bruce, W. R. (1972). "Endotoxin-induced sensitivity of hematopoietic stem cells to chemotherapeutic agents." *Ser. Haemat.* 5, 64.

5. Galanos, C., Lüderitz, O., and Westphal, O. (1969). "A new method for the extraction of R lipopolysaccharides." *Eur. J. Biochem.* 9, 245.

6. Galanos, C., Rietschel, E. Th., Lüderitz, O., Westphal, O., Kim, Y. B., and Watson, D. W. (1972). "Biological activities of lipid A complexed with bovine serum albumin." *Eur. J. Biochem.* 31, 230.

7. Gilman, A. G. (1970). "A protein binding assay for adenosine 3',5'-cyclic monophosphate." *Proc. Nat. Acad. Sci. U.S.A.* 67, 305.

8. Inbar, M., Ben-Bassat, H., and Sachs, L. (1972). "Inhibition of ascites tumor development by concanavalin A." *Internat. J. Cancer* 9, 143.

9. Johnson, R. A. (1972). Assay of cyclic AMP. In "Methods in Cyclic Nucleotide Research" (M. Chasin, ed.) Marcel Dekker, Inc. New York, p. 1.

10. MacGregor, R. R., Sheagren, J. N., and Wolff, S. M. (1969). "Endotoxin-induced modification of *Plasmodium berghei* infection in mice." *J. Immunol.* 102, 131.

11. Marecki, N. M., and Bradley, S. G. (1973). "Enhanced toxicity for mice of combinations of bacterial endotoxin with antitumor drugs." *Antimicrob. Ag. Chemother.* 3, 599.

12. Mayo, J. G., Laster, W. R., Jr., Andrews, C. M., and Schabel, F. M., Jr. (1972). "Success and failure in the treatment of solid tumors. III. 'Cure' of metastatic Lewis lung carcinoma with methyl-CCNU (NSC-95441) and surgery - chemotherapy." *Cancer Chemother. Rept.* (1) 56, 183.

13. Peavy, D. L., Shands, J. W., Jr., Adler, W. H., and Smith, R. T. (1973). "Mitogenicity of bacterial endotoxins: characterization of the mitogenic principle." *J. Immunol.* 111, 352.

14. Reinhold, R., and Fine, J. (1972). "A technique for quantitative measurement of endotoxin in human plasma." *Proc. Soc. Exp. Biol. Med.* 137, 334.

15. Rose, W. C. (1973). "Interaction of bacterial toxins in the toxicity of chemotherapeutic agents." *CRC Critical Rev. Toxicol.* 3, 159.

16. Rose W. C., and Bradley, S. G. (1971). "Enhanced toxicity for mice of combinations of antibiotics with *Escherichia coli* cells or *Salmonella typhosa* endotoxin." *Infect. Immun.* 4, 550.

17. Stobo, J. D., Rosenthal, A. S., and Paul, W. E. (1972). "Functional heterogeneity of murine lymphoid cells. I. Responsiveness to and surface binding of concanavalin A and phytohemagglutinin." *J. Immunol.* 108, 1.

18. Waravdekar, V. S., and Saslaw, L. D. (1959). "A sensitive colorimetric method for the estimation of 2-deoxy sugars with the use of the malonaldehyde-thiobarbituric acid reaction." *J. Biol. Chem.* 234, 1945.

19. Zubrod, C. G. (1974). "Present status of cancer chemotherapy." *Life Sciences* 14, 809.

17

EFFECT OF CONCANAVALIN A AND PHYTOHEMAGGLUTININ ON THE
MODIFICATION OF IMMUNOGENICITY OF CANINE KIDNEY
ALLOGRAFTS*

Luis H. Toledo-Pereyra, Clive O. Callender,
Prasanta K. Ray, John S. Najarian, and
Richard L. Simmons
University of Minnesota
Minneapolis, Minnesota, U. S. A.

ABSTRACT

Mongrel dog kidneys were allografted to unrelated nephrectomized recipients which were then treated with subimmunosuppressive doses of azathioprine (2.5 mg/kg/day). Dog kidneys treated *in vitro* with perfusates containing concanavalin A (Con A) or phytohemagglutinin-P (PHA) survived as long as 60 days (mean 39.8 ± 4.3) after transplantation, whereas normal kidneys survived less than 16 days. The optimal prolongation was achieved by perfusing the kidneys with 500 ml Ringer's lactate containing 25 mg/L Con A, 25 4°C. Lesser effects were achieved with higher or lower concentrations of Con A, or with perfusions carried out at 25°C. Most evidence suggests that Con A and PHA bind to cell surfaces and interfere with the perception of the graft antigens by the host.

I. INTRODUCTION

Prolongation of organ allografts has generally required the use of systemic immunosuppressive agents in order to modify host reactivity to the histocompatibility antigens on the graft. It is thought that introduction of the graft antigens at time of maximal immunologic unresponsiveness leads to a state of partial immunologic tolerance to graft antigen. The opposite approach, i.e., that of

---

*Supported by United States Public Health Service Grants #AM 13083 and #CA 11605 and Grant #DRG 1186 from the Damon Runyon Memorial Fund for Cancer Research.

altering the grafted tissue so that the histocompatibility antigens
of the tissue are masked, and thus not recognized by the host, has
received comparatively little attention (Billingham et al., 1951;
Bonmassar et al., 1966; Callender et al., 1973a,b, 1974; Dukes and
Blocker, 1952; Jolley et al., 1961; Klaue and Jolley, 1971a,b, 1972;
Raju et al., 1969). We have previously demonstrated that the plant
lectins concanavalin A (Con A) and phytohemagglutinin-M (PHA) mask
strong histocompatibility antigens on mouse lymphoid cells when
these agents are bound to the cell surfaces (Ray and Simmons, 1973).
The maximum obliteration of the antigen was seen at concentrations
of Con A greater than 25 µg/ml and of PHA greater than 500 µg/ml.
Similarly we have shown that perfusing kidney grafts with antilymphocyte globulin (ALG) led to prolongation of such kidneys in mildly immunosuppressed dogs (Callender et al., 1973a,b). These results suggest that antigen modification might supplement modest levels of immunosuppression in the prolongation of renal allografts. The present
study was designed to determine if Con A or PHA would mask the histocompatibility antigens on kidney grafts and allow for their prolongation in moderately immunosuppressed dogs.

II. MATERIALS AND METHODS

One hundred and four dogs (weight 16 to 23 kg) received unrelated kidneys by standard operative techniques. After opening the
vascular clamps, 20 mg of furosamide were given intravenously. Bi-
lateral nephrectomy was then performed immediately. Serum creati-
nines were recorded daily. The onset of rejection was defined as
the day on which the creatinine was elevated 2.0 mg% above normal.
Post mortem examinations with kidney histology were performed in all
cases.

The dogs were randomly divided into 13 groups of eight dogs
each (Table I). Most recipient dogs were given azathioprine 2.5
mg/kg/day (I.V. for 2 days, then orally) from the time of grafting
until death. Three groups of dogs (Groups I, III, X) were not given
azathioprine. Immediately after nephrectomy, the kidneys of most
groups were flushed with Ringer's lactate solution (500 ml, 4°C,
procaine 1 gm/L, heparin 10,000 U/L) containing various concentrations of Con A (Calbiochem, San Diego, California) or PHA (Difco
Laboratories, Detroit, Michigan). The kidneys of Groups I, II, and
IX were flushed with the regular perfusate without lectins. The
kidneys in Group VII were perfused with 500 ml of Ringer's containing Con A (25 mg/L), and then perfused a second time with Ringer's
without Con A. The kidneys of Group VIII were flushed at 25°C. Con
A (2.5 mg) was infused intra-arterially into the renal arteries immediately after transplantation of the kidneys in Group IX. Animals
in Group X were given 25 mg/L Con A in the perfusate, but 5 mg was
given intravenously after grafting. Groups XI and XII received kidneys that were given PHA during hypothermic pulsatile perfusion

TABLE I

EFFECT OF RENAL ALLOGRAFT PRETREATMENT ON REJECTION AND SURVIVAL OF RECIPIENT DOGS

Group	Graft Treatment	Recipient Treatment	Mean Onset of Rejection (days ± SE)	p values	Mean Survival (days ± SE)	p values
I	Ringer's	None	6.8 ± 0.5	---	11.0 ± 0.7	---
II	Ringer's	Azathioprine**	8.5 ± 0.7	*	12.5 ± 0.7	*
III	Con A (25 mg/L) in Ringer's	None	8.5 ± 0.9	>0.5	12.8 ± 0.8	>0.5
IV	Con A (12.5 mg/L) in Ringer's	Azathioprine**	19.8 ± 4.8	<0.05	26.6 ± 5.7	<0.05
V	Con A (25 mg/L) in Ringer's	Azathioprine**	34.3 ± 5.6	<0.001	39.8 ± 4.3	<0.001
VI	Con A (50 mg/L) in Ringer's	Azathioprine**	22.2 ± 2.5	<0.001	27.7 ± 2.8	<0.001
VII	Con A (25 mg/L) followed by Ringer's flush	Azathioprine**	10.1 ± 1.0	>0.1	15.1 ± 1.5	>0.1
VIII	Con A (25 mg/L 25° C) in Ringer's	Azathioprine**	9.7 ± 1.0	>0.2	19.6 ± 2.2	>0.05
IX	Ringer's	Con A (2.5 mg) into renal artery + Azathioprine**				
X	Con A (25 mg/L) in Ringer's	Con A (5 mg) IV: No azathioprine	7.8 ± 1.2	>0.5	11.5 ± 1.3	>0.2
XI	PHA (10 mg/L) in plasma	Azathioprine**	8.3 ± 1.0	>0.5	14.1 ± 1.2	>0.2
XII	PHA (20 mg/L) in plasma	Azathioprine**	8.2 ± 1.0	>0.8	13.1 ± 0.9	>0.5
XIII	PHA (160 mg/L) in Ringer's	Azathioprine**	7.1 ± 0.8	>0.1	12.0 ± 0.8	>0.2
			17.9 ± 6.5	>0.2	25.6 ± 5.9	<0.05

*All p values compared to Azathioprine control group (II).
**Azathioprine (2.5 mg/kg/day).

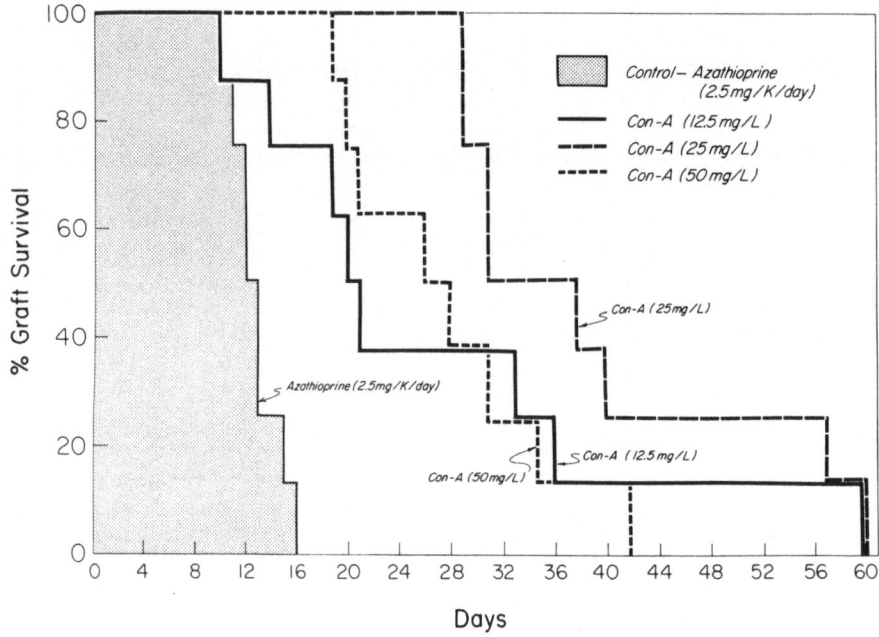

Fig. 1. Effect of Graft Pretreatment with Con A on Renal Allograft Survival. Cumulative survival of dogs receiving renal allografts perfused *in vitro* with concanavalin A (Con A) in Ringer's lactate. The data here include experimental Groups II, IV, V and VI.

(7°C, pH 7.4, pO_2 200 mm Hg, pulse rate 60 min, systolic pressure 60 mm Hg) using pooled dog cryoprecipitated plasma for 24 hours and one hour respectively. Animals in Group XIII were given PHA (160 mg/L) in the regular perfusate.

Ten additional dogs with obvious technical problems such as anesthesia complications, post-operative hemorrhage or ureteral leakage and those that died with other non-related causes such as distemper and intussusception, in the first five days after transplantation were immediately excluded from analysis.

III. RESULTS

A. General Results

The rejection time and survival of the recipients of all renal allografts are summarized in Table I and Figs. 1 and 2. Dogs given

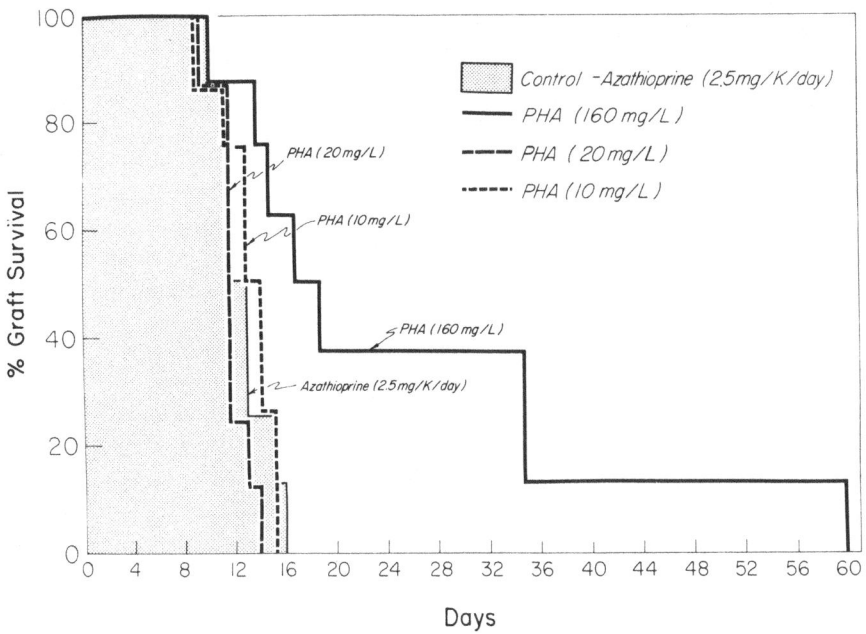

Fig. 2. Effect of Graft Pretreatment with PHA on Renal Allograft Survival. Cumulative survival of dogs receiving renal allografts perfused *in vitro* with phytohemagglutinin-P (PHA) in Ringer's lactate. The data here include experimental Groups II, XI, XII and XIII.

normal kidneys rejected them in less than two weeks even when azathioprine (2.5 mg/kg/day) was given.

B. <u>Concanavalin A Treated Kidneys</u>

Kidneys perfused with Con A in concentrations greater than 12.5 mg/L survived for more prolonged periods in dogs given azathioprine (Groups IV, V, VI) but not in dogs which were not given azathioprine (Group III). The optimal concentration of Con A was 25 mg/L (Group V), higher or lower concentrations having less effect (Groups IV, VI). If the kidneys were perfused with 500 ml Ringer's lactate after optimal Con A perfusion (Group VII) little prolongation was seen. Similarily, optimal concentrations of Con A has less effect if the perfusion was carried out at 25°C (Group VIII) rather than 4°C (Group V).

Two control groups were designed to test if the systemic effect of Con A remaining unbound in the kidney was responsible for the prolongation. Either the systemic Con A (5 mg I.V.) in dogs who had not received azathioprine, or 2.5 mg into the renal artery of dogs given azathioprine had no perceptible immunosuppressive effect (Groups IX, X).

C. Phytohemagglutinin-P (PHA) treated kidneys

PHA in a concentration of 160 mg/L in Ringer's (Group XIII) had a similar effect as Con A. Lesser amounts of PHA in cryoprecipitated plasma perfusate had no effect on graft prolongation (Groups XI, XII).

IV DISCUSSION

The present experiments demonstrate that perfusion of dog kidneys at 4°C with varying concentrations of Con A or PHA will allow moderately immunosuppressed dogs to maintain these grafts for significant periods of time. Ultimate rejection always occurred, but occasional survival as long as 60 days was seen. The doses of azathioprine utilized were not sufficient to produce prolonged survival of kidney grafts by themselves. Similarly, perfusion of kidneys with the optimal concentration of Con A had no effect on dogs who did not receive some systemic immunosuppression. The synergistic effect of minimal immunosuppression and alteration of graft has previously been reported (Callender et al., 1973a,b). We showed that the in vitro perfusion of kidney allografts with horse anti-dog lymphocyte globulin would lead to prolonged survival of dog kidneys only if the dogs also received subimmunosuppressive doses of azathioprine. Neither immunosuppression alone nor treatment of the kidneys with ALG alone achieved graft prolongation. It is likely that all experiments designed to investigate reductions in graft immunogenicity will require the use of some host immunosuppression, since total ablation of graft immunogenicity may be impossible to achieve. This finding may help explain the failure to obtain satisfactory results in previous attempts to reduce graft immunogenicity (Abbott and Pappas, 1970; Callender et al., 1974; Dukes and Blocker, 1952; Grillo and McKhann, 1964; Hellmann and Duke, 1967; Lemperle, 1968; Silverstein et al., 1971).

Con A and PHA were chosen for these experiments because of their ability to bind firmly to tissues. Con A reacts with neutral polysaccharides. It can bind specifically to the C3, C4, and C6 hydroxyl positions on terminal glucopyranose or mannopyranose residues of highly branched polysaccharides (Goldstein et al., 1965; Goldstein and Iyer, 1966). Con A also reacts with mucopolysaccharides, galactan sulfate, glycoproteins, and bacterial lipopolysaccharides (Cifonelli et al., 1956; Manners and Wright, 1962). For PHA, very little information is available in this regard. PHA has been found to bind with nuclei, cytoplasm, or cell surface (Conrad,

1967; Michalowski et al., 1964, 1965; Vassar and Culling, 1964). N-acetyl galactosamine has been suggested to be the constituent of the PHA-combining sites on the cell. Other sugars, such as D-glucose, N-acetyl glucosamine, D-galactose, and D-mannose cannot react with PHA, and D-L-fucose has little tendency to do so (Borberg et al., 1966).

Thus, both Con A and PHA interact with chemical groupings on the cell surface. We have previously shown that both Con A and PHA bound to mouse lymphoid cells will prevent the lysis of these cells by antibody and complement (Ray and Simmons, 1973). In addition, Con A and PHA on the surface of mouse lymphocytes will interfere with the binding of anti-H-2 specific antibody to these cells, as measured by antibody absorption experiments. Thus by analogy, one could expect that Con A and PHA act to mask the immunogenicity of graft antigens by the same mechanism, i.e., by binding to the cell surfaces and interfering with the perception of the antigens on the graft by immune lymphocytes or by antibodies (Jannosy and Greaves, 1971). It is possible, of course, that Con A acts by damaging passenger leukocytes so that the immunogenicity in passenger leukocytes is destroyed (Billingham, 1971; Steinmuller and Hart, 1971; Warden et al., 1971) but other methods of reducing passenger leukocytes have not always been as effective (Freeman et al., 1971).

It is not clear from either these experiments or our previous ones whether these lectins bind to (1) chemical groupings responsible for strong histocompatibility antigens, (2) chemical groupings which are not related to histocompatibility antigenicity but disarrange the cell surface so that the histocompatibility antigenic groupings are distorted, or (3) nonantigenic groupings leading to steric interference with the interaction of antigen and receptors for these antigens on immunoreactive cells. Neither do we know if agglutinin-binding on cell surfaces might induce the cooperative phenomena in cell membranes initiating either *trans* (allosteric) or *cis* effect on the membrane as proposed by Singer and Nicolson (1972) in their fluid mosaic model of the structure of cell membranes. Either of these effects might result in a conformational rearrangement of the membrane structure, and thereby change the functional property of the histocompatibility antigens. There is evidence to suggest that subsequent to binding on the cell surface, lectins may, in fact, be internalized by pinocytosis (Pauli et al., Razavi, 1966). Then a major conformational change might be expected during pinocytosis following the dynamic fluid character of the cell membrane components.

It is clear that these lectins bind with carbohydrate molecules Aub et al., 1965; Burger and Goldberg, 1967; Cifonelli et al., 1956; Inbar and Sachs, 1969; Moscona, 1971; Nakamura and Suzuno, 1965; Naspitz and Richter, 1968; Steck and Wallach, 1965). Although Reisfeld and Kahan (1970) have reported that most of the antigenic determinants of human alloantigens are polypeptide in nature, they admit that it is impossible to rule out that either lipid or carbohydrate moieties express a specific alloantigenic specificity or that they confer a unique protein configuration that determines such a specificity. It appears from the present study that the carbohydrate receptors for Con A and PHA are in some way related or linked to histocompatibility antigens which are recognized by specific alloantibody on the cell surface.

Several aspects of this experiment may ultimately shed light on the mechanism of antigen masking by plant lectins. Apparently washing out the Con A with normal Ringer's lactate removes it, so that the Con A must be bound weakly to cell surfaces. This effect does not seem to be due to the Con A itself acting as an immunosuppressant in the recipient dog (Markowitz et al., 1969) since Con A given intraarterially into the renal artery, or intravenously in amounts that might be found in the kidney after perfusion do not seem to prolong graft survival. In support of this hypothesis is the fact that Con A at 50 mg/L is less effective than at 25 mg/L. If the Con A remaining unbound within the kidney acted synergistically with azathioprine as a systemic immunosuppressive agent, one would expect that the higher concentration of Con A would be more effective than the lesser one. Nevertheless, the failure of Con A to remain bound to the kidney and achieve its masking effect after washing out with Ringer's lactate is puzzling.

One possible mechanism is suggested by consideration of the behavior of polyvalent lectins like Con A and PHA on cell surfaces. When lectins bind to cell surfaces at room temperature, the lectins are found in clusters on the cell membrane (Nicolson, 1973). When lectins bind at 4°C, they are more evenly distributed; when the cells are warmed, clustering takes place presumably due to the fluid nature of the cell membrane. One can hypothesize that clustered Con A on kidney cell surfaces will not mask antigens since perfusion of kidneys with Con A at 25°C will not prolong their survival. The Con A apparently needs to be evenly distributed on cell surfaces to mask antigens. Failure to observe clustering on cell surfaces in the cold may not only reflect the "frozen" status of the cell membrane at 4°C, but also a relatively lower binding affinity of Con A to cell surfaces in the cold so that multiple membrane receptor sites are not pulled together and Con A is easy to wash off.

Why then does transplantation into a warm recipient not interfere with antigen masking? Con A is a polyvalent material which can bind to many types of glycoproteins including those in blood (Chase, 1972; Nakamura et al., 1960; Novogrodosky et al., 1972). We can speculate that rapid warming in the presence of host blood succeeds in binding protein to the evenly distributed Con A on the graft cell surfaces. Such binding may not only insure even distribution of the Con A on the surface but may also further mask the antigens by Con A-protein complexes. These speculations are currently under investigation.

REFERENCES

1. Abbott, W. M., and Pappas, A. M. (1970). "Comparative studies on fresh and preserved skin: Fundamental biological differences in behavior as grafts." *Ann. Surg.* 172, 781.

2. Aub, J. C., Sanford, B. H., and Cote, M. N. (1965). "Studies on reactivity of tumor and normal cells to a wheat germ agglutinin." *Proc. Natl. Acad. Sci. U. S. A.* 54, 396.

3. Billingham, R. E. (1971). "The passenger cell concept in transplantation immunity." *Cell. Immunol.* 2, 1.

4. Billingham, R. E., Krohn, P. L., and Medawar, P. B. (1951). "Effect of locally applied cortisone acetate on survival of skin homografts in rabbits." *Brit. Med. J.* 2, 1049.

5. Bonmassar, E., Francisconi, G., and Manzoni, S. C. (1966). "Chemical deletion of histocompatibility antigens -- Homograft survival of rat skin treated with urethan *in vitro*." *Nature* 209, 1141.

6. Borberg, H., Woodruff, J., Hirschhorn, R., Gesner, B., Miescher, P., and Silber, R. (1966). "Phytohemagglutinin: Inhibition of the agglutinating activity by N-acetyl-D-galactosamine." *Science* 154, 1019.

7. Burger, M. M., and Goldberg, A. R. (1967). "Identification of a tumor-specific determinant on neoplastic cell surfaces." *Proc. Natl. Acad. Sci. U. S. A.* 57, 359.

8. Callender, C. O., Simmons, R. L., Toledo-Pereyra, L. H., Santiago-Delpin, E. A., and Najarian, J. S. (1973a). "Prolongation of kidney allografts perfused by ALG *in vitro*." *Transplantation* 16, 377.

9. Callender, C. O., Santiago-Delpin, E. A., Sutherland, D., Howard, R. J., Toledo-Pereyra, L. H., Olsen, L., Condie, R., Simmons, R. L., and Najarian, J. S. (1973b). "Renal allograft prolongation after graft perfusion with heterologous ALG." *Surg. Forum* XXIV, 306.

10. Callender, C. O., Sutherland, D. E. R., Howard, R. J., Toledo-Pereyra, L. H., and Najarian, J. S. (1974). "Failure of pepsin-digested donor specific alloantibody to prolong renal allograft survival in dogs." *Immunol. Comm.* Submitted for publication.

11. Chase, P. S. (1972). "The effect of human serum fractions on phytohemagglutin- and Concanavalin A-stimulated human lymphocyte cultures." *Cell. Immunol.* **5**, 544.

12. Cifonelli, J. A., Montgomery, R., and Smith, F. (1956). "The reaction of concanavalin A with mucopolysaccharides." *J. Amer. Chem. Soc.* **78**, 2488.

13. Conrad, R. A. (1967). "Autoradiography of leukocytes cultured with tritiated bean extract." *Nature* **214**, 709.

14. Dukes, C. D., and Blocker, T. G. (1952). "Studies on the survival of skin homografts. I. Prolongation of life of full-thickness grafts by the action of streptokinase-streptodornase." *Ann. Surg.* **136**, 999.

15. Freeman, J. S., Chamberlain, E. C., Reemtsma, K., and Steinmuller, D. (1971). "Prolongation of rat heart allografts by donor pretreatment with immunosuppressive agents." *Transplant. Proc.* **3**, 580.

16. Goldstein, I. J., Hollerman, C. E., and Smith, E. E. (1965). "Protein-carbohydrate interaction. II. Inhibition studies on the interaction of concanavalin A with polysaccharides." *Biochemistry* **4**, 876.

17. Goldstein, I. J., and Iyer, R. N. (1966). "Interaction of concanavalin A, a phytohemagglutinin with model substrates." *Biochim. Biophys. Acta* **121**, 197.

18. Grillo, H. C., and McKhann, C. F. (1964). "The acceptance and evolution of dermal homografts freed of viable cells." *Transplantation* **2**, 48.

19. Hellmann, K., and Duke, D. I. (1967). "*In vitro* alteration of skin graft antigenicity." *Transplantation* **5**, 184.

20. Inbar, M., and Sachs, L. (1969). "Interaction of the carbohydrate-binding protein concanavalin A with normal and transformed cells." *Proc. Natl. Acad. Sci. U. S. A.* **63**, 1418.

21. Jannosy, G., and Greaves, M. F. (1971). "Lymphocyte activation. I. Response of T and B lymphocytes to phytomitogens." *Clin. exp. Immunol.* **9**, 483.

22. Jolley, W. B., Hinshaw, D. B., and Peterson, M. (1961). "The effect of RNA on homograft survival." *Surg. Forum* **12**, 99.

23. Klaue, P., and Jolley, W. B. (1971a). "Prolongation of rabbit skin allograft survival following treatment of the skin with

electrophoresis prior to transplantation." *Transplantation* 11, 30.

24. Klaue, P., and Jolley, W. B. (1971b). "The comparative effectiveness of corticosteroids applied topically as pretreatment of rabbit skin allografts." *Surgery* 70, 718.

25. Klaue, P., and Jolley, W. B. (1972). "Prolonged survival of rabbit skin allografts after treatment with locally applied triamcinolone acetonide in combination with small doses of systemic immunosuppressives." *Transplantation* 13, 53.

26. Lemperle, G. (1968). Prolonged survival of skin allografts after incubation with recipient DNA or RNA." *J. Surg. Res.* 8, 511.

27. Manners, D. J., and Wright, A. J. (1962). "Alpha-1, 4-glucosans. Part XIV: The interaction of concanavalin A with glycogen." *J. Chem. Soc. (London)* 4592.

28. Markowitz, H., Person, D. A., Gitnick, G., and Ritts, R. E., Jr. (1969). "Immunosuppressive activity of concanavalin A." *Science* 163, 476.

29. Michalowski, A., Jasinska, J., Brzosko, W. J., and Nowoslawski, A. (1964). "Cellular localization of the mitogenic principle of phytohemagglutinin in leucocyte cultures." *Exp. Cell Res.* 34, 417.

30. Michalowski, A., Jasinska, J., Brzosko, W. J., and Nowoslawski, A. (1965). "Studies on phytohemagglutinin. II. Cellular localization of phytohemagglutinin in white blood cell cultures with immunohistochemical methods." *Exp. Med. Microbiol.* 17, 197.

31. Moscona, A. A. (1971). "Embryonic and neoplastic cell surfaces: Availability of receptor for concanavalin A and wheat germ agglutinin." *Science* 171, 905.

32. Nakamura, S., and Suzuno, R. (1965). "Crystallization of concanavalins A and B and canavalin from Japanese jack beans." *Arch. Biochem. Biophys.* 111, 499.

33. Nakamura, S., Tanaka, K., and Murakawa, S. (1960). "Specific protein of Legumes which reacts with animal proteins." *Nature* 188, 144.

34. Naspitz, C. K., and Richter, M. (1968). "The action of phytohemagglutinin *in vivo* and *in vitro*, a review." *Progr. Allergy* 12, 1.

35. Nicolson, G. L. (1973). "Temperature-dependent mobility of concanavalin A sites on tumour cell surface." *Nature New Biology* 243, 218.

36. Novogrodosky, A., Biniaminov, M., Ramot, B., and Katchalski, E. (1972). "Binding of concanavalin A to rat, normal human and chronic lymphatic leukemia lymphocytes." *Blood* 40, 311.

37. Pauli, R. M., DeSalle, L., Higgins, P., Henderson, E., Norin, A., and Strauss, B. (1973). "Proliferation of stimulated human peripheral blood lymphocytes: Preferential incorporation of concanavalin A by stimulated cells and mitogenic activity." *J. Immunol.* 111, 424.

38. Raju, M. D., Grogan, J. B., and Hardy, J. D. (1969). "Prolonged survival of skin allografts exposed to heterologous antilymphocyte serum *in vitro*." *J. Surg. Res.* 9, 327.

39. Ray, P. K., Jeitz, L., Sarquis, L., and Simmons, R. L. (1972). "Mechanisms of increased immunogenicity of cells treated with concanavalin." *Fed. Proc.* 31, 234.

40. Ray, P. K., and Simmons, R. L. (1973). "Masking of cellular histocompatibility antigens with phytomitogens." *J. Immunol.* 110, 1693.

41. Razavi, L. (1966). "Cytoplasmic localization of phytohaemagglutinin in peripheral white cells." *Nature* 210, 444.

42. Reisfeld, R. A., and Kahan, B. D. (1970). "Transplantation antigens." *Adv. Immunol.* 12, 117.

43. Silverstein, P., Raulston, G. L., Walker, H., Foley, F. D., and Pruitt, B. A. (1971). "Evaluation of formalin-fixed skin as a temporary dressing for granulating wounds." *Surg. Forum* 22, 60.

44. Singer, S. J., and Nicolson, G. L. (1972). "The fluid mosaic model of the structure of cell membranes." *Science* 175, 720.

45. Steck, T. L., and Wallach, D. F. H. (1965). "The binding of kidney-bean phytohemagglutinin by Ehrlich ascites carcinoma." *Biochim. Biophys. Acta* 97, 510.

46. Steinmuller, D., and Hart, E. A. (1971). "Passenger leukocytes and induction of allograft immunity." *Transplant. Proc.* 3, 673.

47. Vassar, P. S., and Culling, C. F. A. (1964). "Cell surface effects of phytohaemagglutinin." *Nature* 202, 610.

48. Warden, G., Reemtsma, K., and Steinmuller, D. (1971). "Survival of rat heart allografts irradiated *in vitro*." *Surg. Forum* XXII, 252.

EVIDENCE FOR CONFORMATIONAL CHANGES IN CONCANAVALIN A UPON BINDING OF SACCHARIDES AS DETERMINED FROM SOLVENT WATER PROTON MAGNETIC RELAXATION RATE DISPERSION MEASUREMENTS*

R. D. Brown, III, C. F. Brewer, and S. H. Koenig
IBM Thomas J. Watson Research Center
Yorktown Heights, New York, U. S. A., and
Albert Einstein College of Medicine
Bronx, New York, U. S. A.

ABSTRACT

In previous studies of the interaction of solvent water molecules with the Mn^{++} ion in Mn-Con A by observation of the dispersion of the spin-lattice relaxation rate (T_1^{-1}) of the solvent water protons over a wide range of magnetic fields (Koenig et al., 1973), we have shown that this rate is dominated by the residence time of a single exchanging water ligand on the Mn^{++} ion. In limited measurements at low fields, we also observed that the binding of α- or β-methyl-D-glucopyranoside to Mn-Con A decreased the relaxation rate by approximately 15 percent. In the present study, we have measured the effects of binding of a series of mono- and oligosaccharides on the solvent water proton relaxation rate over a range of magnetic fields from 5 Oe to 12 KOe and show that the observed decrease in the relaxation rate is due to an increase in the residence time of the single exchanging water ligand. This effect is consistent with a conformational change in the protein upon binding of saccharides. We find that the binding of α- and β-methyl-D-glucopyranoside, α-methyl-D-mannopyranoside and β-(o-iodophenyl)-D-glucopyranoside produce the same increase in residence time and therefore the same conformational change in the protein, whereas galactose and β-(o-iodophenyl)-D-galactopyranoside show no effects. The same reduction in relaxation rate as that caused by the above monosaccharides was ob-

---

*Supported in part by USPHS, NIH grants AI-05336 and CA-16054.

served with the following oligosaccharides: D-maltose, D-maltotriose, D-maltotetraose, o-α-D-mannopyranosyl-(1→2)-D-mannose, o-α-D-mannopyranosyl-(1→2)-o-α-D-mannopyranosyl-(1→2)-D-mannose, o-α-D-mannopyranosyl-(1→2)-o-α-D-mannopyranosyl-(1→2)-o-α-D-mannopyranosyl-(1→2)-D-mannose and melezitose. As observed by Goldstein and co-workers, the first three oligosaccharides have nearly the same affinity constant, whereas the α-(1,2) linked mannans show increasing affinity constants with increasing chain length. Melezitose also shows enhanced binding by a factor of four relative to α-methyl-D-glucopyranoside. The water relaxation data suggest that the above mono- and oligosaccharides bind to Con A by a similar mechanism involving only a single saccharide residue combined with the protein at one time. Determination of the enthalpy (ΔH) and entropy (ΔS) of binding of maltotriose and melezitose indicates that the factor of 8 difference in their affinity constants is due to different ΔS values. This suggests that the greater affinity of melezitose is due to the presence of two glucose residues in the molecule, either of which is capable of binding to the same site on the protein. The increased probability of binding for melezitose results in a larger forward rate constant relative to maltotriose which has only one glucose residue which can bind to the protein. Thus, the greater affinity of the α-(1→2)-mannose oligosaccharides appears to be due to a statistical increase in probability of binding because of the presence of more than one binding residue in the chain and not to an extended binding site on the protein.

Koenig, S. H., Brown, R. D., and Brewer, C. F. (1973). *Proc. Nat. Acad. Sci. U. S. A.* 70, 475.

MAGNETIC RESONANCE STUDIES OF CONCANAVALIN A: LOCATION OF THE BINDING SITE OF METHYL-D-MANNOPYRANOSIDE

Brian H. Barber, Alan Quirt and Jeremy P. Carver
University of Toronto
Toronto, Canada

ABSTRACT

The longitudinal relaxation times of the methyl protons of α-methyl-D-mannopyranoside (α-MeMan) have been measured at 100 MHz in the presence of: (i) demetallized concanavalin A; (ii) a concanavalin A:Zn^{++}:Ca^{++} complex; (iii) a concanavalin A:Mn^{++}:Ca^{++} complex; and, (iv) a concanavalin A:Zn^{++}:Ca^{++}:Gd^{+++} complex. Mn^{++} and Zn^{++} are known to bind to Con A at a site S1; Ca^{++}, at site S2; and, Gd^{+++}, at site S3. Sites S1 and S2 must be occupied before α-MeMan will bind to Con A. The measured relaxation times have been used to calculate the distance from the methyl protons to each of the sites S1 and S3. Regardless of the value of the correlation times for the nuclear-electron dipolar interaction or the value of the α-MeMan exchange rate, the relaxation results are shown to define a very limited set of allowed values for the S1-, and S3- methyl distances. Comparison of these allowed distances with those for the monosaccharide binding sites which have been proposed, indicates that the results are only consistent with the site proposed by Becker *et al.* (*J. Biol. Chem.* [1971], 246, 6123).

THE METAL ION REQUIREMENTS OF CONCANAVALIN A

A. Dean Sherry and A. Newman
Institute for Chemical Sciences
University of Texas at Dallas
Dallas, Texas, U. S. A.

ABSTRACT

Water proton relaxation enhancement studies give evidence for two strong lanthanide ion binding sites (normal transition metal site) and two weaker lanthanide ion binding sites (normal calcium site) per protein monomer. Even though the lanthanide ions compete effectively with a transition metal ion for its site, the saccharide binding sites of concanavalin A are not formed in the absence of a transition metal ion. In the presence of excess Ni^{++}, the lanthanide ions bind primarily in the calcium sites and the protein regains approximately 30% of its saccharide binding activity. The formation of the saccharide site shows the following transition metal ion dependence: $Zn^{++} > Ni^{++} > Co^{++} > Fe^{++} \sim Mn^{++} > Cd^{++}$. In the presence of Mn^{++} and Ca^{++}, the pH profile of concanavalin A binding to Sephadex reflects a dramatic decrease in saccharide binding between pH 5 and 3 (pK=4.3). Electron spin resonance experiments show this loss in saccharide binding activity results from the dissociation of Mn^{++} from the protein.

THE KINETICS OF CELLULAR COMMITMENT DURING STIMULATION OF LYMPHOCYTES BY CONCANAVALIN A

Gary R. Gunther, John L. Wang, and
Gerald M. Edelman
The Rockefeller University
New York, New York, U. S. A.

ABSTRACT

The kinetics of cellular commitment in the stimulation of lymphocytes by concanavalin A (Con A) have been analyzed by measurement of DNA synthesis, autoradiography, and histologic staining techniques. If the competitive inhibitor α-methyl-D-mannoside is introduced into cultures of mouse spleen cells at various times after the addition of Con A, there is a gradual decrease in its capacity to inhibit the lectin-stimulated incorporation of ^{3}H-thymidine. Addition of the saccharide 20 hours after exposure of the cells to Con A has no effect on the level of the cellular response to the lectin. With increasing periods of exposure to Con A, the percentage of blast cells and the percentage of ^{3}H-thymidine-labeled blast cells increased in parallel with the total radioactive thymidine incorporated while the average number of autoradiographic grains per labeled blast cell remained relatively constant. These observations suggest that the rising level of ^{3}H-thymidine incorporation results from an increase in the number of cells that both respond to lectin stimulation and become refractory to inhibition with α-methyl-D-mannoside. Once such cells become committed, they synthesize DNA at a rate independent of the length of exposure to the lectin. The combined data indicate that different cells within a lymphocyte population may require different induction periods to be stimulated. These results may also explain some observations on the inhibition by colchicine and related drugs of the initial events in lymphocyte mitogenesis.

ISOLATION OF A GLYCOPROTEIN RECEPTOR FOR CONCANAVALIN A FROM THE OUTER SURFACE OF MOUSE L CELLS

R. C. Hunt, J. C. Brown, and
C. M. Bullis
University of Virginia Medical School
Charlottesville, Virginia, U. S. A.

ABSTRACT

Two glycoprotein classes having apparent molecular weights of approximately 100,000 (class A) and 50,000 (class B) have been identified as components of the outer plasma membrane surface of mouse L-929 cells by the following criteria: (1) Both are isolated with the smooth membrane fraction of cells lysed by Dounce homogenization. (2) Both are digested when intact cells are treated with trypsin. (3) Both are labeled by two methods, the pyridoxal phosphate-sodium borotritide system and the galactose oxidase-potassium borotritide method, which are designed to introduce radioactive label specifically into cell surface structures. Class A glycoproteins were solubilized for biochemical analysis by dissolving whole cells in 0.3 M lithium diiodosalicyclate and extracting with aqueous phenol; class A glycoproteins were found in the water phase while class B glycoproteins partitioned into the phenol. Further purification of one class A glycoprotein was accomplished by affinity chromatography on a column of concanavalin A-Sepharose. This Con A-binding glycoprotein appears to be a single species of glycopolypeptide chain. The N-terminal amino acid is valine and its cleavage with cyanogen bromide yields five distinct oligopeptides all of which contain carbohydrate moieties.

ELECTRON MICROSCOPIC STUDY ON INTERACTION OF CONCANAVALIN A WITH MOUSE LYMPHOSARCOMA CELLS IN TISSUE CULTURE AND IN ASCITES FORM

Constance A. Feltkamp
The Netherlands Cancer Institute
Amsterdam, The Netherlands

ABSTRACT

Two cell lines of a spontaneous mouse lymphosarcoma, one in suspension culture, the other in ascites form by $i.p.$ passages, behave differently in several aspects; in contrast to the ascites cells, the *in vitro* cultured cells are not transplantable if injected into untreated syngeneic hosts; they have a high Con A binding capacity and are easily agglutinated by Con A. However, treatment of ascites cells with enzymes or culture *in vitro* increases their agglutinability while reducing transplantability (L. A. Smets, and J. Broekhuysen-Davies [1972] *Europ. J. Cancer* 8, 541). These findings indicate that the ascites cells possibly are covered by a coat which masks Con A binding sites and transplantation antigens.

However, ruthenium red staining of cell coat material did not reveal in the electron microscope a significant difference in thickness of this layer between cultured and ascites cells. Other staining methods (colloidal iron) gave indications of a difference in composition of the cell coat. Therefore experiments were performed in order to study whether in this model other factors in addition to the cell coat might play a role in the difference in Con A binding and agglutinability between both cell types.

Staining of bound Con A with hemocyanin showed that the cultured cells bind more Con A per unit of cell surface. Per cell this difference is even enlarged by the presence on the surface of the cultured cells of many long, slender microvilli. This is an indication of increased mobility of the *in vitro* cells which could favor agglutinability by an increased chance of cell interactions.

In both cell lines, Con A binding did not induce much redistribution of receptors in the plane of the membrane, not even after incubation at 37°C.

Trypsin treatment (high concentrations are necessary) increased the binding of Con A to prefixed ascites cells. Observations ot not prefixed trypsinized ascites cells showed, however, that the capacity of redistribution of the receptors in the plane of the membrane was also influenced: many cells showed patching or capping. The quantity of detected Con A bound to trypsinized cells without prefixation seemed to decrease during this redistribution. Moreover, the mobility of the cell surface was increased by the appearance of microvilli and extrusions. Thus, the higher agglutinability of trypsinized ascites cells need not be due exclusively to an unmasking of Con A binding sites.

The present study does not exclude the role of a cell coat in inhibiting the ascites cells to bind Con A and to agglutinate, but stresses the idea that redistribution of Con A receptors in the plane of the membrane and mobility of the cell surface in the form of microvilli may play a role in the process of agglutination.

THE EFFECT OF GLUTARALDEHYDE FIXATION ON THE AGGLUTINATION OF HUMAN ERYTHROCYTES BY CONCANAVALIN A AND SOYBEAN AGGLUTININ

J. A. Gordon and M. D. Marquardt
University of Colorado Medical Center
Denver, Colorado, U. S. A.

ABSTRACT

In order to evaluate the role of membrane fluidity in agglutination following trypsinization, we have investigated the effect of glutaraldehyde fixation on the agglutination of human erythrocytes by concanavalin A and soybean agglutinin. Human erythrocytes were washed with phosphate buffered saline of pH 7.4 (PBS). Crystalline trypsin was used for trypsinization. Fixation was performed with 2.5% redistilled glutaraldehyde in PBS for 2, 4, and 6 hours at room temperature.

The absolute agglutinability of erythrocytes under static conditions was determined in microtiter plates with 1% erythrocyte suspensions in PBS and serial dilutions of lectins. Sedimentation patterns were read macroscopically after 3 hours at room temperature.

Fixation increased the agglutinability of the erythrocytes by both concanavalin A and soybean agglutinin. The increase of agglutinability produced by fixation was not as great as that produced by trypsinization. Fixation of trypsinized erythrocytes did not further increase their agglutinability, but trypsinization of fixed erythrocytes did increase their agglutinability. All agglutination was specifically inhibited by the respective lectin-specific sugars.

When the kinetics of concanavalin A-mediated agglutination were examined by electronically counting particles remaining in solution, no agglutination of either trypsinized-fixed or fixed-trypsinized erythrocytes could be detected within 4 hours. However, when an equal volume of trypsinized-unfixed erythrocytes was mixed with trypsinized-fixed or fixed-trypsinized erythrocytes, the kinetics of the mixed agglutination reaction were identical to the kinetics of trypsinized-unfixed erythrocytes alone. More than half of the erythrocytes were agglutinated, and mixed cell agglutination was confirmed by microscopic observation. Apparently the shear forces produced by agitation of the reaction mixtures during the kinetic experiment are sufficient to prevent or disrupt agglutination of fixed erythrocytes by concanavalin A. In contrast, agglutination of fixed erythrocytes by soybean agglutinin was much less sensitive to disruption by shearing.

Several interpretations of these results may be made. Either glutaraldehyde fixation does not inhibit the mobility of erythrocyte lectin receptors or receptor mobility is unnecessary for lectin-mediated agglutination, as both fixed and unfixed erythrocytes are agglutinable. If fixation restricts receptor mobility, increased receptor mobility cannot account for increased erythrocyte agglutinability following trypsinization, as agglutination of fixed erythrocytes is also increased after they are trypsinized. The strength and stability but not the extent of agglutination of fixed erythrocytes apparently is decreased with respect to unfixed erythrocytes.

ALTERED NET CATION TRANSFER ACROSS THE EHRLICH MOUSE
ASCITES TUMOR CELL DURING EXPOSURE TO CONCANAVALIN A

Felice Aull, Martin S. Nachbar, and
  Joel D. Oppenheim
New York University Medical Center
New York, New York, U. S. A.

ABSTRACT

Ehrlich ascites tumor cells normally accumulate potassium and extrude sodium against electrochemical gradients and maintain gradients for these cations in the steady state. In the presence of concanavalin A (Con A), however, marked alterations in cell sodium and potassium content occur. When exposed to Con A concentrations of 20-105 µg/ml, there was a biphasic response in 11 of 16 experiments; initially cell sodium content increased rapidly to levels which were up to three times higher than in control cells. Potassium content fell concomitantly to values as low as one third of control levels. Then, after 10-2- minutes in Con A, sodium extrusion began, but was prevented when ouabain at 10^{-3} M was also present in the medium. In the remaining 5 experiments with Con A, normal net cation transport was impaired, but the response was not biphasic.

The tumor cells were agglutinated by Con A at all concentrations which altered cation transfer. However, the effect on cation transfer was not the result of agglutination alone since in seven experiments with soybean agglutinin (SBA) either no change (two experiments) or much smaller changes occurred, in spite of the fact that SBA agglutinated the cells to the same extent as Con A.

Preliminary evidence shows that the major Con A receptor of the Ehrlich ascites cell membrane is a glycoprotein of about 120,000 molecular weight. Whether the alterations of cation transfer observed reflect changes in the configuration of this membrane constituent is at this point a matter of speculation.

EFFECTS OF CON A ON FROG NERVE AND MUSCLE

R. J. Person
University of Oklahoma Health Sciences Center
Oklahoma City, Oklahoma, U. S. A.

ABSTRACT

Several parameters of the A-alpha fiber compound action potential recorded from paired *Rana pipiens* sciatic nerves were monitored for periods of up to 24 hours. No significant changes in these parameters were observed within 4 hours after the addition of 1 mg per ml Con A (Sigma, Grade III), when compared with the simultaneously monitored control nerve dissected from the same animal. However, in some experiments in which nerve viability was maintained for 24 hours, A-alpha peak amplitude was significantly decreased. In a limited number of experiments, sartorius muscle fiber resting potentials were recorded intracellularly. Resting potentials of fibers incubated at room temperature in Ringer's with 50 µg per ml Con A added were significantly higher than those recorded from the paired control muscle. It is concluded that Con A may increase permeabilities for both sodium and potassium across nerve and muscle membranes.

MODIFICATION OF THE SURFACE CHARACTERISTICS OF DEVELOPING HEMOPOIETIC CELLS FROM NORMAL HUMAN BONE MARROW REVEALED ULTRASTRUCTURALLY BY THE CONCANAVALIN A-PEROXIDASE-DIAMINOBENZIDINE TECHNIQUE

G. Adolph Ackerman and S. D. Waksal
The Ohio State University
Columbus, Ohio, U. S. A.

ABSTRACT

The concanavalin A-peroxidase-diaminobenzidine (CAPD) technique has been employed as an ultrastructural chemical probe directed toward defining possible modifications in the surface characteristics of hemopoietic cells during normal cell maturation and specialization.

Glutaraldehyde-fixed cell suspensions were exposed to the CAPD sequence as described by Bernhard and Avrameas (*Exptl. Cell Res.*, [1971] 64,232). The concentration of Con A (Sigma Chemical Co., St. Louis, Mo.) employed was 0.1 mg/ml phosphate buffer, pH 7.4. In the control experiments Con A or peroxidase was omitted and α-methyl-D-mannoside was added to the incubation media. The thickness of surface CAPD reactivity was measured in nearly 500 cells in differing morphological categories each sectioned near the cell center and the data were subjected to statistical evaluation.

All nucleated marrow cells showed surface CAPD reactivity and the degree of surface reactivity was dependent upon cell type and the morphological stage of cell maturation. Marked surface staining occurred with primitive hemopoietic cells, viz., myeloblasts and erythroblasts, and with lymphocytes, monocytes, macrophages, and platelets. In contrast, mature neutrophils, eosinophils, and basophils showed limited surface reactivity; erythrocytes were non-reactive or revealed only traces of surface CAPD staining. In the neutrophilic series, surface CAPD reactivity decreased progressively from the myeloblast to myelocyte stage; minimal changes in surface CAPD staining was noted during subsequent neutrophil maturation. Developing nucleated erythrocytic cells showed only minimal difference in surface CAPD staining until the late normoblast stage when a marked decrease in CAPD reactivity developed and continued throughout the reticulocyte phase of erythrocyte development. Polarization of CAPD reactivity was observed along the surface of the reticulocytes and normoblasts. Absence of marrow erythrocyte CAPD reactivity with glutaraldehyde-fixed cells is consistent and reproducible; further work will be required to explain this phenomenon.

These observations indicate that the binding capacity and available carbohydrate residues associated with the cell surfaces are dependent upon cell type. The cell surfaces also are extensively modified during the process of hemopoietic cell differentiation. Membrane changes noted during granulo- and erythropoiesis with the CAPD reaction are distinct from the more uniform surface staining patterns noted with certain anionic binding compounds, viz., ruthenium red and Thorotrast. Thus, although we have noted some variation in the extent of surface reactivity of cells with similar morphological characteristics, Con A appears to serve as a useful tool for exploring the relationship of the cell surface to the differentiation process of hemopoietic cells.

EFFECTS OF SUCCINYL-CON A ON THE GROWTH OF NORMAL AND TRANSFORMED MOUSE FIBROBLASTS

David A. Hilborn and Ian S. Trowbridge
The Salk Institute
San Diego, California, U. S. A.

ABSTRACT

Extensive derivatization of concanavalin A (Con A) with succinic anhydride produces a derivative which has a binding specificity similar to that of the native lectin but which is less toxic and which does not readily agglutinate cells. We have studied the effects of succinyl-Con A on the growth of virally transformed and normal 3T3 mouse fibroblasts to test the hypothesis that covering of the Con A binding sites on the surface of transformed cells restores the cells to normal growth. Binding studies showed that succinyl-Con A binds almost as effectively to the same sites on transformed cells as the native lectin and conditions were established under which 90% of the maximum number of succinyl-Con A molecules were bound to transformed cells. Under these conditions succinyl-Con A causes a small decrease in the rate of growth and maximum cell density of both normal and transformed cells. However, there was no change in the growth characteristics of either type of cell. SV3T3 cells treated with succinyl-Con A grew to a high cell density and did not arrest in G_1 phase as shown by microfluorimetric analysis, whereas 3T3 cells stopped growing at a characteristically low cell density and arrested in G_1 phase. From these results we conclude that it is unlikely that covering Con A binding sites on the surface of transformed cells is alone sufficient to restore normal growth.

ENDOTOXIN-LIKE ACTIVITIES IN CONCANAVALIN A PREPARATIONS

Kenneth W. Brunson and Dennis W. Watson
University of Minnesota Medical School
Minneapolis, Minnesota, U. S. A.

ABSTRACT

Concanavalin A preparations obtained from commercial sources were found to have varying amounts of pyrogen contamination. Employing different dilutions of the lectin preparations, a dose-dependent febrile response in American Dutch rabbits was observed which was strikingly similar to that obtained by the use of gram-negative bacterial endotoxins. If animals were made immune to the pyrogenic effect of endotoxin, those same animals were also immune when cross tested with concanavalin A contaminated with pyrogen. Furthermore, using an intravenous immunization regimen commonly employed to immunize rabbits against endotoxin, the contaminated concanavalin A preparations would also immunize. Evidence that the contaminating material in the lectin preparations was endotoxin was further provided by positive reactions in the *Limulus* coagulation technique, a test generally considered to be specific for endotoxin.

We believe the above findings to be particularly significant regarding the biological activity of concanavalin A preparations, in view of the recognized properties of endotoxin in tumor cell inhibition, B cell (lymphocyte) mitogenicity, and adjuvanticity.

CONCANAVALIN A INDUCED INFLAMMATION*

W. Thomas Shier
The Salk Institute
San Diego, California, U. S. A.

ABSTRACT

A large number of biological activities of concanavalin A (Con A) has been described and studied in detail, but one of the least studied of its activities is the induction of an inflammatory response. As little as 10 µg of Con A injected into the footpad of a mouse induces an intense inflammatory response 6 to 8 hours after injection. Two other lectins, wheat germ agglutinin and pea lectin, have been shown to induce inflammatory responses of similar duration and intensity. Since wheat germ agglutinin is not mitogenic for lymphocytes, it is unlikely that mitogenicity for lymphocytes plays a significant role in Con A induced inflammation. The inflammatory response to these lectins was inhibited by the anti-inflammatory drugs acetylsalicylic acid, indomethacin and hydrocortisone. The response is sufficiently intense to provide a reproducible assay for the inflammatory response and for anti-inflammatory drugs without requiring complex measuring apparatus.

The inflammatory response induced by Con A is prolonged, with detectable edema 72 hrs. after injection into the footpad of a mouse. This phenomenon probably results from the unusually slow rate of elimination of Con A from the site of injection. ^{125}I-Con A required 48 times longer than ^{125}I-bovine serum albumin to undergo 90% elimination from the site of injection in mouse footpads. A possible application of this combination of properties of Con A (i.e., a potent inflammogen that is retained for an unusually long period at the site

---

*Supported in part by the Theodore Gildred Foundation and grants AI 10265 and CA 14195 from the N. I. H.

of injection) was investigated further--the production of an arthritis model disease by a single injection of Con A in rabbit knee joints. The resulting synovitis bore some similarities to rheumatoid arthritis.

ENHANCED IMMUNOGENICITY OF CON A COATED EL-4 LEUKEMIA CELLS

W. John Martin, E. Esber, and
J. R. Wunderlich
National Institutes of Health
Bethesda, Maryland, U. S. A.

ABSTRACT

The immune response of C57BL/6 mice to irradiated EL-4 leukemia cells was analysed. EL-4 inoculated mice possessed lymphoid cells reactive with EL-4 in a colony inhibition assay. Lymphocytes active in a 4 hr ^{51}Cr release cytotoxicity assay were not present in either once or twice immunized mice. Cytotoxic lymphocytes were, however, readily generated by *in vitro* tissue culture of spleen cells of EL-4 inoculated mice. Mice inoculated with Con A coated irradiated El-4 developed low levels of lymphocyte mediated cytotoxicity. Injection of irradiated but uncoated EL-4 cells into Con A coated EL-4 primed mice resulted in the development of high levels of cytotoxic lymphocyte activity. Con A coated EL-4 were shown to be more readily phagocytized, to bind less anti-H-2^b antibody, to show reduced *in vitro* protein synthesis, and to induce marked proliferation of syngeneic spleen cells. The mechanism whereby Con A coated El-4 more effectively primes the immune system for subsequent *in vitro* development of cytotoxic lymphocytes is being investigated.

LIST OF CONTRIBUTORS

G. A. Ackerman, *Department of Anatomy, Ohio State University, Columbus, Ohio, 43210, U. S. A.*

Ateeq Ahmad, *Christian Medical College Hospital, Vellore, India*

Felice Aull, *Department of Physiology, New York University Medical Center, New York, New York, 10016, U. S. A.*

B. K. Bachhawat, *Christian Medical College Hospital, Vellore, India*

K. A. Balasubramanian, *Christian Medical College Hospital, Vellore, India*

Brian H. Barber, *University of Toronto, Toronto, Ontario, Canada*

Joseph W. Becker, *Department of Biochemistry, Rockefeller University, New York, New York, 10021, U. S. A.*

Richard D. Berlin, *Department of Physiology, University of Connecticut Health Center, Farmington, Connecticut, 06032, U. S. A.*

Subal Bishayee, *Albert Einstein College of Medicine, Bronx, New York, 10461, U. S. A.*

J. S. Bond, *Department of Biochemistry, Virginia Commonwealth University, Richmond, Virginia, 23298, U. S. A.*

S. G. Bradley, *Department of Microbiology, Virginia Commonwealth University, Richmond, Virginia, 23298, U. S. A.*

Curtis F. Brewer, *Department of Pharmacology, Albert Einstein College of Medicine, Bronx, New York, 10461, U. S. A.*

J. C. Brown, *Department of Microbiology, University of Virginia Medical School, Charlottesville, Virginia 22901, U. S. A.*

R. D. Brown, III, *IBM Thomas J. Watson Research Center, Yorktown Heights, New York, 10598, U. S. A.*

Kenneth W. Brunson, *Department of Microbiology, University of Minnesota Medical School, Minneapolis, Minnesota, 55455, U. S. A.*

C. M. Bullis, *Department of Microbiology, University of Virginia Medical School, Charlottesville, Virginia, 22901, U. S. A.*

Max M. Burger, *Department of Biochemistry, Biocenter of the University of Basel, Basel, Switzerland*

Clive O. Callender, *Transplant Service, Freedman's Hospital, Howard University, Washington, D. C., 20001, U. S. A.*

Jeremy P. Carver, *University of Toronto, Toronto, Ontario, Canada*

Tushar K. Chowdhury, *Department of Physiology and Biophysics, University of Oklahoma Health Sciences Center, Oklahoma City, Oklahoma, 73190, U. S. A.*

J. G. Collard, *Departments of Experimental Cytology and Electron Microscopy, The Netherlands Cancer Institute, Amsterdam, The Netherlands*

Bruce A. Cunningham, *Department of Biochemistry, Rockefeller University, New York, New York, 10021, U. S. A.*

Gerald M. Edelman, *Department of Biochemistry, Rockefeller University, New York, New York 10021, U. S. A.*

E. Esber, *Immunology Branch, National Cancer Institute, NIH, Bethesda, Maryland 20014, U. S. A.*

Constance A. Feltkamp, *Department of Electron Microscopy, The Netherlands Cancer Institute, Amsterdam, The Netherlands*

Sten Friberg, Jr., *Department of Tumour Biology, Karolinska Institute, Stockholm, Sweden*

I. J. Goldstein, *Department of Biological Chemistry, University of Michigan, Ann Arbor, Michigan, 48104, U. S. A.*

J. A. Gordon, *Department of Pathology, University of Colorado Medical Center, Denver, Colorado, 80220, U. S. A.*

A. P. Grollman, *Albert Einstein College of Medicine, Bronx, New York, 10461, U. S. A.*

Gary R. Gunther, *Rockefeller University, New York, New York, 10021, U. S. A.*

LIST OF CONTRIBUTORS

Sten Hammarström, *Department of Immunology, Wenner-Gren Institute, Stockholm, Sweden*

David A. Hilborn, *Salk Institute, San Diego, California, 92112, U. S. A.*

R. C. Hunt, *Department of Microbology, University of Virginia Medical School, Charlottesville, Virginia, 22901, U. S. A.*

D. T. John, *Department of Microbiology, Virginia Commonwealth University, Richmond, Virginia, 23298, U. S. A.*

M. Jug, *Departments of Biochemistry and General Pathology, University of Trieste, Trieste, Italy*

S. H. Koenig, *IBM Thomas J. Watson Research Center, Yorktown Heights, New York, 10598, U. S. A.*

Myron A. Leon, *Saint Luke's Hospital, Cleveland, Ohio, 44104, U. S. A.*

Ralphael J. Mannino, Jr., *Department of Biochemistry, Biocenter of the University of Basel, Basel, Switzerland*

D. M. Marcus, *Albert Einstein College of Medicine, Bronx, New York, 10461, U. S. A.*

N. M. Marecki, *Department of Microbiology, Virginia Commonwealth University, Richmond, Virginia, 23298, U. S. A.*

M. D. Marquardt, *Department of Pathology, University of Colorado Medical Center, Denver, Colorado, 80220, U. S. A.*

W. John Martin, *Immunology Branch, National Cancer Institute, NIH, Bethesda, Maryland, 20614, U. S. A.*

G. H. McKenzie, *Russell Grimwade School of Biochemistry, University of Melbourne, Victoria, Australia*

N. Miani, *Department of Biochemistry and General Pathology, University of Trieste, Trieste, Italy*

A. E. Munson, *Department of Pharmacology, Virginia Commonwealth University, Richmond, Virginia, 23298, U. S. A.*

Martin S. Nachbar, *Department of Physiology, New York University Medical Center, New York, New York, 10016, U. S. A.*

John S. Najarian, *Department of Surgery, University of Minnesota, Minneapolis, Minnesota, 55455, U. S. A.*

A. Newman, *Institute for Chemical Sciences, University of Texas at Dallas, Dallas, Texas, 75230, U. S. A.*

Garth L. Nicolson, *Cancer Council and Electron Microscopy Laboratories, Salk Institute, San Diego, California, 92112, U. S. A.*

Joel D. Oppenheim, *Department of Physiology, New York University Medical Center, New York, New York, 10016, U. S. A.*

Arthur B. Pardee, *Program in Biochemical Sciences, Princeton University, Princeton, New Jersey, 08540, U. S. A.*

R. J. Person, *Department of Physiology and Biophysics, University of Oklahoma Health Sciences Center, Oklahoma City, Oklahoma, 73190, U. S. A.*

S. K. Podder, *Department of Biochemistry, Indian Institute of Science, Bangalore, India*

George Poste, *Department of Experimental Pathology, Roswell Park Memorial Institute, Buffalo, New York, 14203, U. S. A.*

Alan Quirt, *University of Toronto, Toronto, Ontario, Canada*

Prasanta K. Ray, *Bhabha Atomic Research Center, Trombay, Bombay, India*

George N. Reeke, Jr., *Department of Biochemistry, Rockefeller University, New York, New York, 10021, U. S. A.*

D. Romeo, *Departments of Biochemistry and General Pathology, University of Trieste, Trieste, Italy*

W. H. Sawyer, *Russell Grimwade School of Biochemistry, University of Melbourne, Victoria, Australia*

A. Dean Sherry, *Institute for Chemical Sciences, University of Texas at Dallas, Dallas, Texas, 75230, U. S. A.*

W. Thomas Shier, *Salk Institute, San Diego, California, 92112, U. S. A.*

Richard L. Simmons, *Department of Surgery, University of Minnesota, Minneapolis, Minnesota, 55455, U. S. A.*

L. A. Smets, *Departments of Experimental Cytology and Electron Microscopy, The Netherlands Cancer Institute, Amsterdam, The Netherlands*

M. R. Soranzo, *Departments of Biochemistry and General Pathology, University of Trieste, Trieste, Italy*

LIST OF CONTRIBUTORS

H. Sternlicht, *Department of Pharmacology, Albert Einstein College of Medicine, Bronx, New York, 10461, U. S. A.*

Avadhesha Surolia, *Christian Medical College Hospital, Vellore, India*

J. H. M. Temmink, *Departments of Experimental Cytology and Electron Microscopy, The Netherlands Cancer Institute, Amsterdam, The Netherlands*

D. Thambi-Dorai, *Christian Medical College Hospital, Vellore, India*

Luis H. Toledo-Pereyra, *Department of Surgery, University of Minnesota, Minneapolis, Minnesota, 55455, U. S. A.*

Ian S. Trowbridge, *Salk Institute, San Diego, California, 92112, U. S. A.*

S. D. Waksal, *Department of Anatomy, Ohio State University, Columbus, Ohio, 43210, U. S. A.*

John L. Wang, *Department of Biochemistry, Rockefeller University, New York, New York 10021, U. S. A.*

Dennis W. Watson, *Department of Microbiology, University of Minnesota Medical School, Minneapolis, Minnesota, 55455, U. S. A.*

A. Kurt Weiss, *Department of Physiology and Biophysics, University of Oklahoma Health Sciences Center, Oklahoma City, Oklahoma, 73190, U. S. A.*

J. R. Wunderlich, *Immunology Branch, National Cancer Institute, NIH, Bethesda, Maryland, 20014, U. S. A.*

Ichiro Yahara, *Department of Biochemistry, Rockefeller University, New York, New York, 10021, U. S. A.*

G. Zabucchi, *Departments of Biochemistry and General Pathology, University of Trieste, Trieste, Italy*

INDEX

A

Acetyl-Con A, 23, 45
Actinomycin, 139
Agglutination, 1, 55, 69, 73, 84-85, 101, 109, 117 ff, 153 ff, 173-174, 178, 187, 207 ff, 221 ff, 245 ff, 333-337, 343
Allograft, 2, 309, 311 ff
Amino acid transport, 190, 194-197, 202
Anti-Con A agents, 24-25, 124-126, 156, 161, 174, 261 ff, 274
Antigenic determinants, 128, 246-247, 258, 275, 285, 309 ff, 345, 349
Ascites tumor, 245 ff, 333, 337
Autolysis, 127
Autoradiography, 209, 329
Azathioprine, 310 ff

B

BHK cells, 121 ff
Bovine kidney cells, 138

C

Cap formation, 6, 14, 24-27, 43, 123, 173, 252 ff, 261, 285, 334
Cell:
  culture, 189, 209, 221
  cycle, 2, 207 ff, 221 ff, 343
  dynamics, 187 ff, 239, 336
  electrical potential, 6, 191 ff, 337
  fusion, 140-142
  growth, 2, 130, 187 ff, 207 ff, 221-222, 306, 343, 349
  mobility 29, 193
  morphology, 2 ff, 144, 164, 187 ff, 208, 221 ff
  nucleus, 143, 188 ff, 314
  rounding, 187 ff, 236
  synchronization, 203, 224 ff
  vacuoles, 143-144, 283 ff
Chick embryo cells, 121 ff
Cinematography, 191
Cluster, 123, 124 ff, 153 ff, 175-176, 202, 285
Colchicine, 15, 27-28, 164, 178 ff, 273, 281, 329
Contact inhibition, 6, 29, 156-158, 188 ff, 209 ff
Contractile protein, 127
Cyclic AMP, 128, 177, 217, 221-224, 274 ff, 292 ff
Cytochalasin B, 27, 164, 273, 281
Cytochemical markers, 239
Cytopathogenicity, 5, 137 ff
Cytophotometry, 213-214, 224 ff
Cytotoxicity, 7, 246, 291 ff
Concanavalin A:
  amino acid sequence, 16-17, 19
  binding, 6, 18 ff, 35 ff, 56, 71 ff, 103, 106, 120 ff, 160, 173, 262, 282, 315, 343
  dissociation, 101 ff, 140, 264
  inhibitor, 40, 42
  molecular weight, 15, 36, 58, 78, 85-88
  precipitation, 101

receptors, 14-15, 25-28, 39-43, 84-85, 123 ff, 137, 140, 153 ff, 173 ff, 202, 209, 221, 245 ff, 261, 316, 336-337
saccharide binding, 1 ff, 15-16, 36 ff, 55 ff, 84, 109-111, 140, 211, 247, 314, 323-327
sedimentation coefficient, 72, 75-78
in solution, 21, 35-36, 69, 80
spectroscopy, 19, 45 ff, 74 ff, 107
structure, 2, 13, 15, 29, 36, 56, 64, 69, 71 ff
subunits, 3, 14 ff, 48, 71 ff, 188-191, 207-211
thermal inactivation, 102
Con A-sepharose, 100

D

Daunomycin, 291 ff
Dibutyryl-cAMP, 164, 221 ff
Divalent cations, 3, 15 ff, 60 ff, 74, 87, 223, 323, 325, 327
DNA synthesis, 203, 217, 224 ff, 297, 329

E

Electrical properties, 6, 187 ff, 249, 339
Electron microscopy, 225, 235-237, 333
Electrophoresis, 79 ff, 100, 129 ff, 211, 247
Endocytosis, 45, 153 ff, 185, 284, 304, 315
Endotoxin, 292, 345
Enthalpy, 111
Erythrocytes, 4, 263 ff, 275, 294, 304, 335, 336, 341, 342

F

Ferritin-labelled Con A, 147, 153 ff
Fibroblast, 155 ff, 178, 184, 187 ff, 207 ff, 221 ff, 343
Fluorescein-Con A, 6, 14, 24, 26, 124, 153-156, 208, 250-255, 274-276
5-Fluorouracil, 293 ff
Furosamide, 310

G

Glutaraldehyde, 335, 341
Glycoprotein, 4, 39, 85, 95 ff, 129-135, 261-263, 284, 331, 337

H

Hemocyanin-Con A, 160, 173 ff, 333
Hemolysis, 265
Hemopoietic cells, 341
D-Hexopyranosides, 15 ff, 35 ff, 55 ff
Histocompatibility, 245, 309 ff, 310
Hormones, 275
Host reaction, 246

I

Immunoresistance, 246, 259
Immunosuppression, 2, 309, 316
Inflammation, 347

K

Kinetics, 58 ff, 85, 102 ff, 323, 329, 335

INDEX

L

Leukocytes, 273 ff, 293, 300, 315
Lipid A, 291 ff
Lumicolchicine, 178, 183
Lymphocytes, 1, 14, 24, 45, 85, 184, 223, 266-268, 275, 292, 329, 333, 347, 349
Lymphoid cells, 14, 29, 250, 310
Lysosomal enzymes, 95 ff, 127 ff, 282, 341

M

Membrane:
  fluidity, 2, 26, 239
  permeability, 189, 198, 203, 285, 339
  resistance, 191, 198 ff
  transport, 2 ff, 180, 187 ff, 337
Metabolism, 273 ff
α-Methyl pyranosides, 3, 15, 45, 56 ff, 80, 105 ff, 142-144, 158, 183, 214, 250, 264, 275, 323, 325, 329
Microelectrodes, 191, 203
Microfilament, 162-164, 284
Micropinocytosis, 173
Microtubules, 6, 27, 162, 173 ff, 202
Microvilli, 232-236, 334
Mithramycin, 291 ff
Mitosis, 6, 15, 25, 128, 203, 207 ff, 228 ff, 292
Muscle, 339

N

Neuraminic acid, 40, 98
Neuraminidase, 132, 140, 248
Nerve, 339
Nuclear magnetic resonance, 4, 20, 55 ff, 323, 325

O

Ouabain, 337

P

Pactamycin, 291 ff
Peroxidase-Con A, 155, 160
Phagocytosis, 2, 84, 173 ff, 273 ff, 349
PHA, 245 ff, 268, 309 ff
Phase contrast microscopy, 192-193
Polymorphonuclear leukocytes, 178, 184, 276 ff
Protein synthesis, 141, 145
Pseudopodia, 176, 232 ff

R

RC_1, 106 ff, 160-162, 183-184
Receptor mobility, 155, 164
Replica technique, 231 ff
RNA synthesis, 140, 266, 305

S

Saturation density, 6, 187 ff, 207 ff, 224
Shear forces, 336
Succinyl-Con A, 7, 14, 23-25, 85, 120, 140, 207 ff, 343
Survival time, 201, 203

T

Theophylline, 226
Thymidine, 15, 196-198, 213, 261, 329
Transformed cells, 1 ff, 55, 117 ff, 153 ff, 173, 187 ff, 208 ff, 221 ff, 245 ff, 291 ff, 343
Tumor transplant, 199-201, 204, 246, 333

V

Vinblastine, 27, 164, 178
Vincristine, 27, 291 ff
Virus infection, 5, 117 ff, 188
Virus replication, 141
Virus transformation, 5, 117, 118
Vitamin A, 137

W

Wheat germ agglutinin, 42, 208

X

X-irradiation, 188, 224, 227, 230
X-ray diffraction, 3, 13 ff, 72, 81